Cognitive Electrophysiology

Cognitive Electrophysiology

H.-J. Heinze   T. F. Münte   G. R. Mangun

Editors

With a Foreword by Michael S. Gazzaniga

Birkhäuser
Boston • Basel • Berlin

H.-J. Heinze
Medizinische Hochschule Hannover
Neurologische Klinik
Konstanty-Gutschow-Straße 8
D-3000 Hannover 61
Germany

George R. Mangun
Center for Neuroscience
University of California, Davis
Davis, California 95616
U.S.A

T. F. Münte
Medizinische Hochschule Hannover
Neurologische Klinik
D-3000 Hannover 61
Germany

Birkhäuser

ISBN 0-8176-3726-5
ISBN 3-7643-3726-5

Typeset by Techset Composition Ltd., Salisbury, United Kingdom
Printed and bound by Quinn Woodbine Inc., Woodbine, NJ.
Printed in the United States of America

9 8 7 6 5 4 3 2 1

Foreword

Michael S. Gazzaniga

The investigation of the human brain and mind involves a myriad of approaches. Cognitive neuroscience has grown out of the appreciation that these approaches have common goals that are separate from other goals in the neural sciences. By identifying cognition as the construct of interest, cognitive neuroscience limits the scope of investigation to higher mental functions, while simultaneously tackling the greatest complexity of creation, the human mind.

The chapters of this collection have their common thread in cognitive neuroscience. They attack the major cognitive processes using functional studies in humans. Indeed, functional measures of human sensation, perception, and cognition are the keystone of much of the neuroscience of cognitive science, and event-related potentials (ERPs) represent a methodological "coming of age" in the study of the intricate temporal characteristics of cognition. Moreover, as the field of cognitive ERPs has matured, the very nature of physiology has undergone a significant revolution. It is no longer sufficient to describe the physiology of non-human primates; one must consider also the detailed knowledge of human brain function and cognition that is now available from functional studies in humans—including the electrophysiological studies in humans described here. Together with functional imaging of the human brain via positron emission tomography (PET) and functional magnetic resonance imaging (fMRI), ERPs fill our quiver with the arrows required to pierce more than the single neuron, but the networks of cognition.

Studies of both normal and neurological patients are addressed in this volume. Heinze, Münte, and Mangun have organized a series of essays that represent a significant contribution to the field of "cognitive electrophysiology." Represented in the present book are chapters contributed by both the major leaders in the field of cognitive electrophysiology and the new generation of young experimentalists trained in the complexities of functional human brain analysis. Several chapters are noteworthy in their contributions to current issues in cognition.

Hillyard and colleagues review interesting recent work in visual attention and provide convincing new data on the mechanisms of early selection and the relationship between signal detection parameters and the underlying physiological mechanisms. Their chapter illustrates the most current innovations in research in spatial attention, and helps us understand the role of early selection mechanisms in terms of perception and performance. In a related vein, the chapter by the editors, Heinze, Münte, and Mangun, investigate global and local processing within hierarchically organized visual stimuli and provide physiological evidence for separations in the mechanisms responsible in each. ERPs to global targets versus local targets can index the interactions and interference of "part and whole" in object analysis. Such electrophysiological evidence helps us to understand the importance of the temporal advantage in ERPs by providing strong evidence as to why one perceptual processing stage may or may not interact with another given the particular task constraints.

Linguistics and neuroscience are elegantly merged in the contribution of Kutas and Kluender, "What Is Who Violating?" The precision with which ERPs can dissect the microstructure of cognition within the current theoretical frameworks of language is illustrated by these authors. Can one conceive of a reductionist approach to the brain that will reveal the layers of processing inherent in human language the way in which modern tools such as these are succeeding? The use of language incongruencies and on-line sentence processing to parse the language systems of the brain into understandable units represent a leap forward in brain-based studies of human language. Their combined linguistic and ERP method is a step forward in physiological studies of language and reflects the appreciation of traditionally conservative disciplines, such as linguistics, of the value of asking cognitive neuroscience questions about language.

Studies of cognitive processes in normal persons can be contrasted with related work in neurological patients. The chapter by Nielsen-Bohlman and Knight illustrates the state-of-the-art in combined electrophysiological and neuropsychological studies of cognition. Utilizing their careful, neurologically sophisticated analyses of focal neurological damage, they attack the question of human memory and lead us into new considerations of memory functions. These studies are not mere rehashing of the wealth of knowledge gained by a century of observation by behavioral neurologists. Rather, utilizing the power of modern anatomical imaging, these and like-minded neurologists and cognitive neuroscientists are actively pursuing a refinement of our understanding of human cortical functioning. By combining this approach with ERPs, Nielsen-Bohlman and Knight lead a new charge into the study of higher mental functions from the neuropsychological perspective.

Though one principle advantage of ERPs is the ability to investigate brain processes directly in humans, ERPs do not limit us to asking questions only in humans. In his comprehensive review, Paller bridges the gap between human studies of memory and investigations in non-human primates. As a leader in the field of animal models of memory using ERP measures, Paller provides a thorough review of the issues and instructs the novice in the intricacies of such combined human and animal research. The result is a sense of both the difficulties and the bounty that such studies present the field of cognitive neuroscience.

In an era in which with each passing year a new imaging method is presented as the new oracle of the human brain, ERPs—and their magnetic counterpart—remain the tool of choice for investigating the millisecond-to-millisecond functioning of the brain during sensation, perception, cognition, and action. In the face of the new "phrenology" of functional imaging, ERPs persist as the only way to view the complex interplay between neural processors that operate in the millisecond time domain. With their growing legions of advocates, ERPs will continue to be the functional imaging analog of the cognitive psychologist's reaction time measure. And with no less impact, the cognitive electrophysiological approach will remain the standard of temporal resolution for the foreseeable future. The present volume represents a concise and current reflection of the state of a dynamic field of research.

Contents

Contributors

U. Altrup Institut für Experimentelle Epilepsieforschung, Universität Münster, Hüfferstraße 68, D-48149 Münster, Germany

Erol Başar Institut für Physiologie, Medizinische Universität Lübeck, Ratzeburger Allee 160, D-2400 Lübeck, Germany

Canan Başar-Eroglu Institute for Medical Psychology, Medizinische Universität Lübeck, Ratzeburger Allee 160, D-2400 Lübeck, Germany

N. Birbaumer Eberhard-Karls-Universität, Medical Psychology and Behavioral Neuroscience, Gartenstraße 29, D-7400 Tübingen, Germany

Daniel Brandeis Neurologie, Universitätsspital, CH-8091 Zürich, Switzerland

K. A. Brookhuis Institute for Experimental and Occupational Psychology, University of Groningen, Grote Kruisstraat 2/1, 9712 TS Groningen, The Netherlands

Tamer Demiralp Institut für Physiologie, Medizinische Universität Lübeck, Ratzeburger Allee 160, D-2400 Lübeck, Germany

Michael C. Doyle Wellcome Brain Research Group, Department of Psychology, University of St. Andrews, St. Andrews, Fife KY16 9JU, United Kingdom

T. Elbert University of Münster, Institute for Experimental Audiology, Kardinal-von-Galen-Ring 10, D-4400 Münster, Germany

Michael S. Gazzaniga   Center for Neuroscience and Department of Neurology, University of California at Davis, Davis, California 95616, USA

Riitta Hari   Low Temperature Laboratory, Helsinki University of Technology, Otakaari 3A, SF-02150 Espoo, Finland

Martin Heil   Department of Psychology, Philipps-Universität, Gutenbergstraße 18, D-35032 Marburg, Germany

H.-J. Heinze   Medizinische Hochschule Hannover, Neurologische Klinik, Konstanty-Gutschow-Straße 8, D-30623 Hannover, Germany

Erwin Hennighausen   Department of Psychology, Philipps-Universität, Gutenbergstraße 18, D-35032 Marburg, Germany

Steven A. Hillyard   Department of Neurosciences, 0608, University of California, San Diego, 9500 Gilman Drive, La Jolla, California 92093-0608, USA

A. Hufschmidt   Neurologische Universitätsklinik Freiburg, Hansastraße 9, D-79104 Freiburg, Germany

Sönke Johannes   Neurologische Klinik, Medizinische Hochschule Hannover, Konstanty-Gutschow-Straße 8, D-30623 Hannover, Germany

Robert Kluender   Department of Linguistics, University of California, San Diego, La Jolla, California 92093-0115, USA

R. T. Knight   Department of Neurology and Center for Neuroscience, University of California at Davis, Veterans Administration Medical Center, 150 Muir Road, Martinez, California 94553, USA

R. Köhling   Institut für Experimentelle Epilepsieforschung, Universität Münster, Hüfferstraße 68, D-48149 Münster, Germany

Marta Kutas   Department of Cognitive Science, C-015, University of California, San Diego, La Jolla, California 92093-0115, USA

Dietrich Lehmann   Neurologie, Universitätsspital, CH-8091 Zürich, Switzerland

Steven J. Luck   Department of Neurosciences, 0608, University of California, San Diego, 9500 Gilman Drive, La Jolla California 92093-0608, USA

A. Lücke   Institut für Physiologie, Universität Münster, Robert-Koch-Straße 27a, D-48149 Münster, Germany

C. H. Lücking   Neurologische Universitätsklinik Freiburg, Hansastraße 9, D-79104 Freiburg, Germany

W. Lutzenberger   University of Tübingen, Institute of Medical Psychology and Behavioral Neurobiology, Gartenstraße 29, D-7400 Tübingen, Germany

George R. Mangun   Center for Neuroscience, University of California, Davis, Davis, California 95616, USA

L. J. M. Mulder   Institute for Experimental and Occupational Psychology, University of Groningen, Grote Kruisstraat 2/1, 9712 TS Groningen, The Netherlands

G. Mulder   Institute for Experimental and Occupational Psychology, Kerklaan 30, NL-9751 NN Haren, The Netherlands

T. F. Münte   Neurologische Klinik, Medizinische Hochschule Hannover, D-30623 Hannover, Germany

L. Nielsen-Bohlman   Department of Neurology and Center for Neuroscience, University of California at Davis, Veterans Administration Medical Center, 150 Muir Road, Martinez, California 94553, USA

J. Niemann   Neurologische Universitätsklinik Freiburg, Hansastraße 9, D-79104 Freiburg, Germany

Ken A. Paller   Center for Functional Imaging, Lawrence Berkeley Laboratory, Mail Stop 55-121, Berkeley, California 94720, USA

Frank Rösler   Department of Psychology, Philipps-Universität Marburg, Gutenbergstraße 18, D-3550 Marburg, Germany

Michael D. Rugg   Wellcome Brain Research Group, Department of Psychology, University of St. Andrews, St. Andrews, Fife KY16 9JU, United Kingdom

Martin Schürmann   Institut für Physiologie, Medizinische Universität Lübeck, Ratzeburger Allee 160, D-2400 Lübeck, Germany

H. G. O. M. Smid   Institute for Experimental and Occupational Psychology, University of Groningen, Kerklaan 30, NL-9751 NN Haren, Groningen, The Netherlands

E.-J. Speckmann   Institut für Physiologie, Universität Münster, Robert-Koch-Straße 27a, D-48149 Münster, Germany

T. Trevorrow   University of Hawaii at Manoa, Department of Psychology, 2430 Campus Road, Honolulu, Hawaii 96822, USA

A. A. Wijers   Institute for Experimental and Occupational Psychology, University of Groningen, Kerklaan 30, NL-9751 NN Haren, Groningen, The Netherlands

T. Winker   Neurologische Universitätsklinik Freiburg, Hansastraße 9, D-79104 Freiburg, Germany

Chapter 1

The Cuing of Attention to Visual Field Locations: Analysis with ERP Recordings

STEVEN A. HILLYARD, STEVEN J. LUCK, AND GEORGE R. MANGUN

Event-related potentials (ERPs) were recorded during the performance of different types of spatial attention tasks in which an advance cue directed subjects to attend to a particular location in the visual field. In the first experiment, the effects of central and peripheral cues were compared. These two cue types produced similar validity effects for most of the target-elicited ERP components; however, the earliest ERP component (P1) failed to show a validity effect under peripheral cuing. In a second experiment employing only central cues, subjects attempted to detect faint luminance targets following valid cues, invalid cues, or neutral cues that provided no information about the location of the subsequent target. Significant costs and benefits of precuing were observed on signal detectability (d'), and these attention effects were associated with amplitude modulations of early ERP components (P1 and N1) that were localized over lateral occipital cortex. These results provide the strongest evidence to date that modulation of short-latency (80- to 200-msec) activity in the visual cortex is specifically associated with changes in perceptual processing.

Introduction

Human observers have a remarkable capability for deploying their attention rapidly to the critical regions of a visual scene to extract the information contained therein. Under most circumstances this involves

Cognitive Electrophysiology
H-J. Heinze, T.F. Münte, and G.R. Mangun, editors
© 1994 Birkhäuser Boston

making eye movements toward the attended location, but attention can also be focused covertly in the absence of any changes in eye position. An expanding research literature has established that stimuli occurring at or near a covertly attended location are processed more efficiently than are stimuli at a distance from the focus of attention. The majority of studies of visual-spatial attention have used cuing or priming paradigms in which an initial stimulus (the cue) indicates the probable location of a subsequent stimulus (the target) that must be detected or discriminated. Targets occurring at precued (i.e., attended) locations show enhanced processing in the form of improved detectability for faint stimuli, improved discriminability of stimulus features and patterns, or speeded motor responses (reviewed in Ericksen and Yeh, 1985; Henderson, 1991; Lambert et al., 1987; Muller and Humphreys, 1991).

There has been a long-standing controversy regarding the extent to which advance cuing leads to an actual improvement in the quality of attended sensory information (Prinzmetal et al., 1986; Shaw, 1984; Sperling, 1984). This is related to the classic "levels of selection" issue—does attention act by controlling the flow of sensory information at a relatively early level of processing, or does it bias higher recognition, decision, or response stages to favor specific inputs? Numerous studies have attempted to distinguish early-sensory versus late-decision processes, but recent late selection theories that invoke rapid switching of location-specific decision processes (e.g., Duncan and Humphreys, 1989) appear difficult to distinguish operationally from early input selection. Thus, behavioral studies appear to have reached an impasse at resolving this level-of-selection issue.

In recent studies we have investigated levels of selection during visual-spatial attention by recording event-related brain potentials (ERPs) from the scalp in normal human subjects using multielectrode arrays. Each successive wave or "component" of the human visual ERP is characterized by a latency that indicates the timing of the underlying neural activity and a topography on the scalp determined by the geometry of the active neural tissue (Hillyard and Picton, 1987). By studying how selective attention affects specific early and late ERP components, conclusions may be drawn about which levels of sensory processing are modulated by attention, which types of sensory information are enhanced or suppressed, and which brain areas participate in the attentional control processes.

Using ERP techniques, we have reported that early sensory-evoked activity (onset latency, 70–100 msec) arising from specific zones of occipital cortex is enhanced by visual-spatial attention (Mangun and Hillyard, 1988, 1990; Mangun, Hillyard, and Luck, 1993). Our initial interpretation is that attention to location produces a spatially restricted modulation of information flow through specific pathways leading from primary visual cortex to higher cortical areas, perhaps analogous to a sensory gain control process. These early ERP modulations have been observed in different types of spatial attention paradigms, including sustained attention to continuous sequences of stimuli and cued attention to probable target locations (Mangun and Hillyard, 1991).

The current study employs ERP techniques to investigate two important issues that have emerged concerning the mechanisms of cued spatial attention. In the first experiment, ERPs are compared in reaction time (RT) tasks that utilize central-symbolic versus peripheral cues to direct attention; the aim is to provide physiological evidence bearing on the question of whether these two types of cues invoke fundamentally different attentional mechanisms. In the second experiment, we extend the ERP method to a task in which cued spatial attention improves the detectability of faint, masked signals. The aim is to test further the proposition that advance cuing of attention facilitates perceptual processing per se, as manifested by changes in early ERP components.

Experiment 1: Central Versus Peripheral Cues

The studies of Posner and colleagues (Posner, 1980; Posner et al., 1978) established that RTs are faster in response to targets that occur at precued visual field locations. Posner et al. provided subjects with an advance visual cue (a directional arrow located at fixation) that indicated the most probable location at which a target stimulus would subsequently occur; speeded responses to the targets were required. Three types of trials were used: valid trials in which the cue correctly indicated the location of the subsequent target; invalid trials in which the cue indicated an incorrect location; and neutral trials in which the cue provided no information about target location. Relative to the neutral cue condition, reaction times (RTs) were found to be faster for validly cued targets (benefits) and slower for invalidly cued targets (costs).

The speeded RTs to validly cued stimuli were interpreted by Posner and associates as the result of an enhanced efficiency of sensory processing for attended-location stimuli, that is, an improvement in the quality of their perceptual representations (e.g., Posner, 1980). This view has been supported by signal detection studies that found improved detectability for targets at attended locations (Bashinski and Bacharach, 1980; Downing, 1988; Hawkins et al., 1990; Müller and Humphreys, 1991). Using a different approach, Hawkins, Shafto, and Richardson (1988) also provided evidence that spatial priming affects perceptual processing. They varied target luminance in a spatial priming task and found that the RT difference for valid versus invalid targets increased with decreasing target luminance. Such an interaction suggests that the attention and luminance effects are produced at similar, early levels of visual processing.

Contrasting with the perceptual facilitation hypothesis is the view that RT benefits during spatial priming are the result of changes in postperceptual factors such as decision bias (Shaw, 1984; Sperling, 1984). According to this hypothesis, the quality of the perceptual representation remains unchanged by the cuing of attention, but the subject's willingness to respond to an event (i.e., decision criterion) is altered. If we assume that information about a stimulus accumulates over time and that a response is generated when the amount of accumulated information reaches the subject's criterion level for reporting target occurrence, adopting a lower criterion for validly cued targets could result in faster RTs to those events.

Recordings of ERPs may help to resolve this controversy, because the early components (P1 at 90–120 msec and N1 at 150–190 msec) evoked over the occipital scalp appear to reflect modality-specific information processing in the visual pathways that is sensitive to the direction of attention (e.g., Eason, 1981; Hillyard and Münte, 1984; Mangun and Hillyard, 1990). If spatial priming results in a facilitation of early perceptual processing, then validly cued stimuli would be expected to elicit greater sensory evoked activity than invalidly cued stimuli. We reported the results of such an experiment in Mangun and Hillyard (1991): Following a central arrow cue, a bar stimulus was flashed to either the same visual field as indicated by the arrow (0.75 probability, valid trial) or the opposite field (0.25 probability, invalid trial). In one experiment, subjects were required to respond to the onset of the target bar whether it was validly or invalidly cued (simple RT). In subsequent experiments, subjects were asked to make one of two alternative responses depending on the height of the bar (discriminative RT). ERPs were recorded to the right and left visual field bar stimuli as a function of whether or not they were validly precued.

Mangun and Hillyard reported a validity effect on the early P1 component of the visual ERP in all experimental conditions, with validly cued targets eliciting larger P1 amplitudes than invalidly cued targets. A similar effect was demonstrated by Harter and associates (1989). This modulation of the occipital P1 wave clearly indicates an alteration in early sensory processing as a function of target validity.

The current experiment was aimed at comparing the ERP and behavioral attention effects for different types of attention-directing cues. There is a good deal of evidence that a peripheral cue captures attention in a different manner from a symbolic cue such as a central arrow that points to that peripheral location. Jonides (1981) demonstrated that a peripheral cue adjacent to one position in a ring of letters summoned attention in a more automatic way than did a central-symbolic cue. This automaticity was evident in: (1) strong cue validity effects even when the peripheral cue was not predictive of target position; (2) an inability to ignore the presence of the peripheral cue; and (3) insensitivity of the peripheral cuing effect to a distracting task.

The time course of attentional priming also appears to differ between central and peripheral cues. Posner et al. (1982), for example, showed that a peripheral cue produced an early facilitation of RT for targets that followed at the same location with a stimulus onset asynchrony (SOA) of 50–100 msec, even when that cue signified that an opposite-field location was more likely to contain a target. After several hundred milliseconds, however, the RT facilitation was transferred to the opposite location in accordance with the cue's symbolic meaning. The peripheral cue also had the property of producing an inhibition of RT following the initial facilitation. More recently, Briand and Klein (1987) reported that peripheral cues produced larger validity effects for conjunction targets than for feature targets, but this difference was absent when central-symbolic cues were employed. These authors proposed that exogenous (peripheral) and endogenous (symbolic) cuing engage distinctive attentional systems, with the former being more effective at combining stimulus features into integrated percepts.

A critical problem that arises when comparing central versus peripheral cuing is that the latter will inevitably produce some degree of sensory interaction between cue and target because of their proximity on the receptor surface and in the sensory pathways. Posner et al. (1982) argued that the automaticity and afterinhibition seen with peripheral cues were not a simple consequence of sensory interaction, because similar effects could be produced by dimming as well as by brightening the peripheral location. This observation does not completely rule out contributions from sensory interaction factors, however, because the retina is well endowed with ganglion cells that respond to stimulus offsets. Thus, the fact that the peripheral cue and the target activate common (or neighboring) regions of the sensory pathways could result in sensory interactions that may be confounded with attention affects.

These observed differences between the central and peripheral cuing of attention in degree of automaticity, time course, inhibitory aftereffects, and feature conjunction processing are strongly suggestive of separate mechanisms of attentional orienting. To gain further information on this proposed dichotomy, the current experiment compared visual ERPs elicited by targets that were cued by either central arrow or peripheral cues. To the extent that separable neural systems mediate these two forms of attentional cuing, we would expect to observe qualitatively different validity effects on target-elicited ERPs.

Method

SUBJECTS. Twelve healthy, right-handed persons aged 18 to 30 years participated as paid volunteers. Participants had normal or corrected-to-normal vision. Three subjects were disqualified when the computer-averaged electro-oculogram (EOG) revealed small but systematic eye movements in the direction of the cued location. Thus, nine subjects (7 males) were included in the final analyses.

Stimuli

Central Cue Condition

Trials began with an arrow flashed at fixation that pointed on a random basis to the left or right visual field (0.50 probability each). The cue indicated the most likely (0.75 probability; valid trials) side of presentation for the subsequent target stimulus. The target stimuli were short (1.9 × 0.7 degrees) or tall (2.1 × 0.7 degrees) vertical bars flashed in the left or right visual field 6.4 degrees lateral to fixation. The target locations were each demarcated by four continuously present dots that formed the corners of a vertical rectangle (1.5 × 1.1 degrees) centered on the target's position. The cuing arrow and target bars were flashed for durations of 34 and 50 msec, respectively. The interval from cue onset to target onset (SOA) varied randomly from 600 to 800 msec, with intertrial intervals of 1800 msec. All stimuli were presented white-on-black on a video monitor under microcomputer control.

Peripheral Cue Conditions

Stimuli and procedures were the same as described previously for the central cue condition except as specified here. Trials began with a peripheral cue that consisted of a brief displacement of the dots that marked one of the target locations. The four dots were extinguished and replaced for 50 msec by four dots that formed a new outline rectangle with the dimensions 0.5 × 1.1 degrees. The original dots were then restored, giving the appearance that the continuously present marker dots had jumped toward and then away from each other in the vertical dimension. When viewing these dots during the experiment, some subjects reported the sensation of movement of the dots while others reported that the dots flickered. This peripheral cue occurred in the same visual field as the subsequent target bar 75% of the time (valid trials) and in the opposite visual field 25% of the time (invalid trials). The cue-target SOAs were varied in a manner identical to the central cue condition.

Procedure. During all conditions, subjects maintained eye fixation upon a central point and were required to press one button with the left hand to short bars and another button with the right hand to tall bars, as quickly and accurately as possible on every trial. Subjects were informed of the probabilities of the valid and invalid trial types and were told to make use of this information to maximize their performance. Sixteen 2.5-min runs in each of the two conditions (64 trials/run) were presented in two recording sessions. Presentation was in blocks of 4 runs of each condition, counterbalanced for order effects within and across subjects. Training consisted of 2 runs per condition before the start of the experiment and 1 practice run of the new condition between each block.

ELECTROPHYSIOLOGICAL RECORDING. The electroencephalogram (EEG) was recorded from electrodes located at C3, C4, T5, T6, O1, O2, Oz, left mastoid, and six additional nonstandard sites, as described in Mangun and Hillyard (1990) (P1, P2, PL, PR, OL, OR). All these electrodes were referenced to an electrode placed over the right mastoid; an average mastoid reference was algebraically computed offline. Eye position was monitored via recording of the horizontal EOG. Signals were amplified with a bandpass of 0.01–100 Hz (1/2 amplitude cutoffs), digitized online at a sampling rate of 250 Hz, and stored on magnetic tape. Computerized artifact rejection was used during ERP averaging to remove epochs during which eye movements to the priming cue or target stimulus occurred.

Repeated-measures analyses of variance (ANOVA) were performed on the RT results and on mean amplitude measures of the major ERP components to the targets within time windows specified in Tables 1 and 2. The

Table 1. Statistical analysis of target validity effect on ERP amplitudes for the central cue condition, collapsed across visual field and hemisphere of recording

ERP component	Site	Window (msec)	F Ratio[a]	p Value
P1	01/02	95–135	6.5	.034
	0L/0R	95–135	10.5	.012
	T5/T6	95–135	14.2	.006
	PL/PR	95–135	4.9	n.s.[b]
	P1/P2	95–135	2.7	n.s.
	C3/C4	95–135	0.0	n.s.
N1	01/02	150–200	11.9	.009
	0L/0R	150–200	10.4	.012
	T5/T6	150–200	10.1	.013
	PL/PR	140–190	6.7	.033
	P1/P2	140–190	11.1	.010
	C3/C4	130–170	8.1	.022
N2	01/02	250–300	25.5	.001
	0L/0R	250–300	18.8	.002
	T5/T6	250–300	9.7	.01
	PL/PR	260–310	21.8	.002
	P1/P2	260–310	37.3	.001
	C3/C4	260–320	37.3	.001
P3	01/02	300–400	44.3	.001
	0L/0R	300–400	43.9	.001
	T5/T6	300–400	49.8	.001
	PL/PR	300–400	60.4	.001
	P1/P2	300–400	37.8	.001
	C3/C4	300–400	44.9	.001

[a] F(1, 8), main effect of validity.
[b] n.s., not significant.

Table 2. Statistical analysis of target validity effect on ERP amplitudes for the peripheral cue condition, collapsed across visual field and hemisphere of recording

ERP component	Site	Window (msec)	F Ratio[a]	p Value
P1	01/02	95–135	0.2	n.s.[b]
	0L/0R	95–135	0.6	n.s.
	T5/T6	95–135	1.4	n.s.
	PL/PR	95–135	1.3	n.s.
	P1/P2	95–135	0.3	n.s.
	C3/C4	95–135	0.4	n.s.
N1	01/02	150–200	20.8	.002
	0L/0R	150–200	21.9	.002
	T5/T6	150–200	25.5	.001
	PL/PR	140–190	7.9	.023
	P1/P2	140–190	10.2	.013
	C3/C4	130–170	1.1	n.s.
N2	01/02	250–300	11.8	.009
	0L/0R	250–300	11.6	.009
	T5/T6	250–300	13.4	.006
	PL/PR	260–310	23.1	.001
	P1/P2	260–310	24.9	.001
	C3/C4	260–320	44.8	.001
P3	01/02	300–400	4.9	n.s.
	0L/0R	300–400	10.2	.01
	T5/T6	300–400	16.0	.004
	PL/PR	300–400	19.0	.002
	P1/P2	300–400	12.6	.007
	C3/C4	300–400	22.5	.001

[a] $F(1, 8)$, main effect of validity.
[b] n.s., not significant.

baseline for ERP measurements was the mean voltage of the 100-msec pretarget interval. ANOVA factors for the target ERP analyses were bar height (tall vs. short), visual field of stimulus (left vs. right), validity (valid vs. invalid), and hemisphere of recording (left vs. right). Significance levels were corrected using the Greenhouse–Geisser epsilon coefficient (Jennings and Wood, 1976).

Results and Discussion

BEHAVIORAL PERFORMANCE. Reaction times were significantly faster to validly cued targets than invalidly cued targets in both central ($p < .001$) and peripheral ($p < .001$) cue conditions. For central cues, RTs averaged

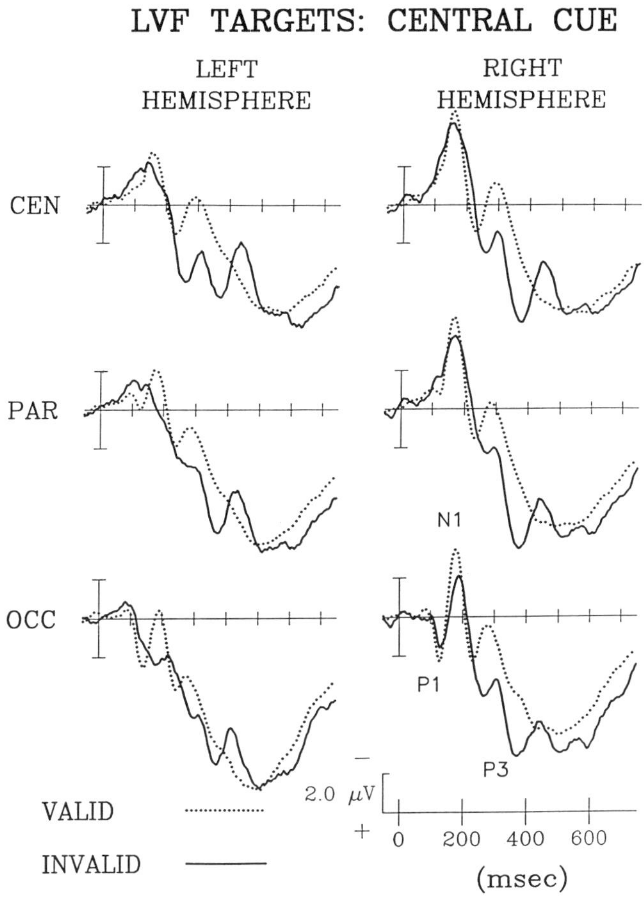

Figure 1.   Grand average ERPs to target stimuli in left visual field (LVF) in central cuing condition of Experiment 1. Recording sites over left and right hemispheres are as follows: CEN (C3/C4), PAR (PL/PR), OCC (OL/OR). Validly precued targets (*dotted lines*) are overlain by invalidly precued targets (*solid lines*).

520 msec for valid trials and 562 msec for invalid trials; for peripheral cues, these values were 537 and 573 msec, respectively. Neither the valid–invalid difference in RT nor the overall RTs differed significantly between cue conditions.

ERPs: CENTRAL CUE CONDITION. Grand average ERPs for the nine subjects to left and right visual field target stimuli are presented in Figures 1 and 2, collapsed over tall and short stimuli. The principal effects of cue validity on the ERPs were amplitude enhancements of the P1 and N1 components for validly cued stimuli relative to invalidly cued stimuli. Significant validity

RVF TARGETS: CENTRAL CUE

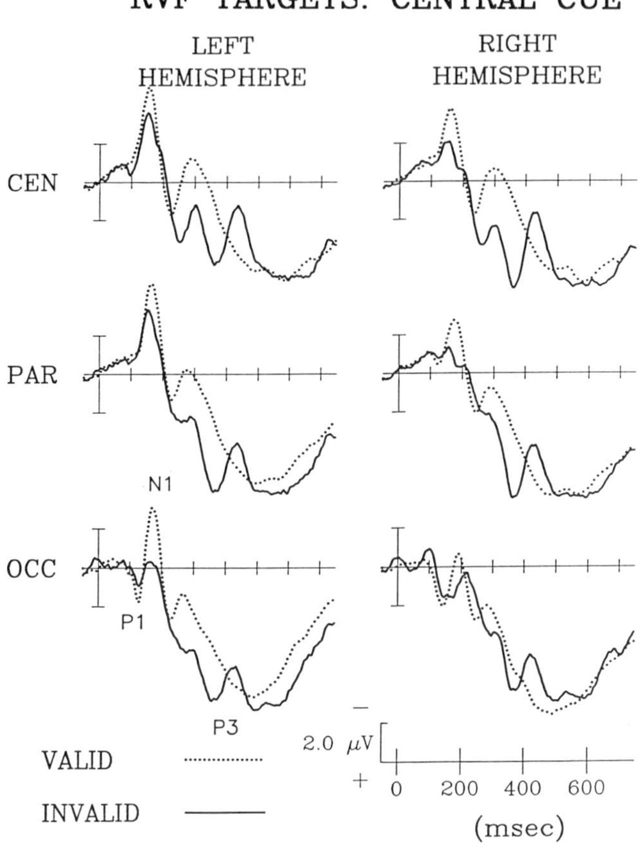

Figure 2. Grand average ERPs to target stimuli in right visual field (RVF) in central cuing condition of Experiment 1; recording sites and abbreviations as in Figure 1.

effects for the P1 component were observed at lateral occipital sites (OL and OR) and at posterior temporal sites (T5 and T6), as listed in Table 1. The N1 (and N2) effects were significant at occipital, temporal, parietal, and central sites. The P3 component measured in the interval of 300–400 msec was significantly larger for invalidly cued stimuli at all scalp sites, although a longer lasting positivity at 400–700 msec was similar in amplitude for valid and invalid targets. These results replicate the major ERP validity effects reported by Mangun and Hillyard (1991).

ERPs: PERIPHERAL CUE CONDITION. Grand average ERPs to left and right visual field target stimuli in the peripheral cue condition are presented in

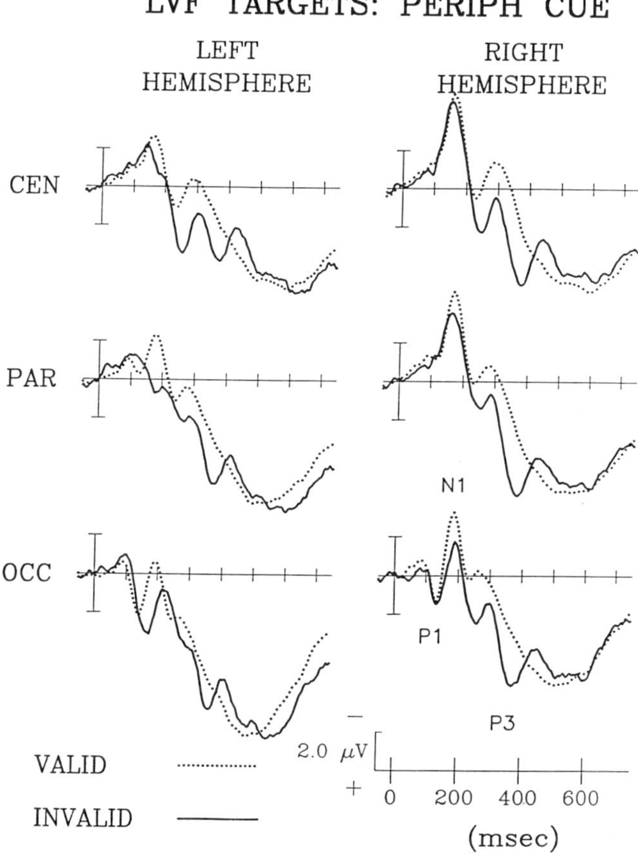

Figure 3.   Grand average ERPs to target stimuli in left visual field (LVF) in peripheral cuing condition of Experiment 1; target sites and abbreviations as in Figure 1.

Figures 3 and 4. In contrast to the results from the central cue condition, the ERP validity effects for peripheral cues were restricted to the N1, N2, and P3 components; no significant effect of cue validity was observed for the P1 component. The N1 and N2 were significantly larger to validly cued stimuli than invalidly cued stimuli at occipital, temporal, parietal, and central scalp sites (see Table 2). The P3 component in the 300- to 400-msec interval was again larger to invalidly cued target stimuli at all scalp sites. These effects were very similar in amplitude and scalp distribution to those of the central cue condition.

Except for the lack of a P1 effect in the peripheral cue condition, the target ERPs and validity effects were highly similar under central and

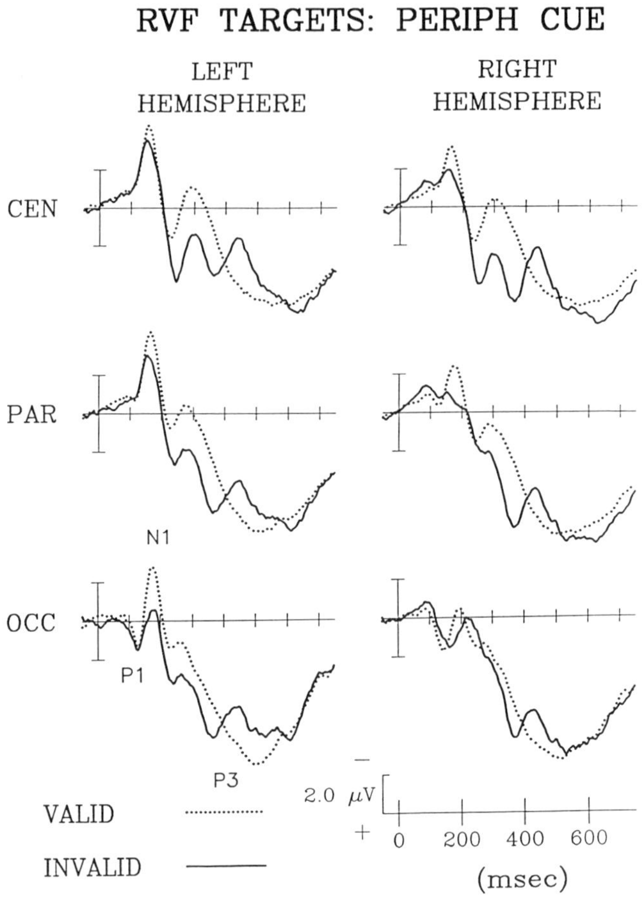

Figure 4.   Grand average ERPs to target stimuli in right visual field (RVF) in peripheral cuing condition of Experiment 1; target sites and abbreviations as in Figure 2.

peripheral cuing. This suggests that, to a large degree, central and peripheral cues affect sensory processing via common or partially overlapping brain systems, at least at the relatively long cue–target intervals that were used here. Such a similarity is in accordance with the comparability of the validity effects on RT seen under central and peripheral cuing in the present experiment and with the results of Warner et al. (1990), who found that the attentional processes activated by central and peripheral cues had much in common. Similar attention effects have also been observed in sustained attention paradigms (e.g., Mangun and Hillyard, 1988, 1990) and in a visual search paradigm (Luck, Fan, and Hillyard, 1993). These similarities suggest that common attentional subsystems are shared across these varied paradigms, although there are clearly some differences corresponding to the particular attentional demands of each task.

Despite the overall similarity of the effects of central and peripheral cuing, the absence of a P1 validity effect in the peripheral cue condition may indicate that these two cue types differ in their capacity to invoke the earliest form of attentional selection. This finding is consistent with behavioral evidence indicating that central and peripheral cues may have different effects on some aspects of perceptual processing (e.g., Briand and Klein, 1987; Hawkins et al., 1990). Although this interpretation seems plausible, other factors may also explain the absence of a P1 validity effect in the peripheral cue condition. In particular, the peripheral cue itself may have caused a fatigue or refractoriness in the visual neurons coding the area of visual space stimulated by the cue. On valid trials, the target stimuli would have been processed by many of the same neurons that were presumably fatigued by the preceding cue, and this may have produced a decrement in P1 amplitude for validly cued targets compared to invalidly cued targets. Such a refractoriness-induced decrement may have offset an attention-induced enhancement of P1 amplitude, thus producing the appearance of no difference between the P1 components elicited by validly and invalidly cued targets. Given the 600- to 800-msec delay between cue and target, this refractoriness/cancelation hypothesis is somewhat unlikely, but additional research is required to rule it out conclusively.

Experiment 2: ERPs During the Detection of Faint Signals

Does the advance cuing of attention actually improve the quality of sensory information arising from cued locations, or do the well-documented validity effects on RT result from postsensory factors such as changes in decision or response bias? To investigate this question further, it is preferable to study the effects of attention on measures of perceptual quality *per se* rather than on RT, which is subject to a host of nonperceptual influences. To this end, a number of studies have examined the effects of attention on the detection of faint signals in the framework of signal detection theory, with increases in d' at the attended location being taken as evidence of improved sensory quality. An initial group of experiments obtained conflicting results on the question of whether or not d' was enhanced for threshold-level targets at precued locations (Bashinski and Bacharach, 1980; Muller and Findlay, 1987; Shaw, 1984) but more recent studies employing a "postcue" design have obtained consistent evidence in favor of increased detectability within the focus of attention (Downing, 1988; Hawkins et al., 1990; Müller and Humphreys, 1991).

In the postcue design used by Hawkins et al. (1990), each trial began with a precue (e.g., a central arrow) that directed the subject's attention to one of four alternative stimulus locations. After a brief delay, target information (consisting of either a small, faint luminance target or a target-free blank interval) was presented at one of the locations, followed by masks at all locations. After a 500-msec delay, a postcue occurred at the same location as the target information, directing the subject to make a target present/absent decision at the postcued location. Subjects were advised to attend to the location indicated by the precue because that site was the most likely to contain the target information and hence to be postcued. The postcue method simplifies the calculation of d' values for target detectability at valid and invalid locations, because subjects report the presence or absence of the target for a specific location on each trial. In designs that lack a postcue, it is difficult to know to which location false alarms should be attributed.

Hawkins et al. interpreted their finding of improved target detectability at precued locations as reflecting facilitated sensory processing for attended-location inputs, perhaps achieved by means of a sensory gain control type of mechanism. There are certain postsensory mechanisms that might also account for these results, however. For example, because of the delay that separated the target interval from the postcue, it is possible that subjects preferentially rehearsed the target information at the precued location, resulting in greater accuracy when that location was postcued (i.e., on valid trials). Conversely, decisions regarding target presence/absence on invalid trials would be based on relatively decayed information in sensory memory, resulting in lower d' values.

The current study used a variant of the paradigm described by Hawkins et al. (1990) in which the delay between the target information and the postcue was eliminated to minimize memory decay effects. Each trial began with an initial cue (central arrow) that pointed to the location at which a target presence/absence decision would most probably be required. That is, on 75% of the trials (valid trials) the precued location was also postcued, and on 25% of the trials (invalid trials) the postcue was presented at a different location. The postcue was a bright, 200-msec flash that was presented at one of the four locations and also served as a mask to terminate target processing. The subject's task was to make a decision about target presence/absence only at the location where the postcue/mask appeared. The design also included a "neutral" condition in which the initial cue consisted of four arrows and each location was equally likely to be postcued. Because the mask and postcue were combined into a single stimulus that was presented immediately on target offset, this design allows the decision to begin immediately on both valid and invalid trials, thus ruling out differences in sensitivity caused by differential memory decay.

Although this design rules out differences in detection sensitivity between valid and invalid trials due to memory decay, it does not rule out certain decision-level models of attention (e.g., Sperling, 1984). For example, consider a limited-capacity decision process that receives information from all four locations, but allows the information from the precued location to be more heavily weighted because it is more likely to be used. When the mask/probe arrives, the flow of sensory information into this mechanism terminates, and the decision is then based upon the weighted noise or signal + noise information that has accumulated at each location. If the decision mechanism has some internal noise, then the total signal-to-noise ratio at this stage will be greater for a location that has a higher initial weight, thus leading to a higher d' for valid trials.

This explanation is actually quite similar to the sensory gain control hypothesis favored by Hawkins et al. (1990), except that the gain control operates at the transition between the sensory and decision stages rather than within the sensory processing stage. In fact, these explanations may be indistinguishable, because it is not clear whether this selective biasing of topographically organized visual inputs should be considered a "sensory" or a "decision" process. Because of this ambiguity, the current study recasts the issue into more concrete terms by attempting to specify the latency at which stimulus processing is first modulated by attention and the brain region in which this modulation occurs. Thus, instead of asking whether attention affects processing at the sensory level or the decision level, we ask whether attention affects processing early or late in the visual system, in terms of both latency and neuroanatomic location in the visual pathways.

To examine the latency and brain location of these attention effects, ERPs were recorded from 12 scalp sites in response to the target–mask complex. Because the targets contained negligible stimulus energy, these ERPs were elicited primarily by the mask/postcue stimulus and reflected the state of the sensory pathways at the postcued location as the target (or nontarget) was being discriminated.

Methods

SUBJECTS. The subjects in this experiment were 10 right-handed, neurologically normal college student volunteers between 18 and 27 years old who were paid for their participation. All subjects had normal or corrected-to-normal visual acuity.

STIMULI AND PROCEDURE. Stimuli were presented on a microcomputer-controlled video display at a distance of 100 cm. A white fixation dot and a set of four location markers constructed from red dotted lines were continuously present (see Fig. 5 for stimulus dimensions). Each trial began with a cue stimulus, either a single arrow that pointed from fixation toward one of the four location markers or a set of four arrows that pointed toward all four markers simultaneously. After a delay of 200–500 msec (randomly varied with a rectangular distribution), a white, single-pixel target dot could appear at the center of one of the location markers for 50 msec. This target dot was present on 50% of trials; a 50-msec blank interval was used on the remaining trials. Immediately after this brief target or blank interval, a mask stimulus

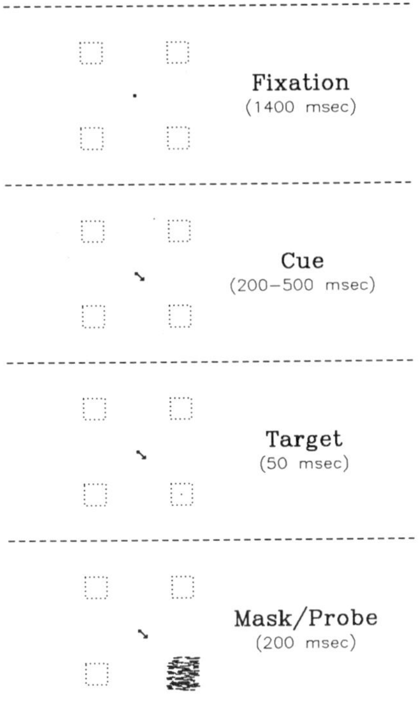

Figure 5.   Sequence and durations of stimulus events for each trial of Experiment 2. Each trial began with fixation interval followed by central arrow cue, brief luminance target (or absence of target), and mask/probe stimulus.

was flashed for 200 msec. The mask consisted of a set of horizontal line segments that were randomly arranged within a 1.7 × 1.7 degree square area, centered over one of the location markers; the mask always contained a line segment that covered the position of the target dot. A 1400-msec intertrial interval followed mask offset.

In addition to terminating the iconic image of the target, the mask was also used as a postcue that instructed the subject to make the target present/absent decision at the location of the mask and to ignore all information from the other locations. The target information always appeared at the masked location. On 80% of trials the cue pointed toward a single location, and on these trials the mask appeared at the cued location with 0.75 probability (valid trials) and at each of the remaining locations with 0.083 probability (invalid trials). On the remaining 20% of trials, the cue pointed toward all four locations, and on these trials the mask (and target, if present) could appear with equal probability at any of the four locations (neutral trials). It should be emphasized that the probability of target presence at the postcued location remained at 0.50 for all these types of trials. Subjects were informed of these probabilities and were instructed to attend to the cued location to maximize their signal detection performance.

Subjects were instructed to respond as accurately as possible and told that speed was irrelevant. They were required to maintain fixation during performance of the task and were aware that their eye movements were being monitored by the experimenter. Before the recording began, subjects were given extensive practice with the task. Practice continued until a stable, intermediate level of performance (d' $\simeq$ 1.5) was reached. Target luminance was periodically adjusted over the course of the experiment to maintain this level of accuracy.

After the practice period, data were recorded over a series of 14 blocks of 120 trials, each lasting approximately 5 min including a 30-sec rest period in the middle. A total of 1344 valid trials, 336 invalid trials, and 336 neutral trials were given over the course of a session. The order of trials and the interval between cue and target varied randomly within and across trial blocks.

RECORDING AND ANALYSIS. The EEG was recorded from electrodes mounted in an elastic cap located at standard scalp positions of the International 10/20 System (F3, F4, C3, C4, P3, P4, 01, 02, T5, T6) and at a nonstandard pair of lateral occipital electrodes halfway between the standard occipital and posterior temporal sites (named OL and OR). These sites and the right mastoid were recorded relative to a left mastoid reference, and the ERP waveforms were algebraically rereferenced to the average of the left and right mastoids. The horizontal EOG was recorded by a pair of electrodes located lateral to each eye, and an electrode beneath the left eye was used to monitor vertical eye movements and eyeblinks. These signals were amplified with a bandpass of 0.01–100 Hz, digitized at 250 Hz, and stored on magnetic tape for later analysis. Trials with eye movements or blinks were automatically excluded from the averages and also from the signal detection analysis.

Due to the short interval between the cue and the mask, the ERP elicited by the cue continued after the onset of the ERP elicited by the mask, and the resulting overlap of these waveforms distorted them both. The ADJAR (adjacent response) filter, described by Woldorff (1993), was used to estimate the overlap between these waveforms and to remove it, thereby providing estimates of the true waveforms. A digital implementation of a single-pole causal high-pass filter with a half-amplitude cutoff of 1.2 Hz was used to remove most of the low-frequency content from the average waveforms before the ADJAR filter was applied.

Three ERP components were measured from the mask-elicited ERP waveforms over the following latency ranges: P1 (60–100 msec), anterior N1 (100–140 msec), and posterior N1 (140–180 msec). The amplitude of each component was quantified as the mean voltage within these windows, relative to the prestimulus baseline voltage. The ERP measures were analyzed in repeated measures ANOVAs, and all *p* values reported here reflect application of the Greenhouse–Geisser correction for nonsphericity. For each component, an omnibus ANOVA was computed using five factors: validity (valid, invalid, or neutral); horizontal mask position (left or right); vertical mask position (upper or lower); electrode location (in the anterior-posterior dimension), and hemisphere (left or right). Whenever the validity factor was significant, further comparisons were conducted comparing valid with invalid, valid with neutral, and invalid with neutral waveforms.

Results and Discussion

SIGNAL DETECTION PERFORMANCE. Target detection sensitivity was quantified using the standard d' measure (Green and Swets, 1966). Target detectability was highest on valid trials ($d' = 1.59$) and lowest on invalid trials ($d' = 1.13$), with neutral trials intermediate ($d' = 1.38$). Both costs and benefits were highly reliable (both, $p < .01$). Decision criteria (beta) did not differ significantly among the three types of trials.

Detection sensitivity was greater in the left visual field than in the right visual field, and greater for the upper locations than for the lower locations, but these effects were independent of cue validity. There were significant main effects of validity ($p < .0001$) and horizontal position ($p < .001$), but none of the interactions reached significance ($p > .20$). These effects were approximately equal in magnitude to the effects reported by Hawkins et al. (1990), suggesting that the presence of a delay between the target and the probe is relatively inconsequential.

ERP WAVEFORMS. The mask-elicited ERP waveforms elicited on valid, invalid, and neutral trials are compared in Figure 6, which displays the waveforms recorded at lateral occipital electrode sites contralateral and ipsilateral to the position of the mask. All stimuli elicited a P1 component, peaking at about 100 msec, and a posterior N1 component, peaking at about 150 msec, both of which were largest at the lateral occipital electrode sites. These components peaked earlier at contralateral scalp sites than at ipsilateral scalp sites, consistent with the direct flow of visual information into the contralateral hemisphere followed by transfer into the ipsilateral hemisphere via the corpus callosum (Mangun et al., 1991). The posterior N1 component was substantially larger at contralateral electrodes than at ipsilateral electrodes, but the P1 component was somewhat larger ipsilaterally, probably because it was partially canceled out at contralateral sites by the larger contralateral N1 component that overlapped it.

As shown in the top row of Figure 6, the posterior P1 and N1 components were larger on valid trials than on invalid trials, replicating the effects of spatial attention that have been observed in previous studies using speeded response tasks (Mangun and Hillyard, 1991). Surprisingly, however, the neutral trials did not produce intermediate amplitudes for these compo-

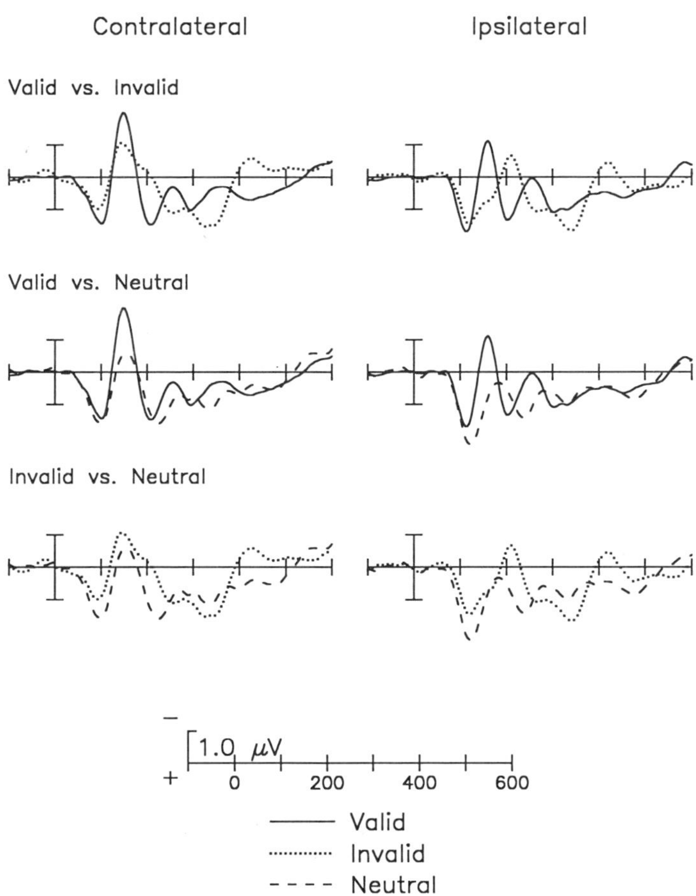

Figure 6.   Grand average ERPs to mask/probe stimulus from 10 subjects in Experiment 2, recorded from lateral occipital (OL and OR) sites. ERPs are collapsed over left and right hemisphere sites and four stimulus positions; left and right columns show recordings from sites contralateral and ipsilateral to visual field of stimulus presentation, respectively. Waveforms associated with valid, invalid, and neutral target positions are overlapped in pairs corresponding to attentional costs and benefits.

nents. Instead, the P1 component was approximately the same for neutral and valid trials (Fig. 6, middle row) and was suppressed on invalid trials (bottom row). The posterior N1 component, in contrast, was larger for valid trials and was suppressed on both invalid and neutral trials. These data suggest the existence of separate mechanisms for attentional "costs," reflected

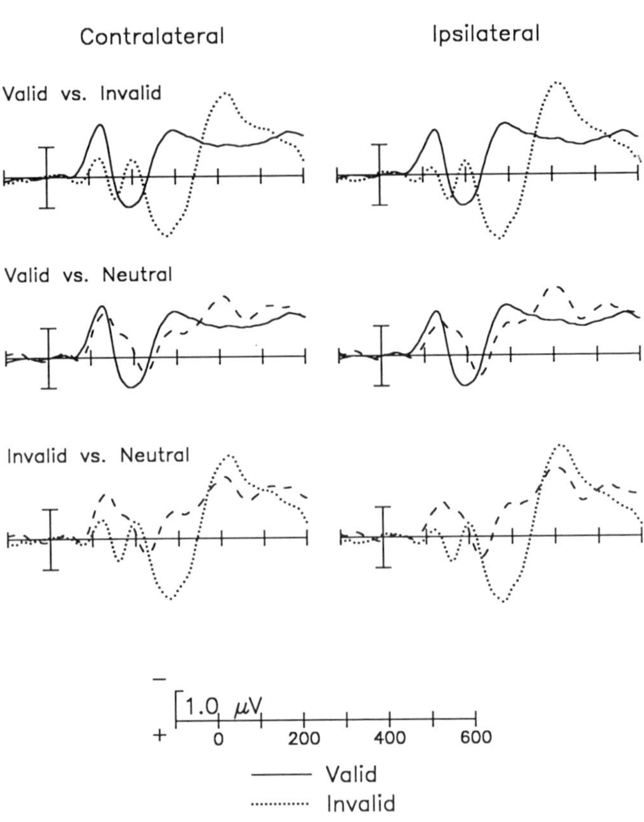

Figure 7.   Grand average ERPs to mask/probe stimulus from 10 subjects in Experiment 2, recorded from frontal (F3/F4) electrode sites. ERP and waveform characteristics as in Figure 6.

in the suppression of the P1 component on invalid trials relative to neutral trials, and attentional "benefits," reflected in the enhancement of the posterior N1 component on valid trials relative to neutral trials.

An anterior N1 component, peaking about 30 msec before the posterior N1, was observed at the frontal and central scalp sites (Fig. 7). This component had approximately equal amplitudes and latencies at the ipsilateral and contralateral sites. Like the P1 wave, the anterior N1 component was suppressed for invalid trials relative to neutral trials, but it was also slightly enhanced for valid trials relative to neutral trials.

The statistical analyses of the validity effects for the P1, anterior N1, and posterior N1 components are summarized in Table 3, which shows

Table 3. Summary of ANOVA significance levels for Experiment 2[a]

Component	Window (msec)	Overall validity effect (p)	Valid versus neutral (p)	Invalid versus neutral (p)
Posterior P1	60–100	.005	n.s.[b]	.01
Anterior N1	100–140	.001	.05	.02
Posterior N1	140–180	.001	.001	n.s.

[a] P1 and posterior N1 components were measured at occipital and temporal electrode sites; anterior N1 component was measured at frontal and central sites.
[b] n.s., not significant.

separate analyses of costs (invalid vs. neutral) and benefits (valid vs. neutral). These analyses support the claim that the P1 reflects mainly costs, the posterior N1 mainly benefits, and the anterior N1 both costs and benefits.

To identify the anatomical sources of these ERP attention effects, topographical maps of source current density (SCD) were created. SCD is

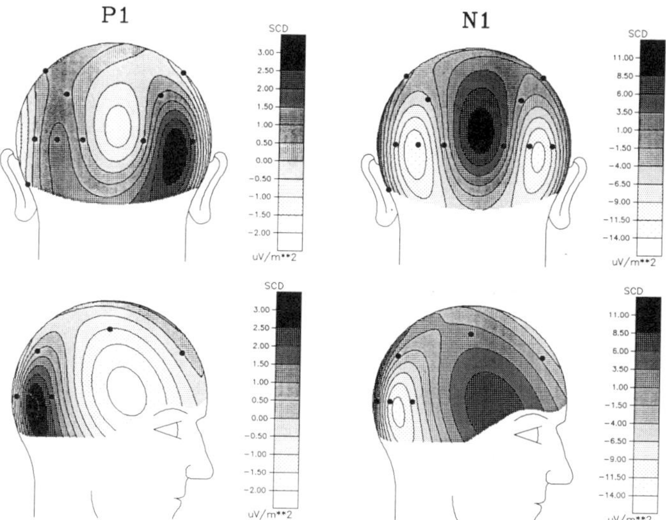

Figure 8. Topographic maps of source current density (SCD) for ERPs in Experiment 2. Left panel corresponds to "attentional costs" for P1 component (measured as mean amplitude between 70 and 100 msec) calculated as difference between neutral and invalid trial ERPs. Right panel corresponds to attentional benefits for N1 component (measured over 140–160 msec), calculated as the difference between valid and neutral trials. Left and right sides of model heads represent ipsilateral and contralateral scalp sites, respectively, with respect to visual field of stimulation. Units of SCD are in microvolts per square meter.

calculated as the second spatial derivative of voltage and yields improved estimates of ERP generator locations by emphasizing superficial cortical generators. The costs and benefits of attention were reflected by the P1 and posterior N1 components, respectively, and SCD maps for these components are shown in Figure 8. Since ERPs were recorded from only 14 scalp sites in this experiment, these maps do not have very high spatial resolution. However, it is clear from these maps that the current flow for the P1 component (shown as dark shading for positive) and the N1 component (shown as light shading for negative) were both maximal over lateral prestriate cortex in the occipital lobe. Thus, spatial attention modulated neural activity in modality-specific visual cortex beginning at approximately 80 msec post-stimulus, strongly suggesting that the increase in luminance sensitivity was associated with facilitated sensory processing.

General Discussion

The present experiments demonstrate that the spatial cuing of attention modulates sensory processing in the posterior visual-cortical pathways in two task situations that have not (to our knowledge) been studied previously using ERP techniques; namely, simple RT with peripheral cues and detection of faint signals with central cues. In both these tasks, stimuli presented to precued (expected) locations elicited enlarged visual evoked potentials in the latency range 80–200 msec. Thus, it appears that the focusing of spatial attention produces amplitude modulations of evoked visual cortical activity, in diverse paradigms, including (1) sustained attention to sequences of unilateral stimuli, (2) sustained attention to sequences of bilateral stimuli, (3) orienting of attention by central or peripheral cues, and (4) visual search for conjunction targets although different patterns of components are affected depending on specific task factors (for reviews, see Harter et al., 1989; Heinze et al., 1990; Luck et al., 1993; Mangun et al., 1993).

The results of Experiment 2 provide the strongest evidence to date that the modulation of short-latency visual cortical activity produced by spatial cuing is specifically associated with actual changes in perceptual processing. The attentional costs and benefits observed in signal detectability (d' values) cannot readily be accounted for by differences in decision criteria, response priming, or memory rehearsal between targets at valid and invalid locations. Thus, the larger P1 and N1 amplitudes evoked by attended-location targets appear to be signs of improved or enhanced sensory processing of inputs occurring in regions of sensory space to which attention has been precued. The fact that these ERP changes appear as amplitude modulations of exogenous components that are also present in unattended (invalid) wave-forms is consistent with a mechanism of sensory "gain control" (Hawkins et al., 1990; Mangun et al., 1993) which is applied selectively to spatio-topically organized sensory pathways.

The current results provide further illustrations of dissociations between the posterior P1 and N1 components under various attentional manipulations (Heinze et al., 1990; Luck et al., 1990; Mangun and Hillyard, 1991). It is clear that these components reflect different aspects of spatial attention, but a clear picture of how to characterize these processes has not yet emerged. The association of P1 suppression with attentional costs in Experiment 2 suggests that this effect may be a sign of an predominately inhibitory process that is applied to inputs coming from unattended locations during focal attention. Similarly, the association of the posterior N1 enhancement with attentional benefits may be a sign of a complementary process that enhances perceptual processing at the attended location but has little direct effect on stimuli presented outside the focus of attention. This fractionation of attention into separate excitatory and inhibitory subcomponent is consistent with behavioral studies showing that attentional costs and benefits can be dissociated (Hawkins et al., 1990). It should be noted, however, that the relationship between the P1 and attentional costs and between the N1 and attentional benefits may depend on the nature of the paradigm used to direct attention (cf. Mangun, this volume; Mangun and Hillyard, 1990).

The scalp distributions of these early attention effects was similar to the distributions observed in previous experiments using both sustained (Mangun and Hillyard, 1990; Mangun et al., 1993) and cued (Mangun and Hillyard, 1991) attention tasks. In these experiments, the augmented P1 activity was associated with a voltage maximum located over the lateral occipital scalp contralateral to the visual field of stimulus presentation. This scalp site overlies prestriate visual cortex, and Mangun et al. (1993) have presented evidence that spatial attention affects visual processing either in the prestriate areas themselves or in the transmission of information from striate to prestriate areas. This accords with evidence from single-cell recordings in monkeys (Colby, 1991; Desimone et al., 1990) showing that spatial attention affects unit responsiveness in prestriate areas of the occipital, temporal, and parietal lobes, but not in striate cortex itself. An important goal for further research will be to identify the anatomic sources of the attention-sensitive human ERP components with the functional divisions of the prestriate visual cortex that have been demarcated in monkey studies.

Returning to the issue of early versus late mechanisms of attention, the present data are in accordance with previous findings (reviewed in Mangun et al., in press) that visual processing is affected by spatial attention as early as 70–80 msec after stimulus delivery (the onset of the P1 wave) and that this most likely occurs in prestriate visual cortex. This narrowing down of the attentional mechanism to early levels of processing both in time and in neuroanatomic location must be taken into account in any comprehensive theory of visual attention.

Acknowledgments. This work has been supported by O.N.R. Contract N00014-89-J-1743 and by grants from N.I.M.H. (MH 25594 and MH 00930), N.I.N.D.S. (NS17778), the Human Frontier Science Program, and the UCSD and UCD McDonnell-Pew Centers for Cognitive Neuroscience.

References

Bashinski HS, Bacharach VR (1980): Enhancement of perceptual sensitivity as the result of selectively attending to spatial locations. *Percept Psychophys* 28:241–248.

Briand HA, Klein RM (1987): Is Posner's "Beam" the same as Treisman's "glue"?: On the relation between visual orienting and feature integration theory. *J Exp Psychol Hum Percept Perform* 13:228–241.

Colby CL (1991): The neuroanatomy and neurophysiology of attention. *J Child Neurol* 6:90–118.

Desimone R, Wessinger M, Thomas L, Schneider W (1990): *Attentional Control of Visual Perception: Cortical and Subcortical Mechanisms.* Cold Spring Harbor Symposium, pp. 963–971. New York: Cold Spring Harbor Press.

Downing CJ (1988): Expectancy and visual-spatial attention: Effects on perceptual quality. *J Exp Psychol Human Percept Perform* 14:188–202.

Duncan J, Humphreys GW (1989): Visual search and stimulus similarity. *Psychol Rev* 96:433–458.

Eason RG (1981): Visual evoked potential correlates of early neural filtering during selective attention. *Bull Psychonom Soc* 18:203–206.

Eriksen CW, Yeh Y (1985): Allocation of attention in the visual field. *J Exp Psychol Hum Percept Perform* 11:583–597.

Green D, Swets J (1966): *Signal detection theory and psychophysics.* New York: Wiley.

Harter MR, Miller SL, Price NJ, LaLonde ME, Keyes AL (1989): Neural processes involved in directing attention. *J Cogn Neurosci* 1:223–237.

Hawkins HL, Shafto MG, Richardson K (1988): Effects of target luminance and cue validity on the latency of visual detection. *Percept Psychophy* 44:484–492.

Hawkins HL, Hillyard SA, Luck SJ, Mouloua M, Downing CJ, Woodward DP (1990): Visual attention modulates signal detectability. *J Exp Psychol Hum Percept Perform* 16:802–811.

Heinze HJ, Luck SJ, Mangun GR, Hillyard SA (1990): Visual event-related potentials index focused attention within bilateral stimulus arrays: I. Evidence for early selection. *Electroencephalogr Clin Neurophysiol* 75:511–527.

Henderson JM (1991): Stimulus discrimination following covert attentional orienting to an exogenous cue. *J Exp Psychol Hum Percept Perform* 17:91–106.

Hillyard SA, Münte TF (1984): Selective attention to color and locational cues: An analysis with event-related brain potentials. *Percept Psychophys* 36:185–198.

Hillyard SA, Picton TW (1987): Electrophysiology of cognition. In: *Handbook of Physiology, Higher Functions of the Nervous System, Section 1: The Nervous System, Vol. V, Higher Functions of the Brain, Part 2*, Plum F, ed., 519–584. Bethesda: American Physiological Society.

Jennings JR, Wood CC (1976): The e-adjustment procedure for repeated-measures analyses of variance. *Psychophysiology* 13:277–278.

Jonides J (1981): Voluntary versus automatic control over the mind's eye's movement. In: *Attention and Performance, IX*, Long JB, Baddeley AD, eds. Hillsdale, New Jersey: Erlbaum.

Lambert A, Spencer E, Mohindra N (1987): Attention by a peripheral display change. *Curr Psychol Res Rev* 6:136–147.

Luck SJ, Fan S, Hillyard SA (1993): Attention-related modulation of sensory-evoked brain activity in a visual search task. *J Cogn Neurosci* 5:188–195.

Luck SJ, Heinze HJ, Mangun GR, Hillyard SA (1990): Visual event-related potentials index focused attention within bilateral stimulus arrays: II. Functional dissociation of P1 and N1 components. *Electroencephalogr Clin Neurophysiol* 75:528–542.

Mangun GR, Hillyard SA (1988): Spatial gradients of visual attention: Behavioral and electrophysiological evidence. *Electroencephalog Clin Neurophysiol* 70:417–428.

Mangun GR, Hillyard SA (1990): Allocation of visual attention to spatial location: Event-related brain potentials and detection performance. *Percept Psychophys* 47:532–550.

Mangun GR, Hillyard SA (1991): Modulation of sensory-evoked brain potentials provide evidence for changes in perceptual processing during visual-spatial priming. *J Exp Psychol Hum Percept Perform* 17:1057–1074.

Mangun GR, Hillyard SA., Luck SJ (1993): Electrocortical substrates of visual selective attention. In: *Attention and Performance XIV*, Meyer D, Kornblum S, eds., 219–243. Cambridge: MIT Press.

Mangun GR, Luck SJ, Gazzaniga MS, Hillyard SA (1991): Electrophysiological measures of interhemispheric transfer of visual information in split brain patients. *Soc Neurosci Abstr* 17:866.

Müller HJ, Findlay JM (1987): Sensitivity and criterion effects in the spatial cuing of visual attention. *Percept Psychophys* 42:383–399.

Müller HJ, Humphreys GW (1991): Luminance increment detection: Capacity-limited or not. *J Exp Psychol Hum Percept Perform* 17:107–124.

Posner MI (1980): Orienting of attention. *Q J Exp Psychol* 32:23–25.

Posner MI, Cohen Y, Rafal RD (1982): Neural systems control of spatial orienting. *Philos Trans R Soc London B Biol Sci* 298:187–198.

Posner MI, Nissen MK, Ogden WC (1978): Attended and unattended processing modes: The role of set for spatial location. In: *Modes of Perceiving and Processing Information*, Pick HL, Saltzman E, eds., pp. 137–157. Hillsdale, New Jersey: Erlbaum.

Prinzmetal W, Presti DE, Posner MI (1986): Does attention affect visual feature integration? *J Exp Psychol Hum Percept Perform* 12:361–369.

Shaw ML (1984): Division of attention among spatial locations: A fundamental difference between detection of letters and detection of luminance increments. In: *Attention and Performance X: Control of Language Processes*, Bouma H, Bouwhis DG, eds., pp. 109–121. Hillsdale, New Jersey: Erlbaum.

Sperling G (1984): A unified theory of attention and signal detection. In: *Varieties of Attention*, Parasuraman R, Davies DR, eds., pp. 103–181. London: Academic Press.

Warner CB, Juola JF, Koshino H (1990): Voluntary allocation versus automatic capture of visual attention. *Percept Psychophys* 48:243–251.

Woldorff M (1993): Distortion of ERP averages due to overlap from temporally adjacent ERPS: Analysis and correction. *Psychophysiology* 30:98–119.

Chapter 2

Selective Visual Attention: Selective Cuing, Selective Cognitive Processing, and Selective Response Processing

G. MULDER, A.A. WIJERS, K.A. BROOKHUIS, H.G.O.M. SMID, AND L.J.M. MULDER

Selective attention systems can be characterized as a collection of brain mechanisms. The function of these mechanisms is the modulation of the impact of external and internal environmental stimuli on overt or covert (e.g., cognitive) behavior of the organism. Many theories of attention assume that the basic function of attentional mechanisms is to protect the brain's limited capacity system from informational overload.

This concept is the basis of the **filter theory** (Broadbent, 1970). Broadbent remarked: "The obvious utility of a selection system is to produce an economy in mechanism. If a complete analysis were performed even on neglected messages, there seems no reason for selection at all." Much research effort has focused on the question where in the system capacity is limited. As a result, many studies of attention aim at the question at what level of processing selection takes place. In the original filter theory, selection must occur before the stage of recognition or categorization. Other theories assume a bottleneck at a late stage and specifically after the level of semantic categorization (Deutsch and Deutsch, 1963). The first conception assumes **early** and the second, **late**, selection.

Kahneman and Treisman (1984) have argued that the long-running controversy over so-called early versus late selection partly results from the use of different paradigms. Two main paradigms can be distinguished: the **filtering** and the **selective-set** paradigm (Kahneman and Treisman, 1984; see

Cognitive Electrophysiology
H-J. Heinze, T.F. Münte, and G.R. Mangun, editors
© 1994 Birkhäuser Boston

Table 1.

Characteristics	Filtering paradigm	Selective-set paradigm
Designs	Selective listening	Search
	Partial report	Priming
Modality	Auditory or visual	Visual
Vocabulary of stimuli	Large	Small
Response choice	Large	Small
Measure	Accuracy	Reaction time (mental chronometry)
Null hypothesis	Perfect early selectivity	Full automaticity
Standard interpretation	Attention prevents or reduces perceptual processing of unattended stimuli	Attention selects and speeds responses to expected stimuli
ERP correlates	Auditory: onset of Nd sensory specific early phase (Nde) later phase (Ndl) P3b on attended targets Visual: enhancements of exogenous components; slow negative shifts	P3b latency and amplitude as a function of display and/or memory load

[a] Adapted from Kahneman and Treisman, 1984.

also Table 1). The first paradigm favors an early selection view of attention (e.g., Broadbent, 1958); the second favors late selection (Deutsch and Deutsch, 1963).

In filtering paradigms, the subject is simultaneously (or sometimes successively) exposed to relevant and irrelevant stimuli. Relevant stimuli in this paradigm control response selection and execution. The property (the selective cue) that distinguishes the relevant from the irrelevant stimuli is usually a simple physical feature. Classical examples are the selective shadowing task invented by Cherry (1953) and the partial report technique introduced by Sperling (1960).

In the selective-set paradigm, the subject's task is to show by a speeded response the detection or recognition of a particular stimulus. Classical examples are the visual or memory search tasks developed by Sternberg (1969) and Shiffrin and Schneider (1977), respectively. Other examples are the studies of Posner (1978) on the costs and benefits of expectations. The response vocabulary in search studies is small, including only "yes" or "no" key presses or sometimes only a "yes" response.

Researchers in cognitive psychophysiology use only one of these two generic paradigms. In the filtering paradigm the onset and extent of selection ("processing" negativity) is the main dependent variable. In the selective-set paradigm, P3b latency and amplitude are usually the variables of interest. In this chapter we mainly review the work on selective visual attention and selective response processing carried out in our laboratory in the period between 1980 and 1991. In this way, we hoped to be able to obtain more information about the nature of "processing negativity" (Näätänen and Michie, 1979). Is processing negativity associated with early selection between relevant and irrelevant information or does it reflect further processing in the attended channel? By definition processing within the attended channel is under attentional control, that is, it is a form of controlled processing (Schneider and Shiffrin, 1977).

In the study of selective attention, a distinction should be made between selective cuing and selective processing. Selective cuing refers to the process (or processes) by which task-relevant information is marked out, by some selection cue, for control of a given response. These two questions are logically independent. One might argue that early selection takes place simply because physical properties can be used as a basis of selection. However, the basis for selecting an item says nothing about the level of processing of that item before selection. It might be that the item has already been identified before it was selected. Strong late selection theories assert that all stimulus items, even unattended ones, are identified before selection. Strong early selection theories hold that only elementary physical properties are extracted preattentively; identification occurs only after attentional selection. Yantis and Johnston (1990) therefore have argued that the approach that has the greatest promise examines the fate of unattended items in a focused-attention task. The combination of filtering and selective-set tasks offers this possibility.

The data obtained in several ERP experiments in our laboratory strongly suggest to us the existence of two different systems: a stimulus evaluation system, and a response activation system. These systems are relatively independent. The structure of this chapter is as follows. The first five sections describe the characteristics of the stimulus evaluation system. The last section deals with the nature of the communication between both systems.

We first review briefly the main findings with the filtering paradigm. In particular, our discussion focuses on the modulation of exogenous components (P1/N1) and their equivalent dipoles and on "processing negativities." We next review our work using the selective-set paradigm. The P3b complex is the main dependent variable. A new paradigm is then introduced: the selective-set **and** filtering paradigm. We next discuss the results obtained with this paradigm and also compare them with those discussed previously. The main question is whether the effects of selective cuing also can be observed in tasks in which the subject has to perform quite complicated cognitive tasks. We conclude with the effects of selective cuing.

We shall try to establish whether there are different types of processing negativities associated with different cognitive processes, with the main emphasis on selective processing. Finally, we shift to processes involved in response control. Is it possible that processes involved in motor preparation and selection use partial information, or do they require full analysis of the input information? Clearly this question again concerns selective processing but now, more specifically, selective response processing.

Most of the experimental data in this chapter have already been published; only data not yet published are extensively discussed.

Selective Cuing in the Filtering Paradigm

Visual Attention: Spatial and Nonspatial Effects

A very robust finding is that attention to a spatial location and attention to nonspatial stimulus attributes are associated with qualitatively different patterns of ERP effects. The current view is that whereas nonspatial attention results in a slow endogenous negativity (much like "processing negativities" in the auditory and somatosensory modalities); spatial attention results in an enhancement of a series of positive and negative deflections. The enhancement is most prominant at posterior electrodes (Hillyard and Mangun, 1986; Näätänen, 1986). The standard interpretation is that spatial attention might involve a modulation of perceptual processing whereas nonspatial attention involves a postperceptual selective process (Näätänen, 1986).

Research on spatial attention has typically used tasks in which stimuli were randomly presented right or left of a fixation point at the horizontal meridian. Usually there is a substantial separation between the point of fixation and the stimulus locations (at least 4 degrees but usually much more). The main finding is an enhancement of the posterior P1 (peaking between 100–160 msec) and the N1 (160–210 msec) components (Eason, 1981; Eason, Harter, and White, 1969; Eason, Oakley, and Flowers, 1983; Hillyard and Münte, 1984; Hillyard, Münte, and Neville, 1985; Hillyard et al., 1984; Mangun and Hillyard, 1987, 1988; Mangun, Hansen, and Hillyard, 1986; Neville and Lawson, 1987; Rugg et al., 1987; Van Voorhis and Hillyard, 1977). However, in several experiments the effect on the P1 component was not reported or failed to reach significance (Eason et al., 1983; Mangun et al., 1986; Van Voorhis, and Hillyard, 1977). Harter, Aine, and Schroeder (1982) obtained an early positivity, but this effect was only significant at central electrodes. It is not clear what factors are favorable for the P1 effect to occur, although Eason (1981) and Eason et al. (1983) suggested that larger effects are obtained when stimuli are less salient.

Spatial attention also influences the ERPs in latency ranges later than the N1 component. However, a variety of patterns of results has been obtained. In some experiments the main finding was that spatial attention also enhanced the amplitudes of the occipital P2 and N2 waves (Eason, 1981; Eason et al., 1969, 1983; Hillyard and Münte, 1984; Hillyard et al., 1984; Mangun and Hillyard, 1987), but in other experiments slow negativities (Harter et al., 1982; Hillyard et al., 1985; Rugg et al., 1987; Wijers et al.,

30   G. Mulder et al.

1989d) and positivities (Neville and Lawson, 1987) have been found. Therefore, these later effects may be more dependent on the nature of the task
(Hillyard et al., 1985).

In research on nonspatial attention the following selection cues have
been investigated: color (Harter and Salmon, 1972; Harter et al., 1982;
Hillyard and Münte, 1984; spatial frequency (checkerboard patterns) (Harter
and Previc, 1978; Kenemans, 1993; Previc and Harter, 1982), orientation
(horizontal versus vertical bars) (Harter and Guido, 1980; Kenemans, Kok,
and Smulders, 1993; Previc and Harter, 1982; Rugg et al., 1987), and contour
(horizontal and vertical bars versus diffuse flashes) (Harter and Guido, 1980).
All these nonspatial selections are associated with negativities in the ERP
with onset latencies of 150 msec or later. In all these cases "processing
negativities," probably endogenous, are present with maximum values at the
occipital derivation (Harter and Aine, 1984; Hillyard and Mangun, 1986;
Näätänen, 1986).

Levels of Selection in Spatial Attention

A filtering paradigm (Wijers et al., 1989d) was designed to investigate the
distribution of attention over visual space. Colored bars (blue or red) were
presented randomly at one of eight different spatial locations. These locations
were arranged as a half-circle around the fixation point (Fig. 1). Subjects
attended to conjunctions of spatial and color cues. In the focused-attention
condition, the subjects were required to respond to stimuli in the target color

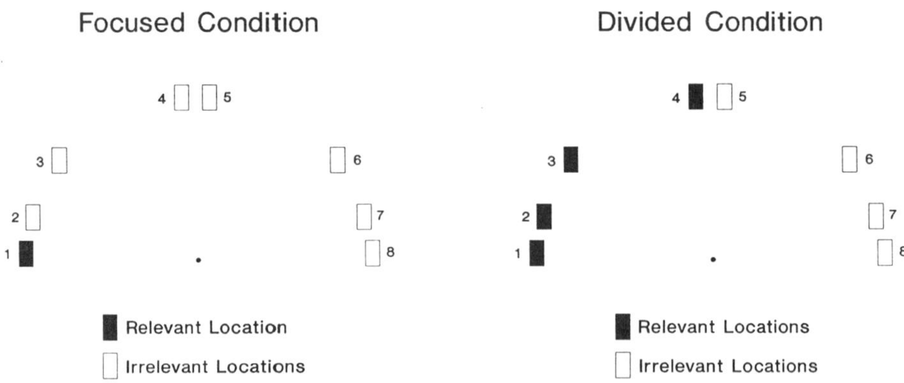

Figure 1.   Subjects fixated the central dot while stimulus bars were presented one
at a time at one indicated position (1–8). Separation between fixation and the stimulus
bars was 3.1° for all stimulus positions. Bars presented at locations 1 and 2 were
separated by 0.15° (shortest distance between the bars; distance between the most
distant corners was 1°). Locations 1 and 3 were separated by 1.23° (most distant
corners 2.1°), 1 and 4 by 3.83° (4.92°), and 4 and 5 by 0.17° (0.87°) (Wijers et al., 1989d).

(either red or blue in different blocks) presented to the most lateral location (location 1 in Fig. 1 is the relevant location). In a divided-attention condition they responded to target bars at all four locations within a visual half-field (locations 1–4 in Fig. 1 are all relevant). In half the blocks the right location(s) was (were) relevant, and in half the blocks the left location(s).

The effect of spatial attention was assessed by comparing the ERP evoked by a stimulus at a particular location when attention was directed at the same visual half-field with the ERP evoked by the same stimulus when attention was directed at the opposite visual half-field.

The half-circle arrangement of stimulus positions was chosen to keep the stimulus–fixation distance equal for all positions. It seemed possible that more attention is allocated to stimuli presented closer to fixation. In this experiment we did not want to consider this question because this requires a more elaborated experiment in which each of the locations is attended in turn.

The rationale for the chosen stimulus positions was as follows: In the focused-attention condition, only information from location 1 was needed to perform the task. If subjects are able to restrict the focus of attention to this location, ERP effects of attention should only be confined to location 1. If attention is a spotlight that is narrowly constricted to an area of approximately 1 degree (Hoffman and Nelson, 1981), locations 1 and 2, which lie within a 1 degree area, would both show a spatial attention effect, but not the other locations. If there is a gradient of attention over a larger visual area (e.g., 5 degrees as estimated by Mangun and Hillyard, 1987), all locations within the same visual half-field as the attended location should exhibit ERP effects of attention, but these effects should be smaller the further away from location 1. Finally, if spatial attention acts by selecting an entire visual half-field (Hughes and Zimba, 1985), stimuli presented at all four locations within the visual half-field containing the relevant location (locations 1–4) should elicit comparable ERP effects and the focused- and divided-attention conditions should not differ. We shall discuss only the results obtained in the focused condition.

In the focused-attention condition, spatial attention resulted in early positivity in the P1 latency range (100–175 msec), followed by a prolonged negativity in the N1, P2, and N2 latency range (175–350 msec) (Fig. 2). These effects generalized to locations in the same visual half-field as the relevant location. The effect of attending and responding to the target color (only present at two adjacent locations) consisted of many different effects. An early anterior positivity, occipital negativity was observed for the relevant location and for locations in the same visual half-field as the relevant location, but not for the locations in the opposite visual field. A later central negativity (N2b) was confined to the relevant location and one location next to it. Finally, a late parietal positivity (P3b) was exclusively evoked by target stimuli at the relevant location (Fig. 3).

To summarize, this study suggested that within the stimulus evaluation

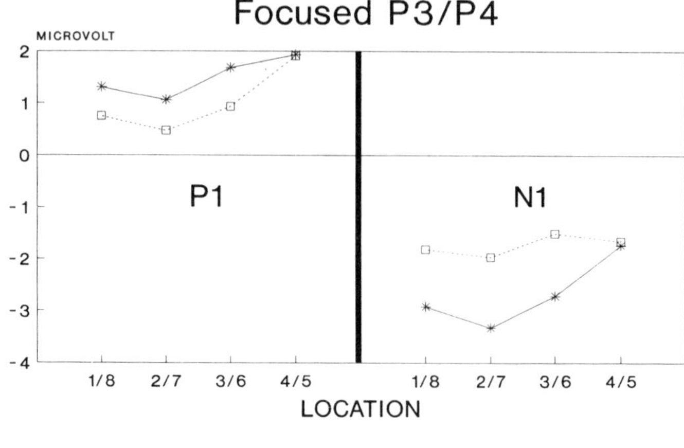

Figure 2. Focused attention conditions. Amplitudes of P1 and N1 at parietal electrodes for each of different spatial conditions, separately for stimuli at relevant location (relevant field, location 1) and locations in same visual half-field as relevant location (relevant field, location 2–4) and for stimuli in the opposite half-field (irrelevant field, locations 5–8). Values are averaged over attend-right and attend-left conditions and over ipsilateral and contralateral electrodes (Wijers et al., 1989d).

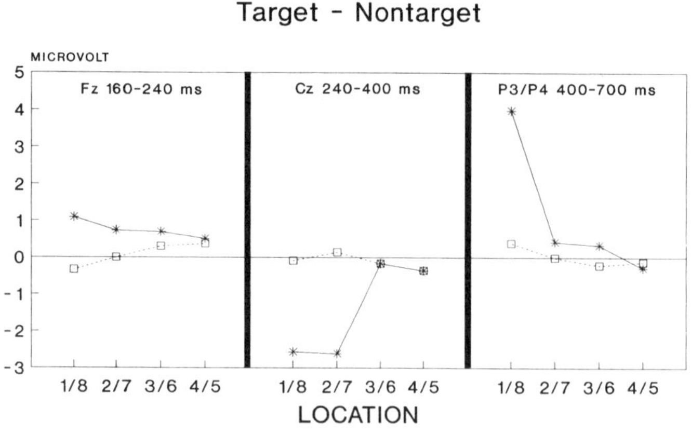

Figure 3. Focused-attention condition. Color selection effects for stimuli presented at relevant location (relevant field, location 1) or at locations in same visual half-field as relevant location (relevant field, location 2–4; see Fig. 1) and for stimuli in the opposite half-field (irrelevant field, locations 5–8). Depicted values were obtained by subtracting mean amplitudes in specified latency ranges for irrelevant color stimuli from mean amplitudes for relevant color stimuli. These values were averaged over attend-right and attend-left conditions and also over attend-blue and attend-red conditions (Wijers et al., 1989d).

system different levels of selection exist: the early stages (indexed by P1, N1, N2) are less selective, and the late stages (indexed by N2b and P3b) are highly selective. We assume that the stimulus evaluation and the response activation system are independent but can communicate. Consequently, it is possible that a lack of selectivity in the initial stages of processing may cause, because of early communication, a similar lack in selective response processing. With early communication we mean that elementary characteristics of a stimulus (e.g., location, color, about 160–240 msec) may be already communicated to the response activation system while identification of the stimulus is not yet complete. When the evaluation system has obtained more selective, detailed, and complete information (about 240–400 msec or later), the response activation system may use this additional information in either inhibiting or in facilitating responses already partly started. Later we discuss the results of some experiments that directly test the possibility of early communication.

Cerebral Localization of Spatial Attention Mechanisms

In a filtering paradigm, we measured the magnetic fields associated with P1 and N1 (Wijers et al., 1992). Five paid subjects received series of stimuli consisting of 192–252 white-on-black checkerboard patterns (consisting of five blocks) presented randomly to the right or the left of a fixation dot on the horizontal meridian. One-sixth of the left-side and one-sixth of the right-side displays were target stimuli, containing an extra dot in the otherwise black part of the checkerboard. The size of the checkerboards was 7.5×5.5 cm, and the distance between the checkerboards and the fixation dot was 2 cm. Subjects were instructed to maintain fixation at the fixation dot while attending to one visual field and counting the number of targets presented in that field. For each measurement position of the magnetometer, subjects received two stimulus series in which they respectively attended the right and left visual fields. The stimuli were presented on a liquid crystal display that was viewed from a distance of about 1.5 m.

The checkerboard patterns were presented for a duration of 300 msec. The interval between stimulus offset and the onset of the next stimulus was randomly chosen as 500, 600, or 700 msec. The (MEG) measurements were carried out in a magnetically shielded room with a three-channel RF-SQUID magnetometer. Each channel had a first-order gradiometer with a baseline of 5 cm and a coil diameter of 4 cm. The distance from the pickup coil to the scalp was about 2.5 cm. The outputs of the squid electronics were bandpassed between 0.1 and 100 Hz and sampled at a rate of 1000/sec. The MEG was recorded at 57 different locations over the posterior and temporal regions of the head. Electrical responses were recorded from five occipital electrodes. Data are discussed for subject A.A.W. For equivalent dipole fitting, we used a program developed by de Munck (1989) and based on a spatiotemporal dipole model. This program localized the activity pattern of a fixed dipole in the 140- to 190-msec (left visual field stimuli) and 150- to

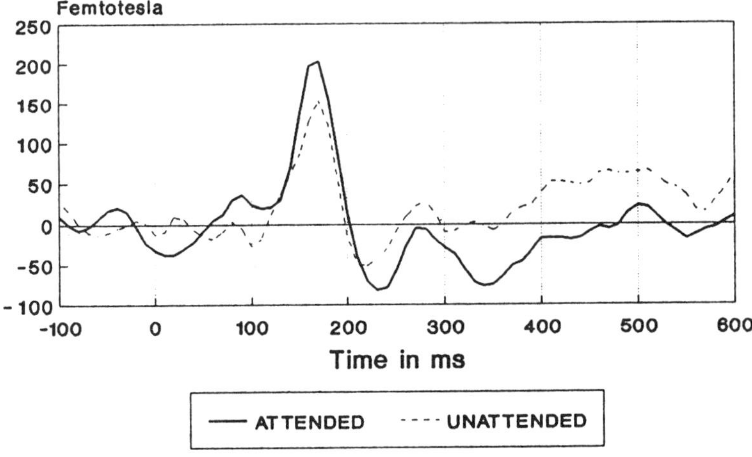

Figure 4.   Magnetic fields associated with attended and unattended visual stimuli (Wijers et al., 1992).

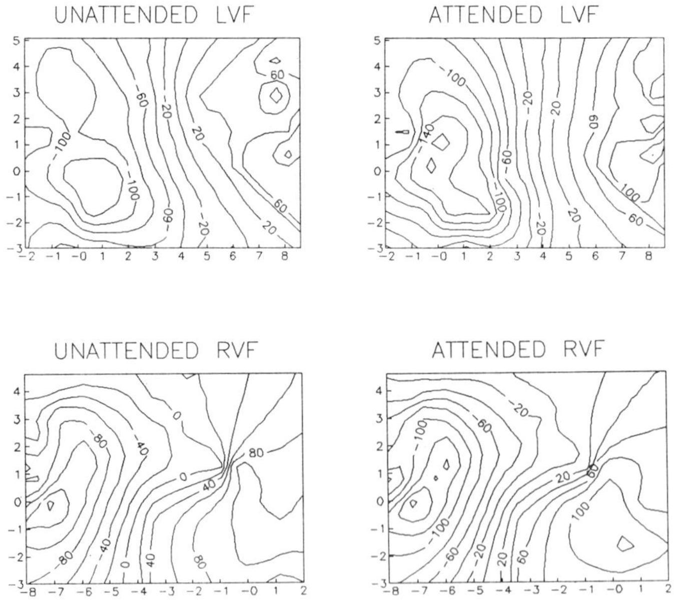

Figure 5.   Magnetic field distributions associated with attended and unattended visual stimuli (Wijers et al., 1992).

200-msec (right visual field stimuli) poststimulus latency ranges. In addition, a single dipole model was fitted to neuromagnetic distributions for individual sample points at 10 msec intervals. A typical time series of magnetic brain responses shows deflections at 170 and 220 msec (Fig. 4).

The electrical responses showed a similar waveform, with a positivity at 170 msec (P1) and a negativity at 220 msec (N1). The latencies of these peaks were later than usual, which is attributed to the slow rise time of the LCD screen. For stimuli presented to the right visual field, the P170-m field direction was positive over the posterior head regions and negative over the left temporal regions. For stimuli presented to the left visual field, an opposite pattern of results was observed (negative at posterior locations, positive at right temporal locations). Similar waveforms were obtained for attended and unattended stimuli, but P170-m amplitude was enhanced for attended stimuli. Figure 5 shows that similar dipolar field maps at 170 msec were obtained for attended and unattended stimuli.

Table 2.

	Latency	X(cm)[a]	Y(cm)	Z(cm)	Q	Fi	Fit (%)
UA LVF[a]	140–190	4.9	2.85	1.17	3347	−.97	83.7
	160	4.14	2.59	2.32	3550	−18.7	81.4
	170	4.37	2.74	1.92	4840	−15.2	89.6
	180	4.9	2.92	1.22	4590	−5.3	91.5
A LVF	140–190	4.33	2.37	1.26	5563	−7.4	83.1
	160	4.41	2.96	2.74	3780	−38.6	79.0
	170	4.05	2.71	2.33	6410	−29.9	85.8
	180	4.0	2.44	1.68	8000	−15.9	90.6
UA RVF	150–200	6.13	−3.27	−.57	2035	−19.3	82.6
	170	5.56	−2.97	−.8	1630	−8.5	73.9
	180	5.12	−2.92	−.97	3180	−16.0	85.0
	190	5.56	−2.98	−.6	3460	−19.3	88.5
A RVF	150–200	5.88	−3.49	−.74	2335	−26.2	84.5
	170	5.72	−3.20	−1.14	2120	−33.7	73.7
	180	5.35	−3.28	−1.2	3500	−32.5	84.3
	190	5.41	−3.32	−.8	3820	−28.1	91.7

[a] Abbreviations: LVF, left visual field; RVF, right visual field; UA, unattended; A, attended; X-axis, from nasion to point about 2 cm above Oz (− is in direction of nasion); Y-axis, between two points 6 cm posterior and 3 cm above the left and right preauricular points (− is to the left); Q, dipole strength in arbitrary units, Fi, orientation.

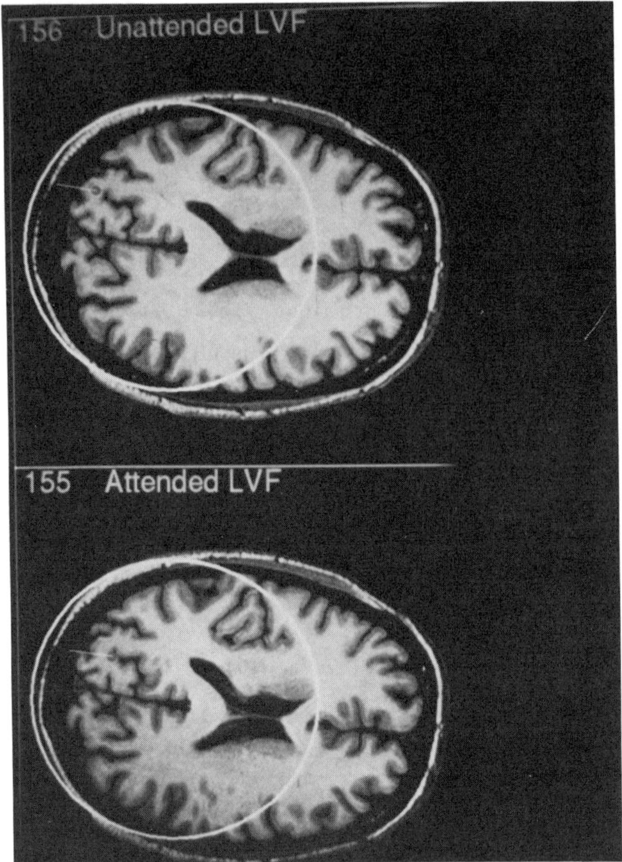

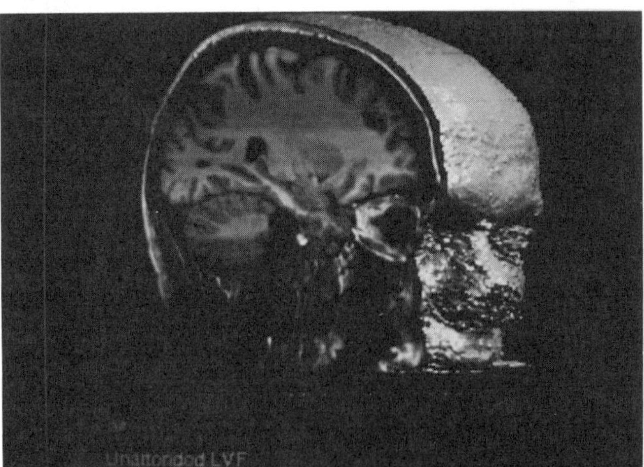

Figure 6.   Localization of equivalent dipole associated with attended and unattended visual stimuli in MRI scan (subject AAW).

Table 2 summarizes the results of dipole modeling. Equivalent dipole sources that were closely similar were computed for the attended and unattended stimuli. In both cases a dipole was localized in the lateral posterior cortex, contralateral to the visual field of stimulus presentation (see also Fig. 6).

Spatiotemporal dipole modeling and modeling of individual sample points yielded comparable results. The dipole parameters remained reasonably constant for several consecutive sample points. Dipole strength was consistently larger for the attended than for the unattended stimuli. For the N220-m, a less consistent pattern of results was obtained. These preliminary results support a "sensory gain" hypothesis of spatial attention. Although the amplitude of the P170-m was enhanced by spatial attention, the equivalent dipole underlying this component was very similar for attended and unattended stimuli. The lateral localization of the dipole is well according to the hypothesis of Mangun et al. (1993, and this volume). These authors suggested that spatial attention modulates the activity of secondary visual brain areas belonging to the ventral projection route from the striate cortex to the inferior temporal cortex. This route is specialized for object recognition.

The Selective-Set Paradigm

The selective-set paradigm has been extensively discussed by Sternberg (1969), Shiffrin and Schneider (1977), and Schneider and Shiffrin (1977). The selective-set paradigm is an example of a binary classification task in which a decision rule partitions a set of stimuli into two exhaustive and mutually exclusive classes or categories. One category is called the positive set. The subject is first required to memorize this set of stimuli, and later, on presentation of a test stimulus, to decide whether it does or does not belong to the memorized positive set. Sternberg (1969) assumed that at least four stages of processing intervene between the presentation of the test stimulus and the overt response. The first stage is stimulus encoding (affected by stimulus quality). The second stage is serial comparison (affected by the size of the positive set). The third stage is binary decision (affected by the nature of the decision, the mean duration being greater for negative than positive decisions); response translation. Finally, the last stage is organization (affected by the relative frequency of response type).

Shiffrin and Schneider extended this paradigm by presenting more than one test stimulus. In this version of the selective set paradigm a display is shown in which the characters are arranged in a square around a central fixation point. The characters consist of either digits, consonants, or random dot masks. If all characters consist of either digits, consonants, or a mixture, display size is four. When display size is less than four, all noncharacter positions are filled with random dot masks. Analysis of response times in these conditions strongly suggested that the subject first chooses a display item and compares it to every memory item in turn. The subject may then choose another display item and continue in the same fashion. Thus, the number of display or memory items determines the number of comparisons the subject has to perform. If load (the product of display size and memory set size) is high, the subject changes from exhaustive to self-terminating search. In both cases the subject uses controlled search, an effortful mode of processing, highly affected by the load of the task. Controlled search is characteristic for *varied mapping* conditions. Varied mapping implies that the positive set items in one trial are negatives in some other trials and vice versa. During *consistent mapping* across trials, the positive set items never occur in the visual display except as targets and the negative set items (called distractors) never become element of the positive set. In consistent mapping conditions automatic detection gradually develops. If automatic detection occurs, performance becomes virtually independent of load, because the subject no longer uses exhaustive or self-terminating search.

We hypothesized that the latency and the amplitude of the P3b component of the ERP is mainly affected by the time required for stimulus evaluation and relatively unaffected by response selection and execution. Consequently, all task variables affecting encoding, serial comparison and binary decision should affect its latency. We developed a method to find the latency of P3b on single trials (Mulder et al., 1984). In addition, we hypothesized that P3b latency is less affected by emphasis on speed, if speed instructions mainly affect response-related processing.

We consistently found in different studies (Brookhuis et al., 1981, 1983; Van Dellen, 1985; see also Brookhuis, 1989) that an increase in display size or memory size increased the latency of P3b. Negative responses were associated with greater P3b latencies.

The distribution of single trial P3b latencies suggested exhaustive search even in conditions in which the behavioral data indicated self-terminating search. An increase in latency of P3b was always accompanied by a decrease in amplitude. However, P3b amplitude was more sensitive than P3b latency. Consistent mapping increased P3b amplitude. If negative responses are more probable than positive responses, the latencies of positive responses increases over those of negative responses, suggesting that relative response frequency also affects the binary decision stage.

All these data strongly suggest that the neurophysiological process producing P3b is contingent on serial comparison and binary decision. The effects of stimulus degradation, a task variable affecting stimulus encoding, and the effects of speed versus accuracy instructions, however, are more complicated and therefore deserve more attention. It was very difficult to

EXPERIMENT 1

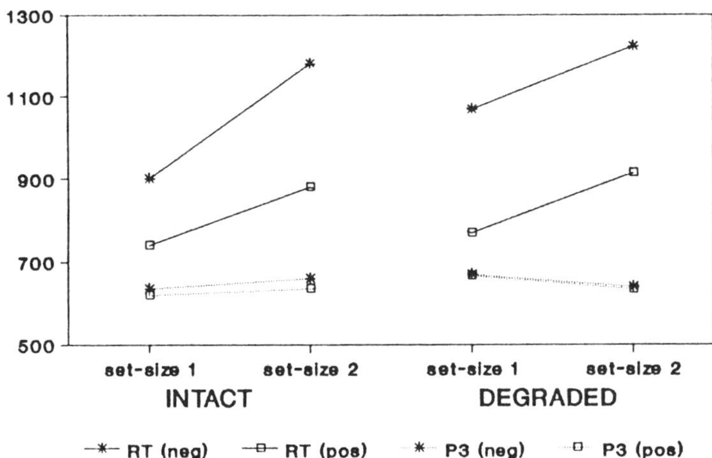

Figure 7. Effects of stimulus degradation and set size on reaction time, shown separately for positive and negative responses (Brookhuis, 1989).

affect stimulus encoding independently from the other processes. We found an interaction of degradation and load on RT, suggesting that when the stimulus was degraded a higher display load did not produce as large an increase in reaction time as it did in the intact condition (Fig. 7) (Brookhuis, 1989). Moreover, when the stimuli were degraded, subjects committed more errors in the higher load conditions. This observation suggests that subjects after a fixed period responded on insufficient information.

Grand averages for the intact and the degraded condition, before the latency adjustment procedure, are shown in Figure 8. A small difference is visible in the amplitude of the P3 complex (the broad positivity ranging from about 300 to 800 msec after the presentation of the imperative signal (at 0 msec). Analysis of variance showed that when stimuli were degraded, latency-corrected P3b amplitudes were lower than in the intact condition. In the degraded conditions the late CNV was larger, suggesting a change in prestimulus response preparation and a readiness to respond faster, but apparently at the cost of more errors.

Manipulating the response speed–accuracy trade-off has a similar prestimulus effect (see Fig. 9). In the conditions emphasizing speed, the late CNV (Contingent Negative Variation) is more pronounced.

Response choice and execution processes after the presentation of the imperative stimulus are accelerated during speed instructions, but again at

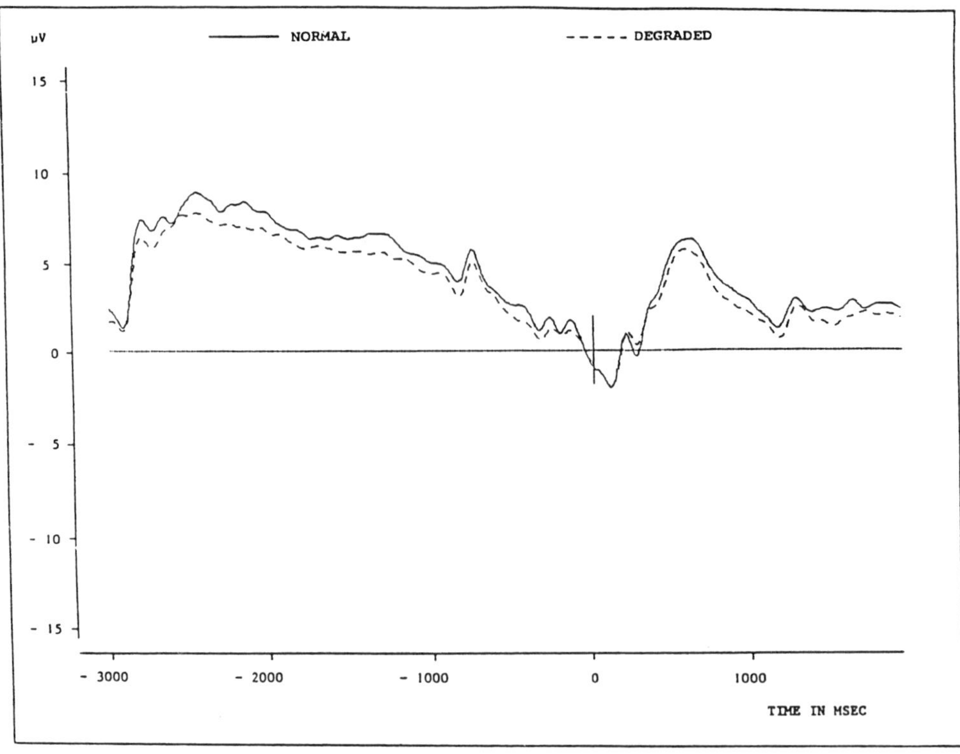

Figure 8. Grand averages obtained in intact and degraded stimulus conditions in visual search task (Brookhuis, 1989).

the expense of accuracy. The P3b component is delayed in the speed condition as compared to the accuracy condition, suggesting that the error information requires additional processing (Fig. 10). All these data suggested to us the existence of at least two independent systems: one (the stimulus evaluation system) in which the stimulus is encoded and compared to internal representations and in which a final decision is made after an exhaustive comparison of the stimulus features with the features of the internal representation. The end of the stimulus evaluation process is indexed by P3b. The second system (the response activation system) is responsible for specific or unspecific response preparation, response selection, and execution. This system can start with preparation before the test stimulus is presented and probably uses incomplete information. "Decisions" in this system are probably available to the first system and increase stimulus evaluation time. This system is more fully discussed later. Task variables such as display load, memory load, response type, probability of response type, stimulus quality etc. predominantly affect both systems. RT signals the end of the processing time in the second system. Unfortunately, from RT

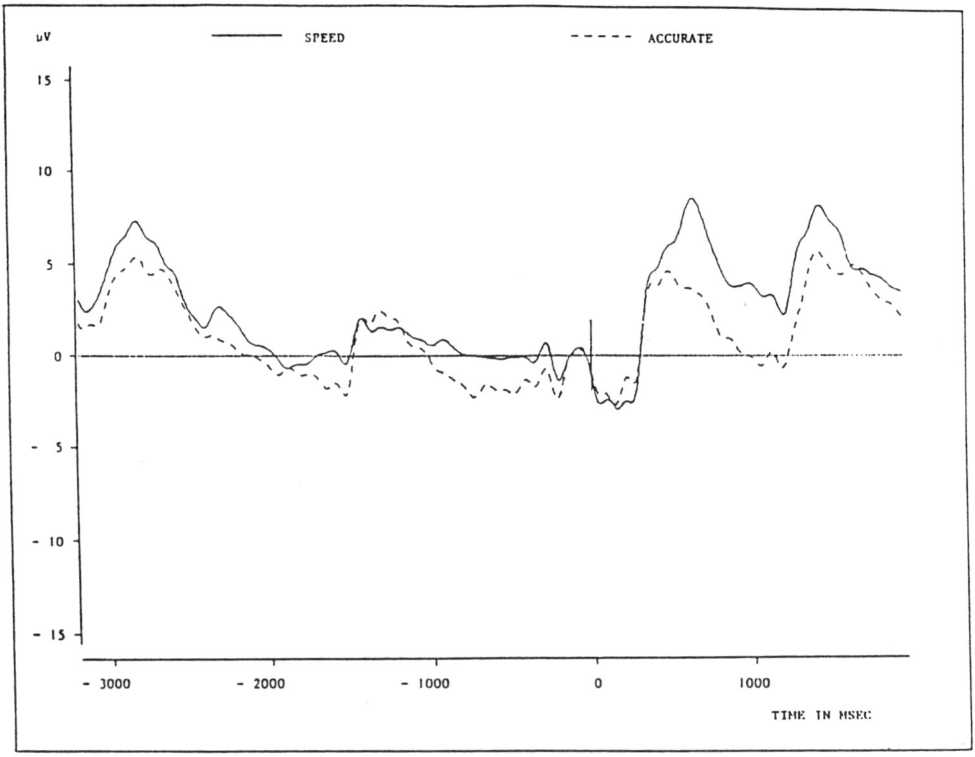

Figure 9. Effects of speed and accuracy instructions on reaction time in visual search task, shown separately for set size and positive and negative responses (Brookhuis, 1989).

EXPERIMENT 2

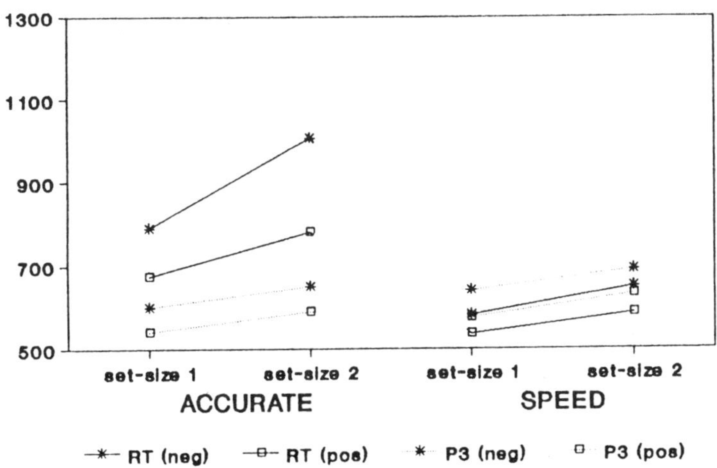

Figure 10. Grand averages obtained in visual search task, shown separately for speed and accuracy instructions (Brookhuis, 1989).

measurements it is not possible to obtain pure estimates of the time required for perceptual and cognitive operations. In addition, speed–accuracy trade-offs predominantly affect the second system. Speed instructions, a speed strategy, or accuracy instructions determine the moment at which the second system starts response selection, preparation, and execution.

A large proportion of RT can be explained by the time required for these processes, and the measure is also highly dependent on the response strategies (Mulder et al., 1984). In the following sections, we further explore the stimulus evaluation system and then briefly discuss recent work on the response activation system.

The Selective Set and Filtering Paradigm

Until now we have discussed results obtained by using either filtering tasks or by using selective set tasks. Here, we explore the fate of unattended information by requiring the subject to further process attended information. First, we attempt to show that electrical brain activity is systematically associated with mental operations such as memory search and rotation. Second, we try to show that this activity is only present in the attended

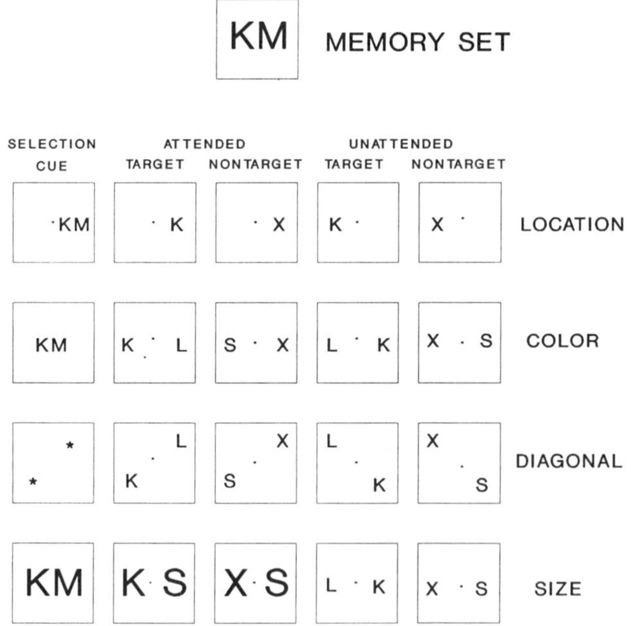

Figure 11.   Stimulus presentation in selective search paradigms with different selection cues.

channel, we have strong evidence for selective processing. Figure 11 shows some examples of these tasks.

In this task, subjects memorize several target letters (the memory set). After that, a random series of two categories of stimulus letters is presented. The categories differ on the basis of simple physical attributes, selective cues (e.g., spatial location, color, size, diagonal of display). Subjects are required to attend to a stimulus category and to detect the occurrence of target letters within the attended category (e.g., the subject should detect target letters presented on the relevant diagonal and ignore the letters presented on the irrelevant diagonal (Okita et al., 1985). In all these paradigms, subjects have to **filter** between relevant and irrelevant information (manipulated by the type of selection cue) and must **search** for target letters (manipulated by the size of the memory set) and **detect** the presence of a target letter.

In case of location and diagonal of display, we are concerned with a special kind of attention: attention allocated to points in visual space (visual spatial attention). Selection cuing in vision may be only effective in the spatial domain. An important question in the research of spatial attention is the bandwidth of the attentional spotlight. When the selective cue is in one representational domain (for example, color), this cross-domain cuing is believed to depend on the shared coding of location (Van der Heijden, 1992). In case of conjunctions we are interested in the hierarchy of processing. In all these task situations, four stimulus categories can be distinguished:

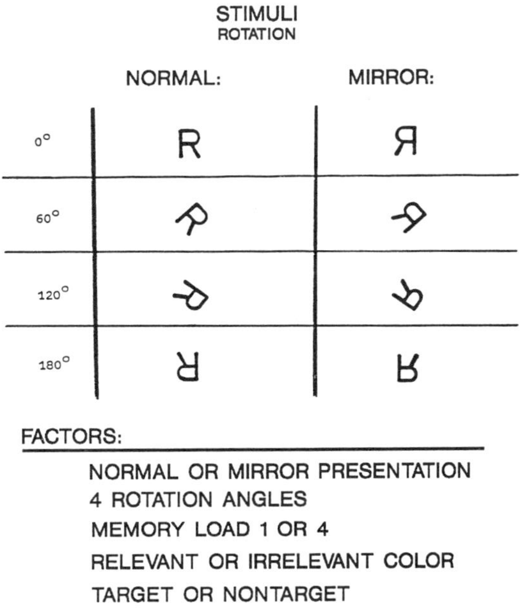

Figure 12.   Relevant and irrelevant stimuli are presented in a normal or mirrored position and at different rotation angles (Wijers et al., 1989c).

relevant targets, relevant nontargets, irrelevant targets, and irrelevant nontargets. In addition, in one experiment selective search was combined with other processes, such as mental rotation. In the latter case, the stimulus letters in each category are presented in one of four different orientations: upright (0 degree), or rotated over 60, 120, or 180 degrees. Half were presented normal and half mirror-reversed. Subjects had to detect target letters from the memory set in the attended color and to decide whether these letters were presented mirror-reversed or not (Fig. 12).

Selecting Cuing in the Selective-Set and Filtering Paradigm

Figure 13 shows the effects of attention to different selection cues in the combined filtering/search paradigm and indexed as difference potentials obtained by subtracting ERPs to unattended stimuli from ERPs to attended stimuli. The vertical axis represents differences in brain activity as a function of attention. Difference potentials are superimposed for several electrodes. Different plots depict separate experiments investigating different selection cues. The arrows show the onset latencies of attention effects as determined by statistical tests.

We can distinguish between two effects of attention: one is dependent on the selection of a specific feature (e.g., color, location) and one is independent of a particular feature. To start with the latter: independent of the type of visual feature that was attended, a late (200–400 msec, Cz maximum) negativity is observed. This component is believed to be identical with the N2b component (Näätänen and Gaillard, 1983), which is thought to reflect a postselection process, namely the covert orienting of attention, a call to the attentional system. Results of control experiments (which are not shown here) suggest that the N2b component is smaller in standard filtering paradigms than in the selective search task, possibly suggesting that more attentional resources are allocated in the latter case. Similar components (and interpretations) have been reported by Harter and Guido (1980) and Rugg et al. (1987).

It has been demonstrated that N2b is sensitive to short-term expectations within stimulus series; the component is larger when a relevant stimulus is preceded by an irrelevant stimulus than when preceded by a relevant stimulus (Wijers et al., 1986b). Apparently subjects expect a repetition of the same class of events: when an irrelevant stimulus is unexpectedly followed by a relevant one, there is a more vigorous attentional orientation (alternatively, subjects could expect alternations of stimuli, and therefore be better prepared for relevant stimuli following irrelevant ones).

We observed feature-specific effects of attention only in the experiments in which a single spatial location or the stimulus color was attended. In the other experiments (diagonal, letter size) only the N2b effect was observed; the onset latency of the attention effects here was about 200–250 msec.

Spatial attention had an early onset, approximately 100 msec, and consisted of an enhancement of the occipital P1 and N1 components, followed by prolonged negativity. When attention was directed at the stimulus color, the onset latency was later, approximately 150 msec, and the effect consisted of negativity at the Oz electrode and positivity at the anterior electrodes. When attention was directed at conjunctions of color and letter size, the early effect resembled that of attention to color only. In two different experiments, we observed that the onset latency of the color selection effect was shorter at Fz than at Oz, suggesting that multiple cortical generators are involved.

Note that the above-mentioned findings are similar to those obtained in standard filtering paradigms (for reviews, see Harter and Aine, 1984; Hillyard et al., 1985). In our experiments, the effects associated with cuing a spatial location or a color were unaffected by task complexity, that is, the amount of required memory search and mental rotation. In this series of experiments we obtained RTs as short as about 450 msec or as long as 1100 msec, yet we obtained very similar onset latencies, amplitudes, and scalp distributions of the color selection effects in these different conditions.

We believe that the feature specificity of the early effects of attention and the independence of task complexity are consistent with the idea that attention has an impact on an early stage of processing, preceding the start of operations in short-term memory.

The qualitatively different patterns of ERP results for spatial and nonspatial selections are now well established and are explained by postulating different brain mechanisms underlying spatial and nonspatial attention. Spatial attention is assumed to be mediated by subcortical gating (Hillyard et al., 1985; Näätänen, 1986) or by the fast, tectopulvinar projection system (Harter and Aine, 1984), while nonspatial attention is thought to be mediated by a postperceptual matching process according to the "attentional trace theory" of Näätänen (1982) or by the slower, geniculostriate projection system according to the "neural-specificity theory" of Harter and Aine (1984). However, the observed absence of early occipital negativities in the case of attention to letter size and diagonal is difficult to explain. For example, the attentional trace theory (Näätänen, 1982) assumes that a common post-perceptual matching process underlies all nonspatial selections (see also Näätänen, 1986). This theory therefore predicts slow endogenous negativities for all nonspatial stimulus attributes. The neural specificity theory (Harter and Aine, 1984) assumes that nonspatial stimulus selections involve the intraperceptual modulation of the neural channels of the geniculostriate system, resulting in occipital negativities for all nonspatial selection cues.

It seems unlikely that the discriminability of the selection cues is a factor that might account for these results. First, all selection cues were highly discriminable. For example, task performances in the color selection and size selection experiments were very similar. Second, discriminability could explain differences in onset latency and amplitude of the selection negativities (e.g., Hansen and Hillyard, 1983) but not an absence of these effects (unless

DIAGONAL

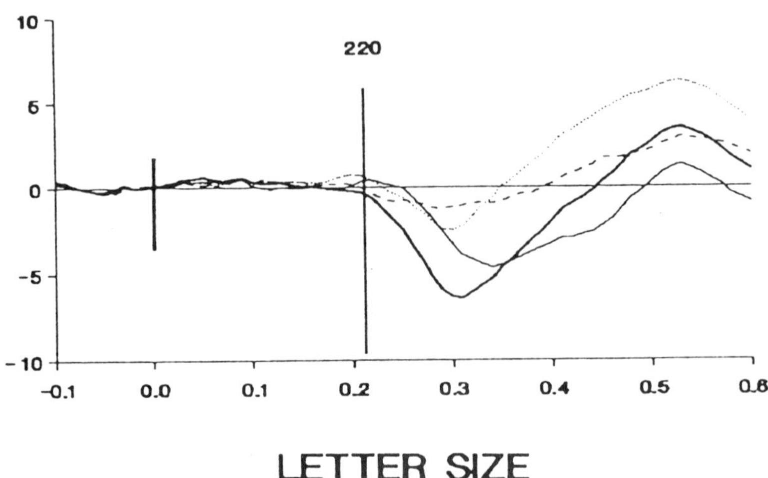

LETTER SIZE

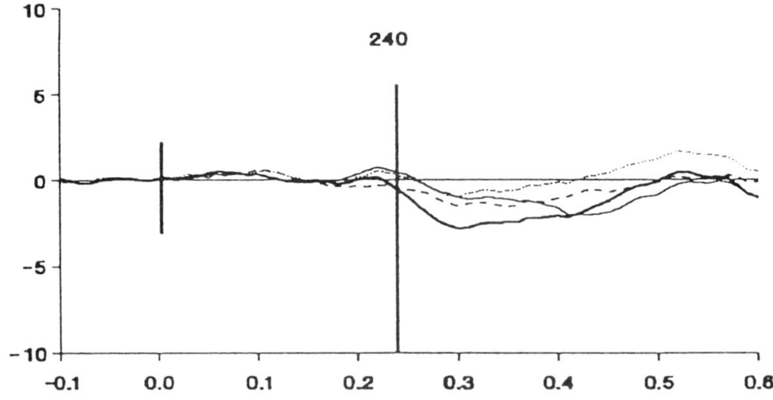

Figure 13.   Effects of selective attention to different visual features (Wijers, 1989). For explanation, see text.

relevant and irrelevant stimuli are so hard to discriminate that they functionally belong to the same input channel (Hillyard and Münte, 1984). As mentioned, this was not true. Third, in a spatial attention filtering paradigm we observed an early selective attention effect, namely, an enhancement of P1 amplitude in a situation in which relevant and irrelevant stimuli were very hard to discriminate (Wijers et al., 1989d).

Only a few nonspatial visual features have been reported. Attention to spatial frequency has been shown to result in occipital negativities (Harter

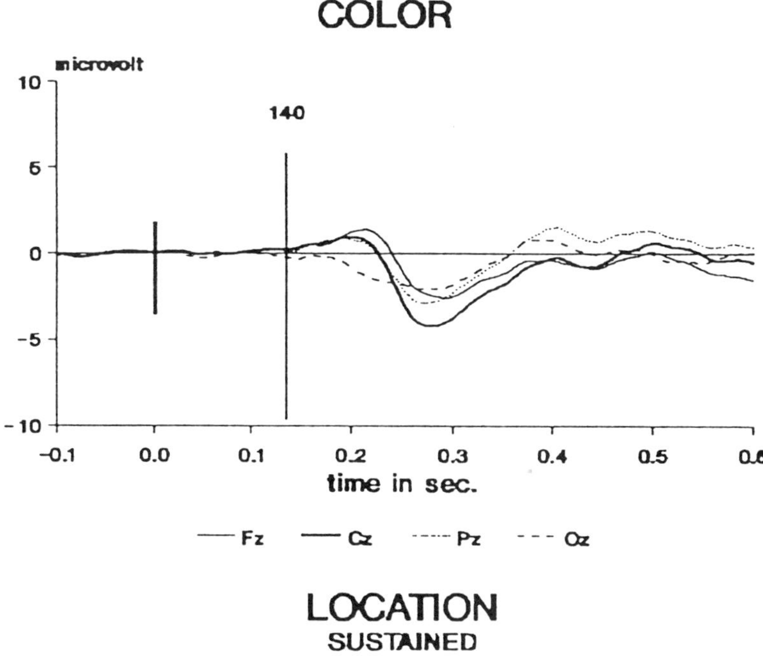

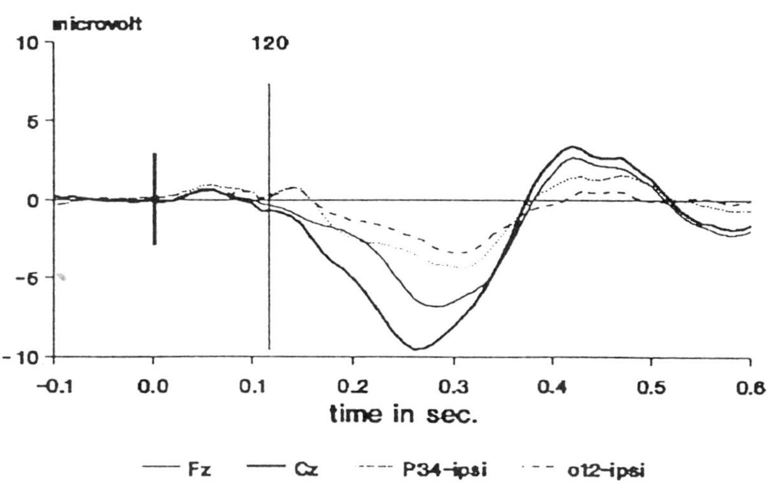

Figure 13.   (*continued*)

and Previc, 1978; Previc and Harter, 1982), but these experiments only
investigated the Oz electrode, leaving the possibility open that the effects
might have been caused by a Cz dominant N2b component. Attention to
orientation has been reported to result in an occipital negativity and frontal
positivity (Harter and Guido, 1980), but Rugg et al. (1987) failed to replcate

such an effect; instead, these authors reported a late anterior negativity, very much comparable to our N2b effect.

Interestingly, the early effect of attention (in the case of attention to location or color) and the later unspecific effect of attention (N2b) are independent, so that stimuli that appear to be rejected in an early stage of processing are attended in a later stage. This is illustrated in Figure 14, which shows the results of an experiment in which subjects attended to combinations of letter size and color. For example, subjects attended to large, red letters while combinations of large or small and red or blue letters were presented. The figure shows the mean ERP amplitude in the 100- to 250-msec range at Fz and in the range 250–400 msec at Cz. The former corresponds with what was denoted earlier as feature specific, and the latter with feature nonspecific (N2b). The early effect of attention shows a hierarchical pattern of results (Hansen and Hillyard, 1983) in which there is an effect of attention to letter size only when the color is relevant. Note that this early effect of letter size was not obtained when subjects attended to letter size only, suggesting that the conjunction of letter size and color is available for selective processing at an early stage, while size as a distinctive feature is only accessible to later stages (for a similar proposal, see Houck and

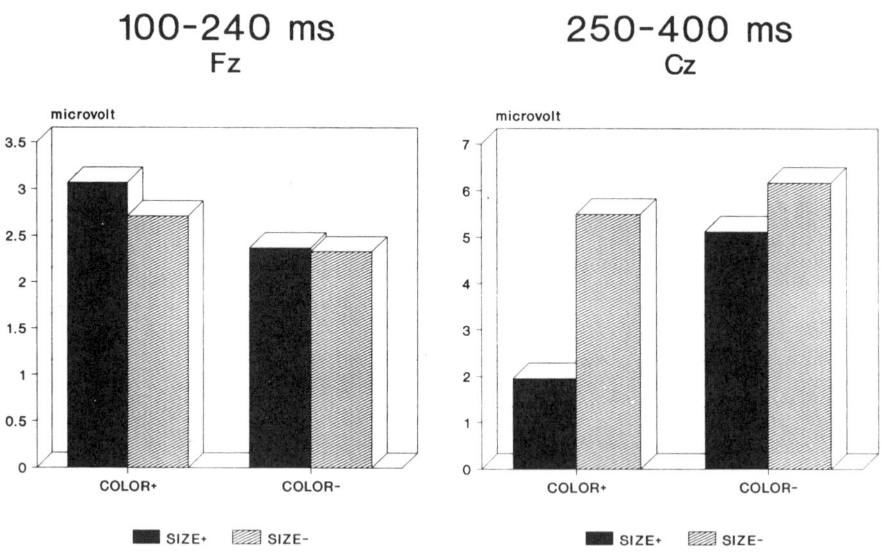

Figure 14.   Mean ERP amplitudes in the 100- to 240-msec interval at Fz and in the 250- to 400-msec interval at Cz for stimuli in the relevant color and letter size (color +, size +), for stimuli in relevant color but not in relevant size (color +, size −), for stimuli in relevant size but not relevant color (color −, size +), and for stimuli in neither relevant color nor relevant size (color −, size −). (Wijers et al., 1989b).

Hoffman, 1986). The later effect of attention, on the other hand, shows a more or less independent organization; that is, effects of the relevant letter size and the relevant color were also observed when the other attribute was irrelevant (however, the effect of both attributes being relevant was larger than the sum of each attribute being relevant alone). Thus, ERPs to stimuli that were only relevant on letter size were indiscriminable from stimuli that were irrelevant on both attributes in an early stage of processing, but showed evidence of being attended in a later stage.

If the early effects reflected a hierarchical all-or-none filtering or feature-matching process, in which first the color of the stimulus is tested and then, dependent on the outcome of this test, its size, it is hard to understand why stimuli that failed the early color test evoked an attentional response later. Therefore, the early effects that are obtained when attention is directed at a color or location seem to reflect an attenuation/amplification of information, for example, a modulating effect of attention on early stimulus representations; this effect does not prevent unattended information from gaining access to later stages of processing. For example, the color of unattended stimuli may be represented less "vividly," but the size of these stimuli may still evoke an orienting reaction later. This shows the important distinction between information representation and information use (e.g., Kahneman and Treisman, 1984). It is an important (although very intricate) problem to decide whether ERP effects of selective attention reflect modulating effects on the creation of stimulus representations, that is, the buildup of information, or alternatively the use of this information, that is, the distribution of information from stimulus representations to other processes. Kahneman and Treisman (1984) suggested that information from stimulus representations could be made selectively available to some mental processes ("agencies of the mind"), but not to others.

Our data may suggest that selective attention to letter size or diagonal has no effect on the accumulation of stimulus information (although the combination of letter size and color does), that later stages of processing select information from early representations. Selective attention to color or location, on the other hand has an effect on the buildup of information. The qualitatively different ERP signs of color and spatial selection suggest that they influence information representations in different ways, involving different brain structures.

It is well known that different visual attributes are represented in different retinotopically organized brain areas (Cowey, 1979, 1985), the visual maps (Nissen, 1985). One might speculate that spatial attention modulates the inflow of information in all the visual maps (e.g., subcortical gating) while attention to color modulates the activity of the color map. However, the dipole localization data obtained with neuromagnetic measurements do not confirm the predictions of a subcortical gating theory. The following conclusions can be drawn regarding the role of selective cuing:

1. Early ERP components related to spatial and color selection are highly similar in standard filtering paradigms and in the much more complicated selective search and mental rotation paradigms that were considered here; that is, at least in these cases generalization across paradigms is possible (Gaillard, 1988).

2. According to Harter and Aine (1984), some visual features seems to have priority over others in selective attention: a spatial location is selected most quickly, followed by color and finally by other attributes. Selection by color and location seem to involve different brain mechanisms, and in addition spatial and color selections also appear to be fundamentally different from selection based on other attributes. However, definite conclusions are premature; more research using dipole localization is needed.

3. Attending to different visual features also involves a common attentional mechanism, which is invoked later. This mechanism, reflected by the N2b, is a postperceptual orienting mechanism that is sensitive to resource allocation, running expectancies, and the general state of the organism (Furda and Otto, 1989; Gunter et al., 1987).

4. The earliest effects of selective attention do not reflect an all-or-none selection or matching process but an attenuating or amplifying operation; in later stages of processing, attention may be oriented to stimuli that are neglected (attenuated) at an earlier stage.

5. Early selective attention effects caused by spatial or color selection were present whether or not the subject was required to further process stimuli in the attended channel, that is, to search and mentally rotate these stimuli (see following).

Selective Cognitive Processing

In the series of experiments considered here, several task variables resulted in attention-dependent effects, suggesting that these variables manipulate stages of processing that are under voluntary control.

Figure 15 shows the effect of memory load (the number of memorized target letters) on the ERPs to attended and unattended nontarget letters (no target was presented) in the experiment in which the relevant input channel was based on color (Wijers et al., 1989b). The more target letters that were memorized, the more comparisons that must be made between the display items and the memorized items. This results in a more extensive, longer duration, serial controlled search process (Shiffrin and Schneider, 1977). Figure 15 shows that a more extensive search process is reflected by increased, prolonged, long latency (in the range of 300 to 600 msec) negative shifts, with a maximum at the Cz electrode. This shift was completely

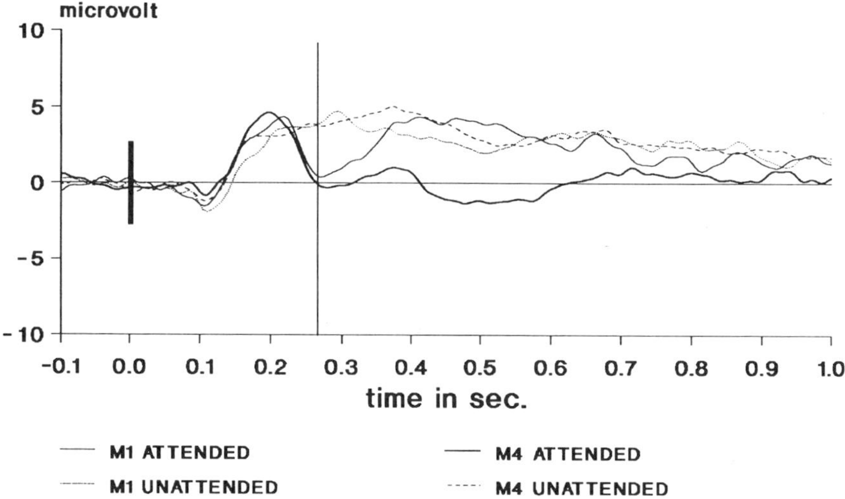

Figure 15.   Grand average ERPs to nontarget stimuli at Cz, superimposed for stimuli in attended and unattended colors and for stimuli in different memory load conditions (memory load 1, M1; memory load 4, M4) (Wijers et al., 1989a).

confined to attended stimuli, showing that the memory search process is under voluntary control. Note that the onset of the memory load effect is later than the effects of attending the stimulus color. Mostly the onset of the memory load effect started at the peak of the N2b component (see earlier). Thus, memory search appears to be under voluntary control, at least in many instances (see following).

Figure 16 shows the design of an experiment in which we investigated the selectivity of display search (Wijers et al., 1987). Subjects searched for the occurrence of one target letter. In condition 1, four letters were presented and all display items were relevant. In condition 2, only two display items were presented, both relevant. The critical condition is a condition in which four display items were presented, but the subjects were instructed to attend to one diagonal of the display, i.e., to a subset of two of the four presented items (see Fig. 16).

Figure 17 shows the mean amplitude of the ERPs in the 300- to 500-msec latency range at the Cz electrode for these three conditions. These were ERPs to nontarget stimuli, in which the target letter was not presented in the display. Figure 17 shows that there is increased negativity in the display load 4 condition as compared to the display load 2 condition. Importantly, the display load 4 condition in which only two elements were defined as

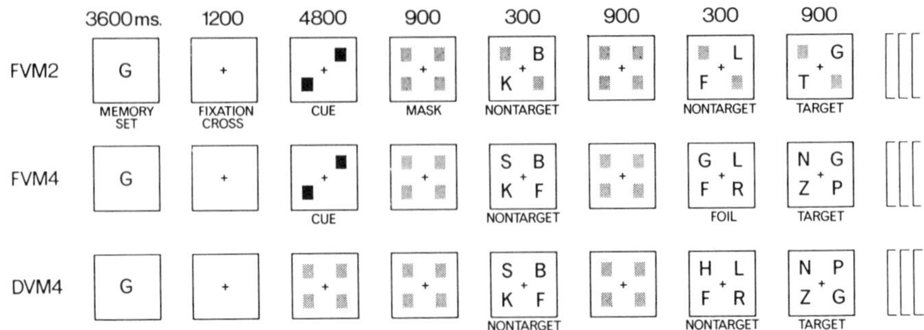

Figure 16.   Filtering and search in different task configurations. First, memory set is presented. Cue indicates which part of display should be attended. Subject's task is to press a button when seeing target in one of two relevant display positions. DVM4 is usual selective-set condition with display load of 4. In FVM2, display load is 2, and relevant and irrelevant stimuli are successively presented. In FVM4, display load is 4, and relevant and irrelevant stimuli are simultaneously presented (Wijers et al., 1987).

being relevant showed about the same amplitude as the display load 2 condition; apparently the subjects effectively confined the display search process to the relevant diagonal, resulting in an effective display load of 2.

300-500 ms Cz

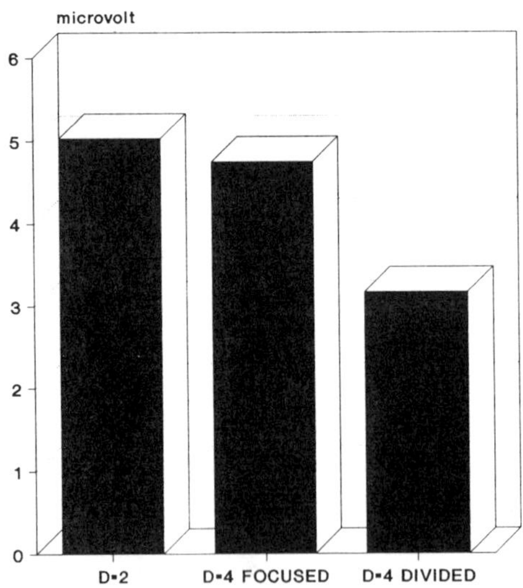

Figure 17.   Mean ERP amplitudes at Cz in the 300- to 500-msec interval in three different conditions: D = 2 (FVM2 in Fig. 16), D = 4 focused (FVM4), and D = 4 divided (DVM4). Note that display load in conditions D = 4 focused and D = 4 divided is 4, but in D = 4 focused, only two display elements, cued by diagonal, are relevant.

Pz
ATTEND COLOR

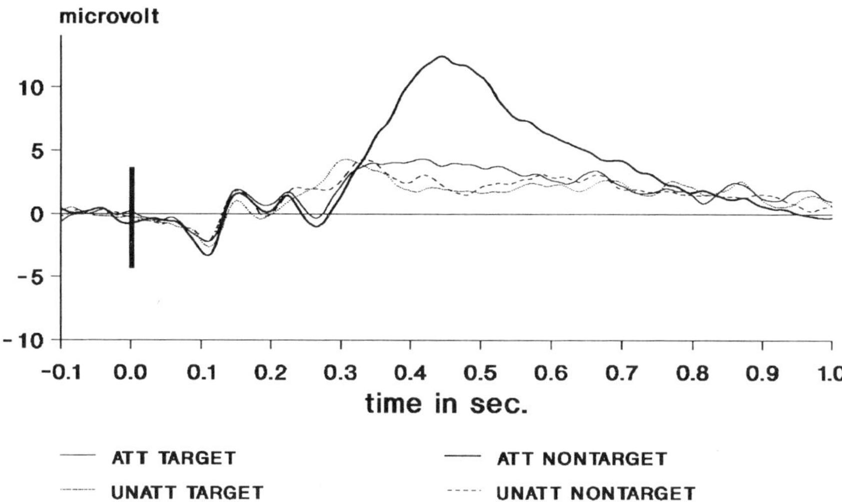

Figure 18.   Grand average ERP (Pz) to stimuli in memory load 1 condition (M1) at Pz, superimposed for target and nontarget stimuli in attended color and for target and nontarget stimuli in unattended color.

Figure 18 shows the effect of presenting target letters in an attended or unattended stimulus color (Wijers et al., 1986b). As can be seen, the difference between nontargets and targets consisted of a late parietal positivity to targets as compared to nontargets. This component was considered to be P3b. This positivity was only observed for attended targets. Unattended targets and nontargets evoked very similar ERPs. The onset latency of target/nontarget differences was about 400 msec.

Until now we have only discussed the effects of visual and memory search on the attended and unattended information. In a recent study, however, we wondered whether other mental operations also produce negativities and whether mental rotation is restricted only to attended information. Figure 12 depicts the stimulus presentation in this experiment. Stimulus letters were presented one by one, half were presented normally and half mirror-reversed; half were red and half were blue. In all these four categories (normal/reversed and blue/red), the letters were presented in upright position or rotated counterclockwise over 60, 120, or 180 degrees. Subjects had to attend to stimuli in one color, detect the occurrence of letters from a memory set (either 1 or 4 in different conditions) in the relevant color, and decide whether relevant target letters were presented normal or reversed, indicated by two different response buttons. The reaction times showed an increase as a function of memory load, and in addition an increase as a function of orientation, being slower the more the letters were rotated from their upright position; memory load and orientation effects were largely independent. The ERP effects of attending stimulus color, memory load, and target detection were as described for the simple selective search task. Attending the stimulus color resulted in an early occipital negativity/ anterior positivity (onset about 150 msec), followed by a Cz maximum N2b component. An increase of memory load resulted in a later, Cz maximum negativity, confined to the attended category. When target letters were presented in the attended color, this evoked a late parietal positivity (P3b). None of these effects were influenced by the orientation of the stimulus letters.

Figure 19 shows the effect of stimulus orientation. The ERPs showed increased negativity as stimulus letters were rotated over larger angles from the upright position. This effect was independent of memory load; it was most pronounced at the descending flank of the P3b component. The latency range was about 400–700 msec. Importantly, this effect was obtained only in the ERPs to attended target stimuli.

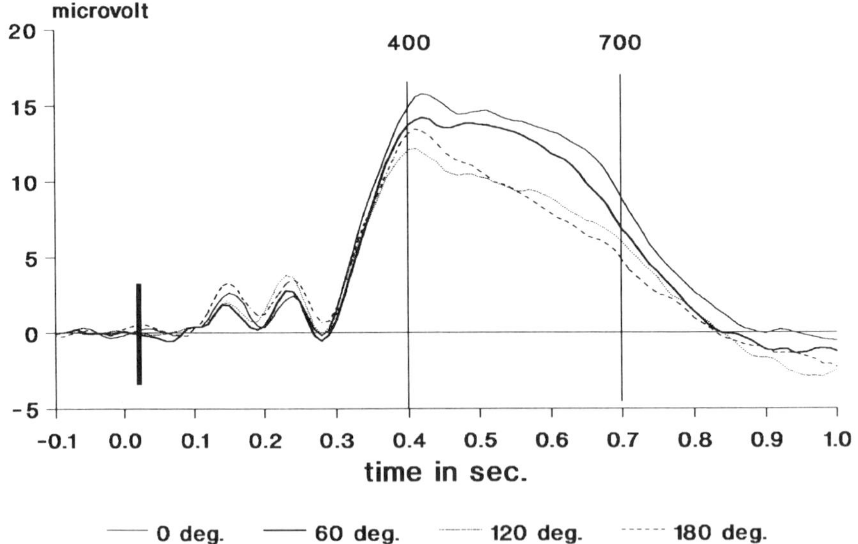

Figure 19.   Grand average ERPs to target stimuli in attended color at Pz, superimposed for four orientations of stimulus letters (0°, 60°, 120°, 180°) averaged over normal and backward presentation modes and over memory load 1 and 4 conditions.

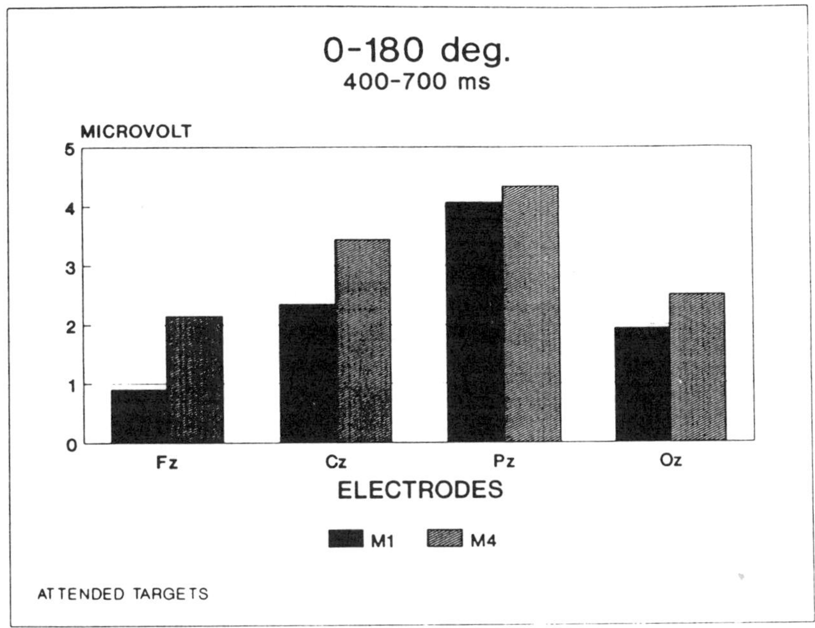

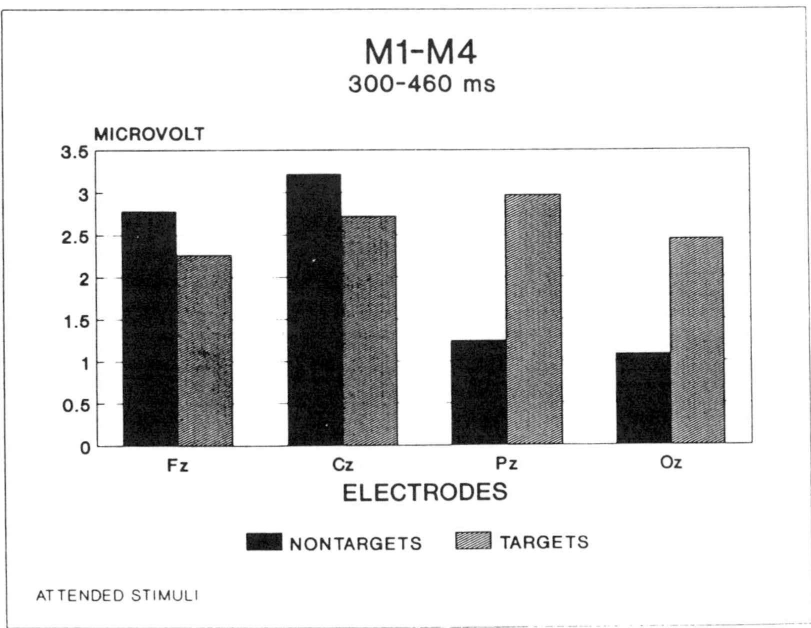

Figure 20.   Scalp distribution of rotation-related negativity (top) computed as average difference between stimuli rotated over 180° as compared to stimuli rotated 0° and separately for M1 and M4. Bottom: Scalp distribution of search negativity computed as the average difference between attended M1 and attended M4 and separately for targets and nontargets.

In Figure 20 the scalp distributions of the memory load and orientation effects are compared. The scalp distribution of the memory load effect is different for targets and nontargets, especially at Pz and Oz. Presumably, for target stimuli a memory load effect on the latency of the P3b component (and also on the single-trial latency jitter of the P3b) enhances the size of the memory load effect at the posterior electrodes. The "true" scalp distribution of the search-related negative component can therefore better be studied by examining the nontarget ERPs. The scalp distribution of the rotation-related negativity is probably less confounded by the P3b, because orientation did not affect P3b latency. Further, the effect of orientation was more strongly present at the parietal electrode than the effect of memory load, which extended more into the anterior regions. Similar rotation related negativities have been reported by Stuss et al. (1983) and Peronnet and Farah (1989).

We have reached the following conclusions concerning selective processing:

1. Increased controlled processing of stimuli belonging to the attended input channel is reflected by late prolonged negativities. Ruchkin et al. (1988) observed an increase of late negativity related to more difficult mental arithmetic. These negativities should be regarded as reflecting further processing of attended inputs (Näätänen, 1982). However, differences in scalp distribution between the effects of memory load and stimulus orientation suggest that the brain mechanisms that generate these negativities may depend on the exact type of controlled processing. In particular, these results suggest that rotation-related negativity and search-related negativity reflect different aspects of working memory perhaps involving visual and symbolic/phonological representations. These two different aspects may correspond to the "visuospatial scratch pad" and the "articulatory loop," respectively, two of the four working memory subcomponents distinguished by Baddeley (1983). The data confirm those obtained by Farah and Peronnet (1989), who also found the effects of mental rotation both largest and most systematic at the parietal, rather than the occipital, recording site. In both cases the data are compatible with the hypothesis that mental rotation is a spatial, rather than visual, form of imagery, and that it is an effortful, attention-demanding process (Hasher and Zacks, 1979).
2. The independence of memory load and orientation confirms both on the behavioral and electrophysiological level an important aspect of the Cooper and Shepard (1973) model, namely, that letter identity can be determined independently of letter orientation and that the mental rotation process is only required to determine whether the letters are of normal or reversed configuration.
3. The sequential onset latencies of the effects related to relevant/irrelevant channel discriminations, memory search, target detection, and mental

rotation suggest that these processes are started in a **serial order**. The finding that later stages of processing are hierarchically dependent on earlier stages (e.g., memory search was confined to the relevant stimulus category; mental rotation was confined to relevant targets) suggests (1) that later stages generally await the completion of earlier stages, and (2) that stages provide discrete codes to the next stage, for example, relevant or irrelevant, target or nontarget. However, these organizational principles are by no means universal. In some conditions the memory search process was initiated nonselectively, and other researchers have obtained evidence for preliminary motor activation (e.g., Smid et al., 1991) (see later section) based on partial stimulus evaluation.

4. An important assumption of the subtraction method (Donders, 1869) is that in switching from one task to another, stages of processing may be inserted or deleted without grossly changing the total structure of information processing. It seems that this assumption is quite valid, at least in the series of experiments considered here. The pattern of ERP results that is normally obtained in standard selective attention paradigms was not altered by inserting a memory search process. In the same vein, the organization of information in the selective search task was not influenced by adding a mental rotation process.

5. Controlled processing is accompanied by late negativities, which probably have different generators depending on working memory structures involved.

Selective Response Processing

In the previous section, we concluded that the negativities associated with further processing indeed were mostly selective: search and rotation were confined to the cued input channel. Selective processing is said to occur. In this section, we ask a similar question but with regard to response processing. To be able to establish whether different levels of response processing are also selective we first must develop an index of selective central response activation and an index of selective peripheral response activation.

A Central Measure of Selective Response Activation:
The Lateralized Readiness Potential

We used the lateralized readiness potential (LRP) as the major index of central response activation. The LRP is derived from the readiness potential (Deecke, Grizinger, and Kornhuber, 1976). The readiness potential (RP) refers to premovement potentials related to voluntary hand movements. The RP is a gradually increasing negative shift, beginning one s or more prior to movement onset. Some variable time before movement onset the RP starts to lateralize, with larger amplitudes above the hemisphere contralateral to the side of the activated response (Kutas and Donchin, 1977, 1980). This lateralization, however, is confounded by other processing and structural asymmetries. These asymmetries are assumed to be equal for left- and right-hand responses. Thus, lateralization of the RP can be corrected for the nonmotor asymmetries by subtracting right-hand asymmetries from left-hand asymmetries, resulting in a difference potential consisting of "pure" motor-related activity (De Jong et al., 1988; Gratton et al., 1988) (also see Fig. 21 for additional explanation). This difference potential is called the LRP.

The logic of the derivation of the LRP implies therefore that it reflects only motor-related activity. De Jong et al. (1988) showed that the LRP is a highly sensitive index of selective motor activation, both in the absence of an overt response (i.e., in the interval between an informative precue and the imperative signal) and preceding the overt response (i.e., after the imperative signal).

Osman et al. (1992) presented evidence that lateralization can occur to a multiattribute NO–GO stimulus, of which one attribute is associated with a response and the other with the inhibition of that response. This lateralization also occurred in the absence of peripheral activation. De Jong et al. (1990) showed that lateralization of the LRP in response to a response-choice

---

Figure 21.   Two subtraction steps are necessary to derive LRP. Monopolar ERPs are recorded from electrode positions C3' and C4' (see subject's head in center). Imaginary ERPs are shown relative to imperative stimulus (IS) at time zero when subject produces left- or right-hand reaction at about 500 msec. On left-hand response trials (left-upper), C4' amplitude is more negative than C3' amplitude. On right-hand response trials (right upper), C3' amplitude is more negative than C4' amplitude. Middle left and right panels: C3'–C4' subtraction potentials left- and right-hand response trials show large difference with opposite sign on left- and right-hand response trials and small difference with equal sign on these trials. Difference with opposite sign reflects pure motor-related activity; difference with equal sign reflects nonmotor activity. To obtain pure motor-related activity, a second subtraction is performed: LRP = RH(C3'–C4') − LH(C3'–C4'), subtracting left-hand asymmetry (middle-left panel) from right-hand asymmetry (middle-right panel) to obtain difference potential depicted in bottom panel. This difference potential, the LRP, now logically only contains pure motor-related activity. Note that asymmetries with equal sign in left- and right-hand response trials (middle two panels) have disappeared due to the second subtraction in the bottom panel. Because of double subtraction, LRP does not reflect right or left response hand activation but differential activation of response hands. This activation can be in the direction of the response side indicated by the stimulus (correct activation, negative amplitude), or in the direction opposite to the response side indicated by the stimulus (incorrect activation, positive amplitude).

DERIVATION OF THE LATERALIZED READINESS POTENTIAL:

THE LRP

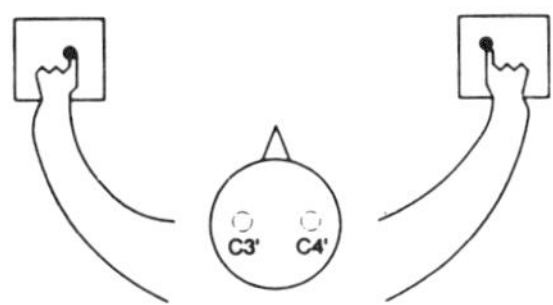

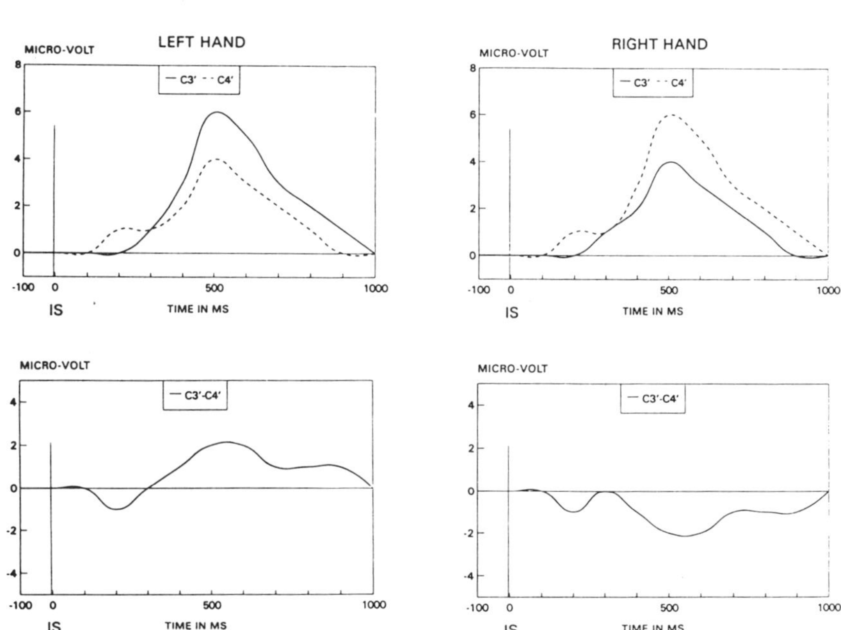

LRP = DIFFERENCE: [RH (C3' - C4')] - [LH (C3' - C4')]

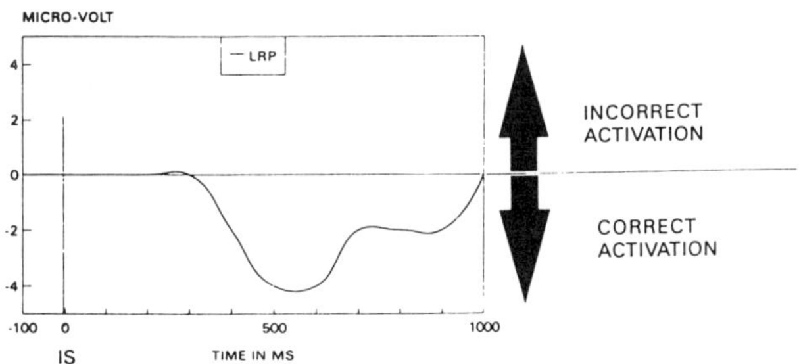

stimulus occurred, but lateralization could be prevented from producing more peripheral activation (of EMG and squeeze activity) when a stop signal followed the choice signal at some critical delay. These results show that the onset of significant lateralization of the LRP can be interpreted as evidence that at that time a response has at least been selected.

An important issue is the onset latency of the LRP. To determine LRP onset latency, first at each sample point, a Wilcoxon ranksum test was performed on the amplitude difference between C3′ and C4′ (i.e., the difference in electrical brain activity above the relevant parts of the motor cortex) on left-hand response trials and C3′–C4′ on right-hand trials. Prestimulus differences were corrected by subtracting a base-line voltage averaged over the 100-msec interval before stimulus onset.

The Wilcoxon tests were performed on a matrix of trials by sample points for each subject and category; thus, a new time series was obtained for each subject and category consisting of Wilcoxon W statistics for the 100 sample points following stimulus onset. The W statistic is used to obtain an index of difference that is less sensitive to noise than the amplitude of the difference (note that the LRPs shown in the figures are still expressed in amplitude differences). Next, we computed the onset latency of the LRP across subjects on the basis of their W statistics by performing a t test at each sample point in each condition in which the null hypothesis of zero average Wilcoxon W statistics across subjects is tested (De Jong et al., 1988, 1990; Smid et al., 1990; Van Dellen et al., 1985). The latency at which the Wilcoxon W statistic starts to differ significantly from zero is called LRP onset ($p < .01$, one-tailed). When the LRP accompanies an overt response, the t statistic is usually found significantly different from zero for more than 200 msec. The t test in this procedure is performed in a one-tailed version because the W statistics are expected to have values in a predefined direction i.e., they should be negative when accompanying a correct response and positive when accompanying an incorrect response (see Fig. 21).

A Peripheral Measure of Selective Response Activation: Electromyogram

A second important measure is the electromyogram (EMG), recorded from the muscles activated before a response (when a button press is required, these muscles involve the flexor digitorum). This measure is important for several reasons. In many choice RT tasks, subjects have to choose between a left- and a right-hand criterion response (e.g., a button press or a criterion squeeze), only one of which is correct. It has been shown that on several trials (10%–40%) the incorrect response muscle is activated although only a correct criterion response was registered (Coles et al., 1985; Gratton et al., 1988; Smid et al., 1990, 1991). These incorrect EMG responses were found to be related to factors as flanker identity in the response priming paradigm designed by Eriksen and Eriksen (1974; see also following) and to response bias. On these trials, there was also lateralization of the LRP toward the incorrect response. If such trials are not discarded from the LRP analysis to establish the onset time of correct lateralization in a particular experimental condition, the onset time may be delayed by incorrect lateralization. Thus, LRP analysis requires EMG analysis. From these observations it also follows that the EMG measure is important because it provides a more sensitive measure of the occurrence of errors than counting the number of incorrect criterion responses. Even more importantly, the EMG onset time can be used as an index of the time at which response processes related to actual execution of the response become manifest at the peripheral level. If no incorrect EMG activation occurs on a trial, this time point can be taken as a more sensitive index for the timing of the ultimate response than criterion RT, because aspects of the choice signal and response bias have been shown to affect the interval between EMG onset and criterion RT (Coles et al., 1985; Smid et al., 1990).

It has been shown (Requin, 1985) that there is a constant time interval between central initiation of a response and onset of EMG activation. Also, Bullock and Grossberg (1988) reviewed evidence that suggests that response production is mediated by two distinct processes. One process, cortically located, "computes" response parameters like direction of movement and the other, more peripherally located, provides for parameters such as force and speed of the response. With a typical chronometric technique as the stop signal task, Osman et al. (1992) and De Jong et al. (1990) presented evidence obtained from the LRP and EMG that supports this model of response production.

It seems, therefore, that it is not unwarranted to interpret EMG onset time as a relative index of response initiation, that is, differences in EMG onset between experimental conditions reflect differences in response initiation times although the absolute time of initiation is unknown.

To summarize: these measures, then, in combination with P3b, a central index of stimulus categorization, and with RT provide powerful tools to observe the chronometric aspects of response production because they show consistent and selective effects of the main variables affecting CRT.

On the Selectivity of Response Processing:
The Case of Multiple Stimulus Displays

In a first experiment (Smid, Mulder, and Mulder, 1990) we applied the choice reaction time paradigm designed by Eriksen and Eriksen (1974). In this paradigm the subject should only respond to either an H or a D (targets) with different hands. Four different types of trials are presented to the subject.

1. No Flanker (NOF) trials. In these trials only one target letter is presented.

2. Compatible Flanker (COM) trials. During these trials the target letter is presented with letters that may call for the same response as the target (e.g., 'DDDDDDD').
3. Neutral Flanker (NEU) trials. During these trials the target is presented together with letters that are not associated with a response (e.g., 'XXXDXXX').
4. Incompatible flanker (INC) trials. In these trials the flankers may call for the opposite response (e.g., 'HHHDHHH').

The continuous flow conception of human information processing assumes that any information in the display associated with a response hand will activate that channel. If a particular stimulus display contains information that activates two different response channels, the concurrent activation of these channels produces mutual inhibition (response competition). This would occur at least on incompatible trials. If, on the other hand, a stimulus array contains redundant information associated with a response channel, it will activate it earlier and to a higher degree. Thus, in compatible displays response facilitation may occur.

We defined four error categories based on the sequence of correct and incorrect EMG activation and incorrect button presses. No-Error trials (NOE) contained only correct EMG activation, followed by a correct button response. Correct-Incorrect-Error trials (CIE) contained two EMG responses: first a correct EMG response that produced a button press, followed by an incorrect EMG response that did not result in a button press. Incorrect-Correct-Error trials (ICE) also contained dual EMG responses, but in the reverse order; first, incorrect EMG activation occurred that did not result in a button press, followed by correct EMG activation that did produce a button press. Finally, the Button-Press-Error category consisted of trials on which an incorrect button press was made (the traditional RT error).

The behavioral and psychophysiological data replicated the findings by Eriksen et al. (1985) and Coles et al. (1985) and provided additional evidence for the mechanisms postulated by the continuous flow conception. RTs to targets alone did not differ from RTs to compatible arrays. The latter were faster than RTs to neutral arrays, which were faster than RTs to incompatible than on neutral trials, and longer on compatible than on target alone trials.

The LRP indicated that incorrect central response activation on incompatible trials, and correct central response activation on compatible trials, both began earlier than on target-neutral trials. More incompatible but less compatible trials than neutral ones exhibited incorrect peripheral (EMG) response activation. Peripheral response execution (the time interval between EMG onset and overt responding) was faster and more accurate on compatible than on target-alone trials, while it was slower and less accurate on compatible than on neutral trials (See Fig. 22). ICE responses were mainly present in INC trials.

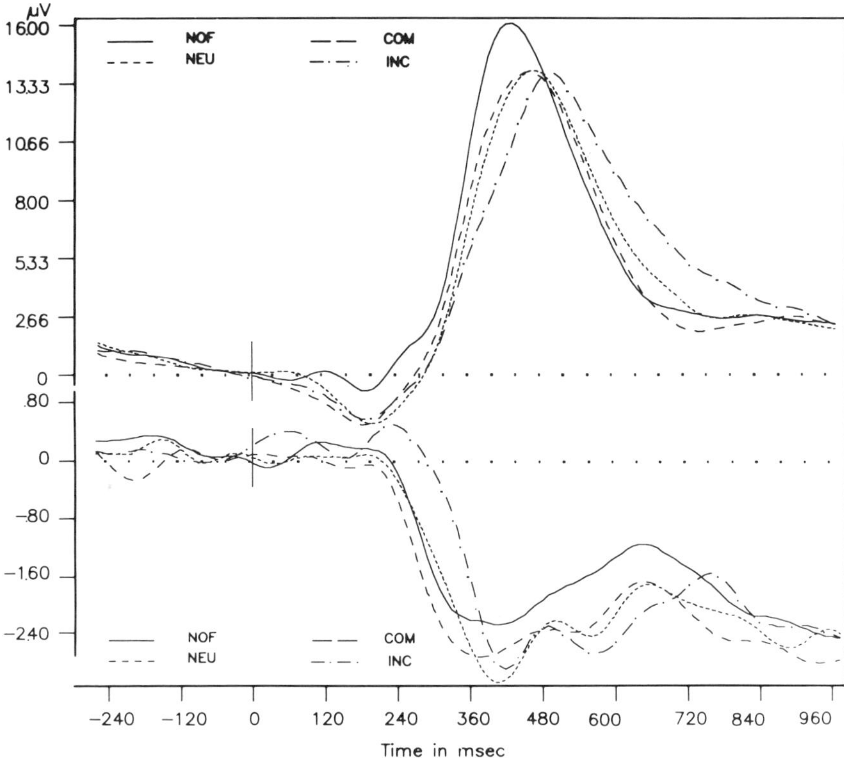

Figure 22.   Waveforms of P3 component (upper panel) and LRP (lower panel) obtained on no-flanker trials (NOF) and trials with compatible flankers (COM), neutral flankers (NEU), and incompatible flankers (INC). Stimulus onset was at time 0. Note that P3 component peaks later in COM and NEU trials than in the no-flanker trials and peaks later in INC than in all other trial types. LRP can change in two directions: upward change (positive amplitudes) indicates incorrect response activation; downward change (negative amplitudes) indicates correct response activation. Note that there is small positive LRP in INC trials (dash-dot waveform) starting significantly around 200 msec post stimulus (the small positive deflection starting at stimulus onset in these trials is an artifact of outliers because it disappeared in the *t* tests (see Electrophysiological Measures). LRP changes in negative direction at about 300 msec post stimulus. Note also that LRP in COM trials (dashed waveform) starts to activate correct response (downward deflection) earlier than in all other trial types. (Amplitude is in microvolts.)

These results indicated that flankers activate their associated response before the target has been identified, and, perhaps as a result of this, that response facilitation and competition occurred. It was concluded that the set of stimulus recognition processes consisting of letter identity analysis and location analysis and the set of response activation processes cannot be regarded as independent stages of processing. Our data strongly suggest that central response activation on the basis of the identity of the flankers had already begun before the target was located and identified. This means that partial output of stimulus recognition processes (i.e., flanker identity) was transmitted and served as input for transformation by response activation processes. It follows, that these sets of processes cannot have a stage relationship.

In the next section we explore whether partial information can be transmitted and used in tasks in which only a single stimulus is presented.

Selective Response Processing: The Case of Single Stimulus Displays

The task we used in this experiment (Smid et al., 1992) is a variation of a task introduced by Miller (1982) that is based on a method of precuing subsets of a set of relevant responses. The subjects had to respond with one of four fingers, the index and middle fingers on the left and right hand. Miller (1982) observed that if two fingers on the same hand are precued, RTs are faster than if two fingers on different hands are precued. This same-hand benefit suggests that response preparation is more efficient if the precue signals two fingers on the same hand. Miller (1982) further reasoned that stimuli that can be coded on two dimensions (e.g., colored letters) might serve both as a precue and as an imperative stimulus, if the possible values of one of the dimensions (e.g., red and green) are easier to discriminate than the possible values of the other dimension (e.g., M and N). However, a necessary condition for this to happen is that partial information about the color dimension (the precue) can be transmitted to contingent motor processes before information about the letter dimension (the imperative stimulus) has been processed. Thus, if two letters in the same color are assigned to two fingers on the same hand, preparation of response hand might begin on the basis of color information before letter information is available. If same colored letters are assigned to two fingers on different hands, no preparation based on color would be possible.

The LRP measure seems perfectly suited to study whether this is the case, because it reflects the differential activation of response hands. If color can be used to discriminate between hands before letter information is available, LRP onset should be earlier than if no stimulus dimension discriminates between hands. We combined this task with a method introduced by Osman et al. (1992) that makes it possible to obtain an LRP in response to partial information in trials in which no overt response occurs. To achieve this, easy discriminable stimulus features (e.g., red and green) must signal the hand with which a response has to be made, whereas features difficult to discriminate (e.g., M and N) indicate whether a response has to be made (the GO response) or not (the NOGO response). If information about color can be used to prepare response hand before information about the letter has been fully processed, the motor system would initially begin to prepare response hand, both on GO and NOGO response trials. Next, after the letter would have been fully processed, the response would either be executed with the prepared hand (on GO trials), or the advance preparation would be cancelled (on NOGO trials). Thus, if on NOGO trials an LRP would occur, it would strongly support the idea that partial information is transmitted and used for preparation of response hand.

We combined the GO/NOGO technique with the four-choice task that was introduced by Miller (1982) in the following way. Three conditions were designed in which four letters appeared that were difficult to discriminate (e.g., the GO letters M and N, and the NOGO letters H and K). These letters appeared in two colors that were easy to discriminate (e.g., red and green). Table 3 shows a sample of these conditions. In one condition, called the "hand:color" condition (the H:C condition in Table 1), the color of the letter signaled two response fingers on the same hand, while the identity of the letter signaled two fingers on different hands or that no response should be made. We reasoned that if color is available before letter identity and is used to initially select the response hand associated with it, an LRP would appear on both GO and NOGO trials. Next, when letter identity comes available and this identity represents a NOGO letter, it would be used to cancel the initial activation of the selected hand. If it represents a GO letter, it would be used to select one of the two fingers on the hand already selected, which would subsequently execute the required response.

In the second condition, called the "hand:letter" condition (the H:L condition in Table 3), the identity of the letter either signaled two response fingers on the same hand or that no response should be made, while the color of the letter signaled two fingers on different hands. On NOGO trials in this condition there is no information which permits discrimination between hands. Therefore, on NOGO trials no LRP can occur. In the third condition, called the "hand:neither" condition (the H:N condition in Table 1), both color and letter identity signaled two responses on different hands, while the GO/NOGO decision again depended on the identity of the letter.

Table 3. Sample mapping of colored letters on responses in four-choice tasks[a]

Hand	GO–LH						GO–RH	
Finger	M	I		NOGO			I	M
H:N	M1	N2	H1	W2	H2	W1	M2	N1
H:L	M1	M2	H1	H2	W1	W2	N2	N1
H:C	M1	N1	H1	W1	W2	H2	N2	M2

[a] Abbreviations/symbols: Digits denote colors (e.g., '1', red letter; '2', green letter); 'M', 'middle finger'; 'I', 'index finger'; 'LH', 'left hand'; 'RH', 'right hand'; 'GO', GO stimuli; 'NOGO', NOGO stimuli; 'H:N', 'Hand:Neither' condition; 'H:L', 'Hand:Letter' condition; 'H:C', 'Hand:Color' condition.

In this case, there is also no information on NOGO trials on the basis of which an LRP can occur.

These latter two conditions should provide baseline controls. The (psycho-physiological) response latencies obtained in the hand:letter and hand:neither conditions would be diagnostic for the occurrence of a same-hand benefit in the hand:color condition. Selection of response hand in these two conditions would not occur before at least letter identity had been processed, so if color can be used to select a response earlier than letter identity there should be a same-hand benefit in LRP onset on the GO trials of the hand:color condition. Because on all trials of all three conditions both the color and the identity of the stimulus letter have to be recognized, P3 latency should not differ across the S–R mapping conditions. This would rule out the possibility that the same-hand benefit might be the result of faster stimulus recognition. In these three mapping conditions, the probability of a NOGO trial was .50.

Figure 23 depicts the latencies of the overt button press response, the onset of the response at the EMG level, the peak of the P3 component, and the onset of the LRP. As can be seen, a clear and statistically significant

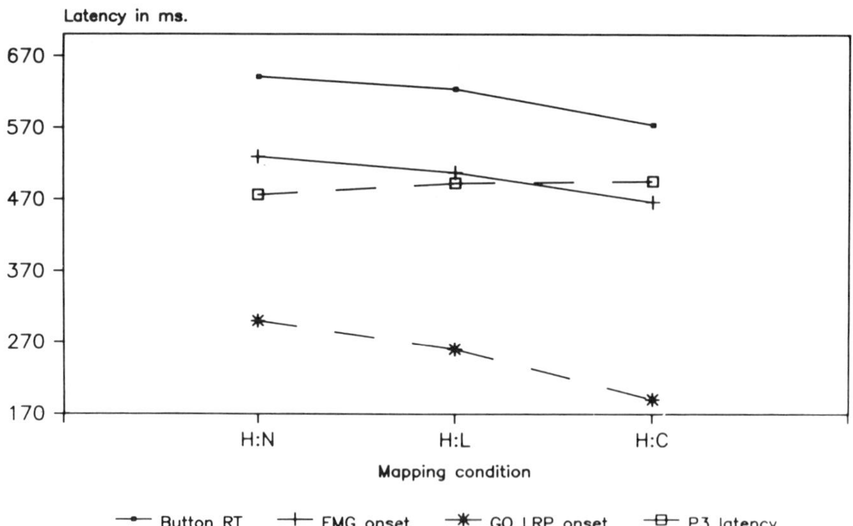

Figure 23.   Latencies in milliseconds of button presses (RT), correct EMG onset (CEMG), onset of lateralization of RP (LRP), and peak of P3 component (P3) in GO trials of three S–R mapping conditions. H:N denotes Hand:Neither; H:L denotes Hand:letter; H:C denotes Hand:Color. Note clear effects of S–R mapping on LRP onset, EMG onset, and RT, but not on P3 latency.

same-hand benefit (70 msec) in the hand:color condition was obtained at the RT level. This result is consistent with the earlier findings of Miller (1982), Proctor and Reeve (1985), and De Jong et al. (1988), which indicate that responses are faster when color can be used to discriminate between hands. Moreover, the EMG and LRP onset latencies also showed a clear same-hand benefit, indicating that the same-hand benefit was present even at the central motor level.

The peak latencies of the P3 component were not affected by S–R assignment, nor by the interaction of processing GO or NOGO stimuli with S-R mapping. These results are consistent with those obtained by De Jong et al. (1988) and suggest that the time at which full information about the stimulus becomes available was not affected by how the four finger responses were assigned to the stimuli. This makes it unlikely that the same-hand benefit obtained when color discriminated between hands is the result of faster stimulus identification.

In Figure 24 are depicted the LRP waveforms obtained in the GO trials of the three S–R mapping conditions together with the LRP obtained in the NOGO trials of the hand:color condition. In this latter condition, LRP waveforms were derived on GO and NOGO trials by subtracting the C3′–C4′ ERPs evoked by the color indicating a right-hand response from those evoked by the color indicating a left-hand response. There is a clear LRP present on NOGO trials in the hand:color condition, which was significantly present from 260 msec until 360 msec post stimulus. This NOGO LRP reached highly significant amplitudes as tested by the combined Wilcoxon–t test [$t(23) > 3.49$; $p < .001$). NOGO trials on which increased EMG activity was observed were essentially absent ($<1\%$) and were removed from the analyses, so that we were sure that the subjects had always processed both stimulus attributes.

Note that if in NOGO trials color and letter identity had been recognized before response activation there would have been no reason for activating the response hand. Thus, the presence of the NOGO LRP strongly suggests that in the hand:color condition partial information concerning the color of the stimulus is transmitted and used by central motor processes before information concerning letter identity is available. Because the LRP represents differential motor activation of the two hands, the LRP data indicate that this partial information affects motor processes. This finding supports the interpretation of Miller (1982) that the same-hand benefit results from advance selection of response hand based on partial information.

Several other conditions were applied in this experiment, which yielded similar and consistent results. The hand:color task was used in three versions, either with 50%, 25%, or no NOGO trials. Whether there were NOGO trials or not did not affect LRP onset to color, but P3 latency was increased and EMG onset delayed when NOGO trials occurred. These results suggest that insertion of NOGO trials (i.e., increasing the number of S–R alternatives) increased the time to identify the letter, but not the color of the letter. When

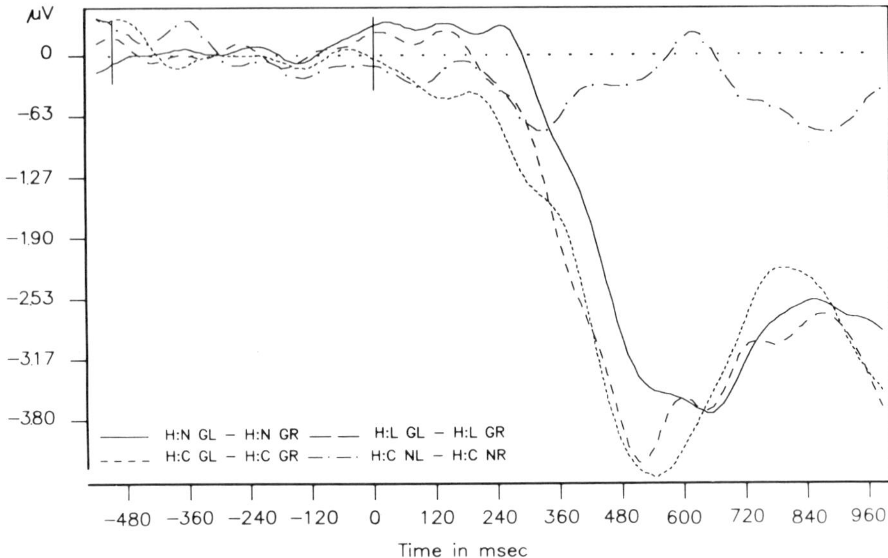

Figure 24.   LRP waveforms obtained in hand:neither (H:N), hand:letter (H:L), and hand:color (H:C) conditions. Stimulus presentation occurred at time 0 (second vertical line). Warning signal occurred 500 msec before onset of stimulus (first vertical line at time − 500). In H:C condition, two LRPs were found, one in GO trials and one in NOGO trials. In GO trials, LRP was obtained by subtracting C3′–C4′ waveforms accompanying right-hand responses from those accompanying left-hand responses (thus, in "H:C GL − H:C GR", 'H:C' denotes hand:color, 'GL' denotes the left-hand responses in GO trials, and 'GR' denotes right-hand responses in GO trials). In NOGO trials, LRP was obtained by subtracting C3′–C4′ waveforms evoked by color signaling right-hand response from those evoked by color signaling left-hand response (thus, in "H:C NL − H:C NR", 'H:C' again denotes the hand: color condition, 'NL' denotes the left-hand color in NOGO trials, and 'NR' denotes right-hand color in NOGO trials). Summarized, "H:C GL − H:C GR" stands for hand:color GO left minus hand:color GO right (dotted waveform); "H:C NL − H:C NR" stands for hand:color NOGO left minus hand:color NOGO right (dash-dot waveform). In same vein, LRPs in other conditions are denoted: "H:N GL − H:N GR" stands for hand:neither GO left minus hand:neither GO right (continuous waveform), and "H:L GL − H:L GR" means hand:letter GO left minus hand:letter GO right (dashed waveform). (Amplitudes are in microvolts.)

the relative frequency of NOGO trials in the hand:color condition was 25%, an LRP was also found on the NOGO trials.

Further, a hand:letter condition was used that was a mirror image of the hand:color condition. In this hand:letter task, one of two letters appeared, each signaling two fingers on one hand (e.g., M signaled two fingers on the left hand and Q signaled two fingers on the right hand). These two letters could appear in one of four colors. Two of these colors signaled that no response had to be made (e.g., white and blue), whereas two other colors each signaled two fingers on different hands (e.g., green signaled both middle fingers and red both index fingers). As in the hand:color condition, an LRP was found in the NOGO trials of this condition. This finding suggests that letter identity can also be used to prepare response hand before full stimulus information has been processed.

In summary, the results of this experiment strongly suggest that perceptual analysis of a single two-dimensional stimulus may transmit information about one dimension (i.e., partial information) before it has finished analyzing the other dimension. They further suggest that motor processes receiving partial information subsequently use this information to prepare a response in advance.

Selective Response Processing: The Case of Stimulus Features

Visual search tasks are used to study how people detect a prespecified piece of information in highly varying visual environments. In such tasks, displays are presented that contain several elements and the subject has to choose one response when a specific target form is present among them and another response when it is absent.

As a starting point for our study (Smid et al., 1991), we used the Duncan and Humphreys theory of visual search (Duncan and Humphreys, 1989). This theory incorporates the main mechanisms of earlier theories and can explain a wide range of experimental data. Duncan and Humphreys proposed three processes enabling the detection of a target among nontargets. One process produces a parallel, hierarchically structured representation of the visual field at several levels of spatial scale. This is accomplished by segmenting and organizing the visual input into structural units according to gestalt principles. Another process matches input descriptions against an internal template of the information searched for. Nonmatching elements lead to a lower, and matching elements to a higher, probability of being selected for further processing. The third process transfers the selected elements to visual short-term memory (VSTM), where an eventual target enters awareness and can become input for cognitive and/or motor processes.

However, as the results in the response priming paradigm of Eriksen and Eriksen (1974) suggested, it might be that a response could already be selected and prepared before search is completed. A positive response might be initially selected and prepared based on features that nontargets currently processed share with the target (e.g., when the target is a Q and nontargets in the display are C's and D's). A negative response might be selected initially on the basis of features that nontargets currently processed do not share with the target(s) (e.g., when the target is Q and the current nontargets are W's and N's).

If preliminary response activation on the basis of features of letter symbols occurs, it would suggest that information transmission between the search process and response activation processes occurs in a continuous fashion and not discretely. That is, response activation would be based initially on information having a very small grain size (Miller, 1989). This information might represent an unfinished percept of the contents of the display. Later, when perceptual analysis has detected and identified a target (or has found no target), response activation would be based on the identity of this target (or on the absence of a target).

To test this prediction of the continuous flow conception, a task was constructed, based on the response competition paradigm (e.g., Eriksen and Eriksen, 1974), which ensured that search would have to be carried out. The search task consisted of detecting a target that could appear in any one of the four corners of an imaginary square with equal probability. On two-thirds of the trials nontargets were present on the other three corners of the square, and on one-third the target appeared alone. Two targets were used, one (e.g., Q) as the target for a left-hand response and another one (e.g., H) as the target for a right-hand response.

When nontargets were present in the display, they had physical features in common with one of the targets, either with the one in the display or with the one not in the display. Thus the nontargets used were neutral with regard to their relevance for a response on the level of a letter, but compatible or incompatible on the level of the features of which they were composed. This procedure produced response-compatible displays (e.g., a Q with nontargets D, G, C), response-incompatible displays (e.g., a Q with nontargets N, W, M), and target-alone displays (e.g., a Q on one of the four corners). A target was always present; thus, there were no target-absent displays.

Several psychophysiological measures were used besides RT. As a relative index for the duration of the set of visual search processes, the peak latency of the P3 component was used. In the current task, duration of visual search consists of the time needed for target detection and identification. Brookhuis et al. (1981) and Van Dellen et al. (1985), using a similar task, showed that P3 latency is a reliable index for the duration of visual search (see also Kramer and Strayer, 1988). Perceptual processing of the target-alone displays consists only of the time needed for target identification, because no nontargets are present. Perceptual processing of compatible and in-compatible displays consists of target detection plus target identification. It was therefore expected that P3 latencies (and of course RTs) would be longer to these latter displays than to target-alone displays.

The P3 latency measure seems to be especially useful here because it allows distinguishing between effects on search duration and effects on response processes. Smid et al. (1990) observed that the presentation of a target with incompatible targets on irrelevant locations delayed both P3 latency and RT, compared to the presentation of a target with letters that were not targets in the task (i.e., neutral letters). The authors attributed the effect on P3 latency to competition among letter identities that are candidates for recognition. The effect on RT was attributed to response competition as was evidenced by speed-accuracy functions and preliminary incorrect response activation produced by the incompatible targets. Because the distractors used in the current study are nontargets [i.e., comparable to the neutral letters in the study by Smid et al. (1990)], it was expected that P3 latencies produced by compatible and incompatible displays would be about equal, while RTs to incompatible displays would be delayed because of response competition.

As an index of the start of selective response preparation the lateralized readiness potential (LRP) was used. If selective response preparation begins before the target has been detected and identified, the compatible and incompatible displays should activate the correct and the incorrect response, respectively, at about the same time as target-alone displays would activate the correct response. This would suggest that there is continuous information transmission between search processes and response processes, starting with information from the physical features of the nontargets that are shared with one of the two targets and ending with information about the target actually present in the display. If selective response activation begins only after a target has been detected, compatible and incompatible displays should activate the correct response only and at a much later time than target-alone displays. As an index for the beginning of peripheral response execution, the start of a reliable EMG was used.

We found that RTs and P3 latencies were longer in trials in which a target with nontargets appeared than in trials in which only a target was presented (Fig. 25). This indicates that on target-alone trials the target only had to be identified, whereas on trials in which nontargets also appeared, additional time was required to detect the target among the nontargets. Thus, visual search delayed both P3 latencies and RTs if nontargets were presented. These results are consistent with previous results presented by Brookhuis et al. (1981), Hoffman, Simons, and Houck (1983), and Van Dellen et al. (1985). In all these studies, P3 latency and RT were found to be positively related to the number of nontargets in a display.

If nontargets were present in the display, they could have features that were either compatible or incompatible with the target in the display. In the upper panel of Figure 26, the P3 waveforms obtained in target alone, compatible and incompatible trials are shown. As can be seen both in figure 4 and 5, P3 latencies to displays with compatible nontargets were not different from those to displays with incompatible nontargets. However, RTs were slower and EMG onsets were later to incompatible displays than to compatible ones. Further, the number of overt button press errors and covert EMG errors produced was also larger on incompatible than on compatible trials.

Significant lateralization of the RP began at about the same time for all display types and concerned the response (correct or incorrect) signaled by most the physical features in the display. On incompatible trials the

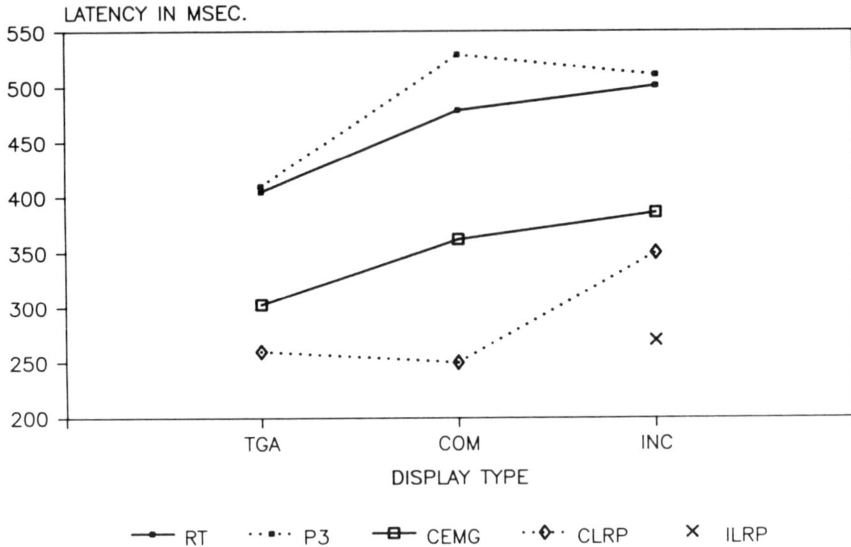

Figure 25.   P3 peak latencies, correct EMG (CEMG) onset, correct LRP (CLRP) onset, incorrect LRP (ILRP) onset latencies, and RTs in milliseconds averaged across no-error trials as function of display type ('TGA' = TarGet Alone; 'COM' = COMpatible; 'INC' = INCompatible displays).

incorrect response channel was activated first at about the same time as the correct channel was activated first at about the same time as the correct channel was activated on compatible and target-alone trials. In Figure 26, activation of the incorrect response in incompatible trials is visible as a small positive LRP (the dotted waveform) between 200 and 320 msec post stimulus. At the same time, a small negative LRP (the dashed waveform) can be seen, indicating activation of the correct response on compatible trials. On both types of trials the waveforms accelerated negatively after 320 msec, indicating increasing activation and execution of the correct response. Note the remarkable similarity of these LRPs with the LRPs obtained in the response priming paradigm discussed earlier.

These results are consistent with and extend the previous findings of Gratton et al. (1988) and Smid et al. (1987, 1990). In combination, the findings presented here confirm that response competition took place. If the incompatible response is activated first, activation of the correct response has to compete with it. This conflict has to be resolved, which delays activation of the correct response and RT in addition to the time needed for visual search.

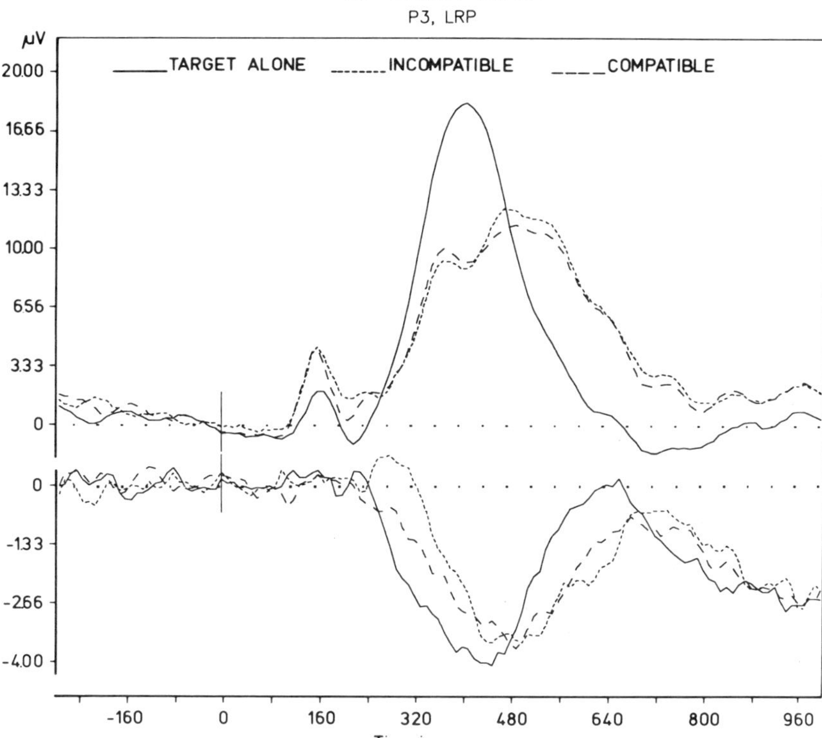

Figure 26.   Stimulus-locked (stimulus onset at time 0) P3 (upper panel) and LRP (lower panel) waveforms averaged across no-error trials as function of display type. Note that P3 waveforms in compatible trials (dashed waveforms) are nearly identical to those in incompatible trials (dotted waveforms). (Amplitudes are in microvolts.)

The LRP results not only suggest that response competition produced delayed EMG onsets and RTs, but also that response preparation began before the target was detected and fully identified. The preliminary character of response preparation follows from two observations. On incompatible trials, central activation of the incorrect response was not accompanied by peripheral (EMG) response activation, but was followed by central activation of the correct response. On compatible trials, activation of the correct response began at the same time as on target-alone trials, but peripheral (EMG) activation of the correct response was postponed and began much later than on target alone trials.

On the basis of these findings we concluded that partial information about nontarget letters was transmitted to, and served as input for, the response activation processes. The findings suggest that letter identification produced output of such a small grain size that it may best be characterized as continuous instead of discrete. This is strong evidence for the continuous flow conception of information processing. It seems reasonable therefore to conclude that the AFM is limited in its application to visual search tasks.

However, one may think about plausible alternative explanations. One alternative could be that a letter that looks like a target activates the representation of that target. Next, this representation sends a discrete output (i.e., a guess) to its associated response, activating it below some criterion level. The problem with this alternative explanation is that the motor system receives no information about the correctness of the output it receives. The result would be that it would always produce an incorrect response when incompatible nontargets are present in the display. This alternative therefore needs to postulate additional assumptions.

Another alternative is that the perceptual output responsible for preliminary response selection does not arise from the letter identification process. It might stem from a global feature analysis process such as proposed by Hoffman (1979) and Treisman and Gelade (1980). This process might send a low-level, discrete output (representing either roundness or angledness of features in the display), which activates a response always below its criterion level (i.e., for example only at the LRP level). Only output from the higher level letter identification process can activate it above its criterion level (e.g., at the EMG level). The problem with this alternative explanation is that we found that the incorrect responses to nontargets are not activated at only one level (e.g., LPR), but also very frequently at other levels (e.g., EMG). Thus, the initial output does not seem to be as discrete as it should be in discrete models.

Our results suggest effects on response selection of elementary products of perceptual processes that are hard to model in a discrete conception of human information processing.

Conclusions

In the studies we have reviewed, several tasks were used, each with its own characteristic stimuli and response requirements. The combination of filtering and selective-set paradigms made it possible to explore the effects of selective cuing and to explore the conditions in which selective processing occured. Cognitive processes such as search and mental rotation in general appeared to be selective: the associated negativities were almost always only present in the attended channel. On the other hand, response processing is not always selective. This became evident if we use indices of central (LRP) and peripheral response activation (EMG) in addition to overt response.

Results obtained in three paradigms indicate that information transmission between perceptual and motor processes consists of at least two steps. The letter identification process even seems to transmit continuous output. This leads to the conclusion that, for a number of important paradigms, it is questionable whether response selection is selective. Response activation can occur quite early, in some conditions about 170 msec, while overt responding occurs about 400 msec later. EMG onset precedes overt responding about 100 msec. A large proportion of overt response errors appear to be related to incorrect motor activation, but these type of errors can never be detected with traditional measures. ERPs, LRPs, and EMGs appear to provide additional and independent information about the time course of human information processing.

Acknowledgments. The authors wish to express their gratitude to Mick Rugg for his valuable comments on an earlier draft of this chapter.

References

Alho K, Sams M, Paavilainen P, Näätänen R (1986): Small pitch separation and the selective-attention effect on the ERP. *Psychophysiology* 23:189–197.

Alho K, Lavikainen J, Reinikainen K, Sams M, Näätänen R (1990): Event-related brain potentials in selective listening to frequent and rare stimuli. *Psychophysiology* 27:73–86.

Alho K, Paavilainen P, Reinikainen K, Sams M, Näätänen R (1986): Separability of different negative components of the event-related potential associated with auditory stimulus processing. *Psychophysiology* 23:613–623.

Alho K, Sams M, Paavilainen P, Reinikainen K, Näätänen R (1989): Event-related brain potentials reflecting processing of relevant and irrelevant stimuli during selective listening. *Psychophysiology* 26:514–528.

Alho K, Donauer N, Paavilainen P, Reinikainen K, Sams M, Näätänen R (1987a): Stimulus selection during auditory spatial attention as expressed by event-related potentials. *Biol Psychol* 24:153–162.

Alho K, Töttölä K, Reinikainen K, Sams M, Näätänen R (1987b): Brain mechanisms of selective listening reflected by event-related potentials. *Electroencephalogr Clin Neurophysiol* 68:458–470.

Baddeley, AD (1983): Working memory. *Phil Trans R Soc London* 302:311–324.

Broadbent DE (1958): *Perception and Communication.* London: Pergamon Press.

Broadbent DE (1970): Stimulus set and response set: Two kinds of selective attention. In: *Attention: Contemporary Theory and Analysis* Mostofsky DI, ed. New York: Appleton-Century-Crofts.

Brookhuis KA (1989): Event-Related Potentials and Information Processing. Doctoral Thesis, University of Groningen.

Brookhuis KA, Mulder G, Mulder LJM, Gloerich ABM, Van Dellen HJ, Van der Meere JJ, Ellermann HH (1981): Late positive components and stimulus evaluation time. *Biol Psychol* 30:107–123.

Brookhuis KA, Mulder G, Mulder LJM, Gloerich ABM (1983): The P3 complex as an index of information processing: The effects of response probability. *Biol Psychol* 17:277–296.

Bullock D, Grossberg S (1988): Neural dynamics of planned arm movements: Emergent invariants and speed accuracy properties during trajectory formation. *Psychol Rev* 1:49–90.

Cherry EC (1953): Some experiments on the recognition of speech with one and two ears. *J Acoust Soc Am* 25:957–979.

Coles MGH, Gratton G, Bashore TR, Eriksen CW, Donchin E (1985): A psychological investigation of the continuous flow model of human information processing. *J Exp Psychol Hum Percept Perform* 11:529–553.

Cooper LA, Shepard RN (1973): Chronometric studies of the rotation of mental images. In: *Visual Information Processing*, Chase WG, ed., pp. 75–176. New York: Academic Press.

Cooper R, Osselton JW, Shaw JC (1980): *EEG Technology*, 3d Ed. London: Butterworths.

Cowey A (1979): Cortical maps and visual perception. *Q J Exp Psychol* 31:1–17.

Cowey A (1985): Aspects of cortical organization related to selective attention and selective impairments of visual perception: A tutorial review. In: *Attention and Performance XI*, Posner MI, Marin O, eds., pp. 41–62. Hillsdale, NJ: Erlbaum.

Deecke L, Grzinger B, Kornhuber HH (1976): Voluntary finger movements in man: Cerebral potentials and theory. *Biol Cybernet* 23:99–119.

De Jong R, Coles MGH, Logan GD, Gratton G (1990): In search of the point of no return: The control of response processes. *J Exp Psychol Hum Percept Perform* 16:164–182.

De Jong R, Wierda M, Mulder G, Mulder LJM (1988): The use of partial information in response preparation. *J Exp Psychol Hum Percept Perform* 14:682–692.

de Munck JC (1989): *A Mathematical and Physical Interpretation of the Electromagnetic Field of the Brain*. Doctoral Thesis, University of Amsterdam.

Deutsch JA, Deutsch D (1963): Attention: Some theoretical considerations. *Psychol Rev* 70:80–90.

Donders FC (1868/1969): On the speed of mental processes. In: *Attention and Performance II*, Koster WG, ed., pp. 412–431 Koster WG, trans. Amsterdam: North-Holland.

Duncan J, Humphreys GW (1989): Visual search and stimulus similarity. *Psychol Rev* 96:433–458.

Eason RG (1981): Visual evoked potential correlates of early neural filtering during selective attention. *Bull Psychonom Soc* 18:203–206.

Eason R, Harter M, White C (1969): Effects of attention and arousal on visually evoked cortical potentials and reaction time in man. *Physiol Behav* 4:283–289.

Eason RG, Oakley M, Flowers L (1983): Central neural influences on the human retina during selective attention. *Physiol Psychol* 11:18–28.

Eriksen B, Eriksen CW (1974): Effects of noise letters upon the identification of a target letter in a nonsearch task. *Percept Psychophys* 16:143–149.

Eriksen CW, Coles MGH, Morris LR, O'Hara WP (1985): An electromyographic examination of response competition. *Bull Psychonom Soc* 23:165–168.

Farah MJ, Peronnet F (1989): Event-related potentials in the study of mental imagery. *J Psychophysiol* 3:99–109.

Gratton G, Coles MGH, Sirevaag E, Eriksen CW, Donchin E (1988): Pre- and poststimulus activation of response channels: A psychophysiological analysis. *J Exp Psychol Hum Percept Perform* 14:331–344.

Gunter TC, Van der Zande RD, Wiethoff M, Mulder G, Mulder LJM (1987): Visual selective attention during meaningful noise and after sleep deprivation. Current Trends in Event-Related Potential Research EEG Supplement 4. Elsevier: Amsterdam.

Hansen JC, Hillyard SA (1983): Selective attention to multidimensional auditory stimuli. *J Exp Psychol Hum Percept Perform* 9:1–19.

Harter MR, Aine CJ (1984): Brain mechanisms of visual selective attention. In: Varieties of Attention Parasuraman R, Davies DR, eds., pp. 293–321. Orlando: Academic Press.

Harter MR, Guido W (1980): Attention to pattern orientation: Negative cortical potentials, reaction time, and the selection process. *Electroencephalogr Clin Neurophysiol* 49:461–475.

Harter MR, Previc FH (1978): Size-specific information channels and selective attention: Visual evoked potential and behavioral measures. *Electroencephalogr Clin Neurophysiol* 45:628–640.

Harter MR, Salmon LE (1972): Intra-modality selective attention and evoked cortical potentials to randomly presented patterns. *Electroencephalogr Clin Neurophysiol* 32:605–613.

Harter MR, Aine CJ, Schroeder C (1982): Hemispheric differences in the neural processing of stimulus location and type: Effects of selective attention on visual evoked potentials. *Neuropsychologia* 20:412–438.

Hillyard SA, Kutas M (1983): Electrophysiology of cognitive processing. *Annu Rev Psychol* 34:33–61.

Hillyard SA, Mangun GR (1986): The neural basis of visual selective attention: A commentary on Harter and Aine. *Biol Psychol* 23:266–279.

Hillyard SA, Münte TF (1984): Selective attention to color and location: An analysis with event-related brain potential. *Percept Psychophys* 36:185–198.

Hillyard SA, Münte TF, Neville HJ (1985): Visual-spatial attention, orienting, and brain physiology. In: *Attention and Performance XI*, Posner MI, Marin OSM, eds., pp. 63–84. Hillsdale, NJ: Erlbaum.

Hillyard SA, Hink RF, Schwent VL, Picton TW (1973): Electrical signs of selective attention in the human brain. *Science* 182:177–180.

Hillyard SA, Simpson GV, Woods DL, Van Voorhis S, Münte T (1984): In: Event-related brain potentials and selective attention to different modalities. *Cortical Integration*, Reinoso-Suarez F, Ajmone-Marsan C, eds., pp. 395–414. New York: Raven Press.

Hoffman JE (1979): A two-stage model of visual search. *Percept Psychophys* 25:319–327.

Hoffman JE, Nelson B (1981): Spatial selectivity in visual search. *Percept Psychophysic* 25:319–327.

Hoffman JE, Simons RF, Houck MR (1983): Event-related potentials during controlled and automatic targets detection. *Psychophysiology* 20:625–632.

Houck MR, Hoffman JE (1986): Conjunction of color and form without attention: evidence from an orientation–contingent color aftereffect. *J Exp Psychol Hum Percept Perf* 12:186–199.

Hughes HC, Zimba LD (1985): Spatial maps of directed visual attention. *J Exp Psychol Hum Percept Perf* 25:319–327.

Kahneman D, Treisman A (1984): Changing views of attention and automaticity. In: *Varieties of Attention*, Parasuraman R, Davies DR, eds., pp. 29–61. London: Academic Press.

Kenemans JL, Kok A, Smulders FTY (1993): Event-related potentials to conjunctions of spatial frequency and orientation as a function of stimulus parameters and response requirements. *Electroencephalography Clin Neurophys* 88:51–63.

Kramer AF, Strayer DL (1988): Assessing the development of automatic processing: An application of dual-task and event-related brain potential methodologies. *Biol Psychol* 26:231–267.

Kramer AF, Sirevaag EJ, Braune R (1987): A psychophysiological assessment of operator workload during simulated flight missions. *Hum Factors* 29:145–160.

Kramer AF, Schneider W, Fisk A, Donchin E (1986): The effects of practice and taskstructure on components of the event-related brain potential. *Psychophysiology* 23:33–47.

Kutas M, Donchin E (1977): The effects of handedness, of responding hand, and of response force on the contralateral dominance of the readiness potential. In: *Attention, Voluntary Contraction and Event-Related Cerebral Potentials*, Desmedt J, ed., pp. 189–210. Basel: Karger.

Kutas M, Donchin E (1980): Preparation to respond as manifested by movement-related brain potential. *Brain Res* 202:95–115.

Mangun GRR, Hillyard SA (1987): The spatial allocation of visual attention as indexed by event-related brain potentials. *Hum Factors* 29:195–211.

Mangun GR, Hillyard SA (1988): Spatial gradients of visual attention: Behavioral and electrophysiological evidence. *Electroencephalogr Clin Neurophysiol* 70:417–428.

Mangun GR, Hillyard SA, Luck SJ (1993): Electrocortical substrates of visual selective attention. Attention and Performance, Kornblum S, Meyer DE, eds., 14:219–243, Erlbaum: Hillsdale, NJ.

Mangun GRR, Hansen JC, Hillyard SA (1986): Electroretinograms reveal no evidence for centrifugal modulation of retinal input during selective attention in man. *Psychophysiology* 23:156–165.

Miller J (1982): Discrete versus continuous stage models of human information processing: In search of partial output. *J Exp Psychol Hum Percept Perform* 8:273–296.

Miller J (1988): Discrete and continuous models of human information processing: Theoretical distinctions and empirical results. *Acta Psychol* 67:191–257.

Mulder G, Gloerich ABM, Brookhuis KA, van Dellen HJ, Mulder LJM (1984): Stage analysis of the reaction process using brain-evoked potentials and reaction time. *Psychol Res* 46:15–32.

Näätänen R (1982): Processing negativity: An evoked-potential reflection of selective attention. *Psychol Bull* 92:605–640.

Näätänen R (1986): The neural-specificity theory of visual selective attention evaluated: A commentary on Harter and Aine. *Biol Psychol* 23:281–295.

Näätänen R, Gaillard AWK (1983): The orienting reflex and the N2 deflection of the event-related potential (ERP). In: *Tutorials in Event-Related Potential Research: Endogenous Components*, Gaillard AWK, Ritler W, eds., pp. 119–141. Amsterdam: North-Holland.

Näätänen R, Michie PT (1979): Early selective attention effects on the evoked potential: A critical review and reinterpretation. *Biol Psychol* 8:81–136.

Neville HJ, Lawson D (1987): Attention to central and peripheral visual space in a movement detection task: an event-related potential and behavioral study. I. Normal hearing adults. *Brain Res* 405:253–267.

Nissen MJ (1985): Accessing features and objects: Is location special? In: *Attention and Performance XI*, Posner MI, Marin O, eds., pp. 205–220. Hillsdale, NJ: Erlbaum.

Okita T, Wijers AA, Mulder G, Mulder LJM (1985): Memory search and visual spatial attention: An event-related brain potential analysis. *Acta Psychol* 60:263–292.

Osman A, Bashore TR, Coles MGH, Donchin E, Meyer DE (1992): On the Transmission of Partial Information: Inferences from Movement–Related Brain Potentials. *J Exp Psychol Hum Percept Perf* 18:217–232.

Peronnet F, Farah MJ (1989): Mental rotation: An event-related potential study with a validated mental rotation task. *Brain Cognit* 9:279–288.

Posner ML (1978): *Chronometric Explorations of Mind.* Hillsdale, NJ: Erlbaum.

Previc FH, Harter MF (1982): Electrophysiological and behavioral indicants of selective attention to multifeature gratings. *Percept Psychophys* 32:465–472.

Proctor RW, Reeve TG (1985): Compatibility effects in the assignment of symbolic stimuli to discrete finger responses. *J Exp Psychol Hum Percept Perform* 11:623–639.

Requin J (1985): Looking forward to moving soon: Ante factum selective processes in motor control. In: *Attention and Performance XI*, Posner MI, Marin OSM, eds. Hillsdale NJ: Erlbaum.

Ruchkin DS, Johnson R, Jr., Mahaffey D, Sutton S (1988): Toward a functional categorization of slow waves. *Psychophysiology* 25:339–353.

Rugg MD, Milner AD, Lines CR, Phalp R (1987): Modulation of visual event-related potentials by spatial and non-spatial visual selective attention. *Neuropsychologia* 25:85–96.

Schneider W, Shiffrin RM (1977): Controlled and automatic human information processing: I. Detection, search, and attention. *Psychol Rev* 84:1–66.

Shiffrin RM, Schneider W (1977): Controlled and automatic human information processing: II. Perceptual learning, automatic attending and a general theory. *Psychol Rev* 84:127–190.

Smid HGOM, Mulder G, Mulder LJM (1987): The continuous flow model revisited: Perceptual and motor aspects. Current Trends in Event-Related Potential Research, EEG Suppl 40, Johnson R, Rohrfbaugh, Parasuraman R, eds., Elsevier: Amsterdam.

Smid HGOM, Mulder G, Mulder LJM (1990): Selective response activation can begin before stimulus recognition is complete: A psychophysiological and error analysis of continuous flow. *Acta Psychol* 74:169–201.

Smid HGOM, Mulder G, Mulder LJM, Brands GJ (1992): A psychophysiological study of the use of partial information in stimulus-response translation. *J Exp Psychol Hum Percept Perform* 18:1101–1119.

Smid HGOM, Lamain W, Hogeboom MM, Mulder G, Mulder LJM (1991): A psychophysiological study of response preparation during visual search. *J Exp Psychol Percept Perform* 17:696–714.

Sternberg S (1969): On the discovery of processing stages. *Acta Psychol* 30:276–315.

Sperling G (1960): The information available in brief visual presentation. Psychological monographs 74:498.

Stuss DT, Sarazin FF, Leech EE, Picton TW (1983): Event-related potentials during naming and mental rotation. *Electroencephalogr Clin Neurophysiol* 56:133–146.

Treisman AM, Gelade G (1980): A feature-integration theory of attention. *Cognit Psychol* 12:97–136.

Van Dellen HJ, Brookhuis KA, Mulder G, Okita T, Mulder LJM (1985): Evoked potential correlates of practice in a visual search task. In: *Clinical and Experimental Neurophysiology*, Papakoustopoulos D, Butler S, Martin I, eds., pp. 132–155. Beckenham: Croom Helm.

Van der Heijden AHC (1992): Selective Attention in Vision. Routledge: London.

Van Voorhis S, Hillyard SA (1977): Visual evoked potentials and selective attention to points in space. *Percept Psychophys* 22:54–62.

Wijers AA (1989): *Visual Selective Attention. An Electrophysiological Approach.* Doctoral Thesis, Groningen.

Wijers AA, Dunajski Z, Peters M, Mulder G (1992): Magnetic brain responses in visual spatial attention. Biomagnetism: Clinical Aspects, Hoke M, Erne SN, Okada YC, Romani GL, eds., pp. 213–216. Elsevier: Amsterdam.

Wijers AA, Mulder G, Okita T, Mulder LJM, Scheffers MK (1989a): Attention to colour: An ERP-analysis of selection, controlled search and motor activation. *Psychophysiology* 26:89–109.

Wijers AA, Mulder G, Okita T, Mulder LJM (1989b): An ERP-study on memory search and selective attention to lettersize and conjunctions of lettersize and color. *Psychophysiology* 26:529–547.

Wijers AA, Mulder G, Otten L, Feenstra S, Mulder LJM (1989c): Brain potentials during selective attention, memory search and mental rotation. *Psychophysiology* 26:452–467.

Wijers AA, Lamain W, Slopsema S, Mulder G, Mulder LJM (1989d): An electrophysiological investigation of the spatial distribution of attention to colored stimuli in focused and divided attention conditions. *Biol Psychol* 29:213–245.

Wijers AA, Okita T, Mulder G, Mulder LJM, Lorist MM, Poiesz R, Scheffers MK (1987): Visual search and spatial attention: ERPs in focussed and divided attention conditions. *Biol Psychol* 25:33–60.

Yantis S, Jonides J (1984): Abrupt visual onsets and selective attention: Evidence from visual search. *J Exp Psychol Hum Percept Perform* 10:601–621.

Yantis S, Jonides J (1990): Abrupt visual onsets and selective attention: Voluntary versus automatica allocation. *J Exp Psychol Hum Percept Perform* 16:121–134.

Yantis S, Johnston JC (1990): On the locus of visual selection: Evidence from focused attention tasks. *J Exp Psychol* 16(1):135–149.

Chapter 3

Orienting Attention in the Visual Fields: An Electrophysiological Analysis

GEORGE R. MANGUN

The mechanisms of attentional orienting in space and the selective processing of visual events were examined in a trial-by-trial cuing task. Subjects were cued by a left- or right-pointing arrow to orient covert attention to a lateral field location. The subjects' task was to discriminate the features of a subsequently flashed target stimulus if at the cued location. Event-related brain potentials (ERPs) were recorded from the scalp in response to both the cue and the target. During the cue–target interval, two effects were observed. The first was in the latency range of 250–350 msec post cue and was manifest as a relative negativity over the parietal-temporal scalp contralateral to the direction of the cue. The second effect was observed only over the right hemisphere between 300 and 500 msec latency; the response to the left cue was more negative than the response to the right cue. The ERPs to the subsequent target stimuli showed enhancements of the sensory-evoked occipital P1 (100–140 msec) and N1 (160–200 msec) peaks when the targets appeared at the cued location. These findings suggest that in response to the attention-directing cues there are lateralized executive orienting mechanisms that enable selective sensory processing during spatial attention.

Introduction

Selective attention to objects or events in the visual world aids in their perception. Such attentional processes may operate overtly, such as in the

Cognitive Electrophysiology
H-J. Heinze, T.F. Münte, and G.R. Mangun, editors
© 1994 Birkhäuser Boston

movement of the eyes to foveate a region of the visual field, or may instead proceed covertly in the absence of eye movements. The process of covert attentional orienting is the topic of this chapter. Evidence from electrophysiological studies in healthy human subjects is presented to investigate the brain processes involved in the orienting of visual attention to relevant spatial locations.

Behavioral studies of attention have shown that selective spatial attention can improve both the detection and discrimination of stimuli at attended locations (Downing, 1988; Eriksen and Yeh, 1985; Hawkins et al., 1990; Hoffman and Nelson, 1981; Jonides, 1981; Müller and Rabbitt, 1989; Posner, Nissen, and Ogden, 1978; Posner, Snyder, and Davidson, 1980; Reinitz, 1990; Van der Heijden et al., 1987). In such tasks, the subjects are typically cued on a trial-by-trial basis to expect a target at a particular spatial location. The target usually appears at the cued location, but occasionally appears at uncued locations. The reaction times or detection scores for a given target are compared as a function of whether or not it had been correctly cued. These studies have established that reaction times to targets at cued locations (valid targets) are faster than those to identical physical stimuli presented to the same uncued retinal locus (e.g., Posner, Nissen, and Ogden, 1978).

Speeded reaction times have been interpreted as resulting from the facilitating effects of focal attention on the encoding or processing of the cued versus uncued targets (Posner, 1980). Although this is a plausible interpretation, it has been appropriately challenged by those who have suggested that speeded reaction times to cued targets might not result from changes in sensory-perceptual processing, but rather from differences in the subject's willingness to respond (i.e., decision/response criterion) to targets occurring at cued versus uncued locations (e.g. Müller and Findlay, 1987; Shaw, 1984; Sperling, 1984; Sperling and Dosher, 1986). It is likely that both mechanisms of improved sensitivity and changes in decision bias are operative in different tasks, and the question has centered on whether changes in sensitivity are ever elicited during spatial cuing/spatial attention and, if so, under what circumstances. Recent experiments support the idea that expectancy-based cuing of target locations leads to changes in sensory-perceptual processing for detection of targets. These studies found improvements for detection sensitivity in the absence of changes in decision criterion during spatial cuing tasks (Downing, 1988; Hawkins et al., 1990).

Electrophysiological Studies of Selection

Electrophysiological studies using cuing paradigms have added to the understanding of the nature of changes in information processing of cued and uncued targets by recording brain responses to the two stimulus conditions. For example, when central field arrow cues indicate the most probable location of a target stimulus, enhancements of ERP components over the occipital scalp have been consistently elicited in response to cued versus uncued targets (Mangun and Hillyard, 1991). Such a finding supports the idea that behavioral improvements reflect changes in the sensory-perceptual processing of the attended stimuli. Moreover, the ERP evidence provides information as to the locus of attentional selectivity in the visual pathways. Specifically, the early P1 attention effect appears to reflect visual information processing within the extrastriate visual cortex (Mangun, Hillyard, and Luck, 1993). These findings are consistent with studies in animals that have shown relative enhancements in the firing rates of single cortical neurons in extrastriate area V4 when the eliciting target is at an attended spatial location (Fischer and Boch, 1985; Moran and Desimone, 1985; Spitzer, Desimone, and Moran, 1988).

The behavioral and electrophysiological experiments in humans, together with the animal studies, indicate that sensory and perceptual processing are indeed affected by selective spatial attention. However, as noted by Mountcastle et al. (1987), evidence of changes in activity in the sensory cortex with spatial attention is only part of the story. In addition, we must ascertain which neural systems were involved in the executive or command processes that resulted in the observed changes in sensory-neural excitability. That is, changes in excitability in the sensory-specific cortex reflect only one of many neural actions that must participate in attentional processing. Thus, precisely which neural systems control the excitability of extrastriate visual neurons during spatial attention remains to be elucidated.

Neuropsychological Studies of Attention

Some evidence concerning the control systems involved in modulating sensory-perceptual processing comes from neuropsychological studies of attentional orienting in brain-damaged patients. Indeed, such work has provided a substantial amount of evidence about the roles of both cortical and subcortical structures during spatial attention. Posner, Rafal, and their colleagues have investigated the role of parietal cortex in attentional orienting (Posner et al., 1984, 1987). Given the role of parietal cortex lesions in neglect syndrome, they reasoned that parietal cortex might play a special role in the orienting of attention to locations in space. Indeed, patients with lesions of parietal cortex unilaterally did show significantly slowed reaction times for validly cued targets in the visual field contralateral to the lesioned hemisphere (see also Petersen, Robinson, and Currie, 1989). This therefore is evidence to support the notion that the ability to shift attention in the direction contralateral to the lesion was somewhat impaired by parietal damage.

The more dramatic finding, however, was that parietal lesion patients showed the greatest reaction time deficits when the visual field ipsilateral to the lesioned hemisphere was cued but the subsequent target was presented in the contralesional direction. In this condition, reaction times to the targets were much slower than for the same target condition in the ipsilesional field. For example, in a patient with *right* parietal damage, a *right* field cue that was followed by a target in the *left* field produced dramatically slowed reaction times. This was termed an "extinction-like" reaction time pattern by Posner et al. On the basis of these findings, they concluded that parietal cortex is involved in the disengagement of attention from its current locus so that it can be shifted to a new region of visual space. *Disengagement* is a hypothetical mental operation that must occur before attention can move from a given locus to another to then *engage* a target stimulus (Posner et al., 1984). The move and engage operations are believed to be controlled by the midbrain and thalamus, respectively (Rafal and Posner, 1987; Rafal et al., 1988). Lesions of the frontal or temporal lobes also produce similar slowing for validly cued contralesional targets, but do not show the extinction-like reaction time (RT) pattern for the contralesional, invalidly cued targets that is seen in parietal patients (Petersen, Robinson, and Currie, 1989).

The study of patients with neurological damage has greatly aided progress in understanding the role of various brain structures in attentional processes. The application of cognitive psychological methods in these groups has also facilitated the dissection of the specific information-processing deficits that result from focal cortical or subcortical lesions. Nonetheless, it is important to be cognizant of the limitations inherent in such approaches. Behavioral measures such as RT are indirect measures of mental processing, and thus inference must be used to relate them to specific psychological or neural events. Measuring such neural processes directly would provide a significant advantage.

Electrophysiological Studies of Orienting

In humans, ERPs can be used to investigate brain activity directly, as noted above. In the case of attentional orienting tasks, ERPs related to the cue-induced orienting of attention can be obtained. Brain electrical activity can be recorded in the cue–target interval and evaluated in the period prior to target onset, even in the absence of motor responses. Harter and his colleagues have utilized this approach in trial-by-trial cuing tasks to examine what they refer to as the "executive" processes that may relate to either the shifting of attention from one location to another, or modulation of sensory structures to affect attentional control over perceptual processing (Harter and Anllo-Vento, 1991; Harter et al., 1989).

In their experiments, Harter and colleagues examined children and adults using the following paradigm. Trials began with attention-directing arrow cues located at fixation that instructed the subject where to attend. After a delay of about 600 msec, a lateral flash stimulus occurred at either the cued or uncued location, with equal probability. Subjects were instructed to make a speeded motor response to the stimulus on the cued side only.

ERPs time-locked to the left- and right-pointing cues were obtained and found to differ significantly over left and right hemisphere scalp regions. They reported that from 200 to 400 msec after the onset of the arrow cue, the hemisphere contralateral to the direction of the arrow cue was more negative in voltage than when it was ipsilateral to the cue direction. Thus, ERPs recorded from left hemisphere sites were more negative following right-arrow cues, but for left-arrow cues the right hemisphere sites had more negative ERPs.[1] This effect was referred to as the "early directing attention negativity" (EDAN).

Following the EDAN effect, in the latency range from 400 to 700 msec after cue onset (at about the time of arrival of the target), the polarity and scalp distribution of the effects changed. In this longer latency range, the posterior scalp regions were more positive over the right hemisphere for left-pointing arrows, whereas right-pointing arrows resulted in greater positivity over the left posterior scalp: this was referred to as the "late directing attention positivity" (LDAP). These effects were interpreted as evidence that in response to an attention-directing cue, executive processes involved in shifting attention to the cued location must be invoked and that the early contralateral negativity was a sign of such processes. The longer latency posterior contralateral positivity is hypothesized to reflect the modulation of cortical excitability in the sensory cortical areas that will process the upcoming target stimulus. The result is larger sensory-evoked potentials to the target when validly precued.

Topographical Voltage and Current Density Mapping in the Study of Orienting and Selection

The goal of the study reviewed here was to replicate the groundbreaking findings of Harter and his colleagues, and to provide additional detail about the scalp distribution of these effects by using dense electrode arrays over the scalp. Such information is useful in attempting to relate scalp recordings to underlying brain processes. Signal detection measures and ERPs to target stimuli were also obtained: These will aid in the evaluation of ERP signs of preparatory/executive processes that follow an attention-directing cue. Presumably, increased cortical excitability, as indexed by the posterior positive shifts following a cue, should lead to increased attentional modulation of the

---

[1] The description of these effects as "negativities" is somewhat arbitrary because they are actually positive polarity effects that are "negative going." The basis for choosing this terminology is in part the physiological data which suggest that increased negativity reflects increased cortical excitability. Thus, presumably activity in the right hemisphere to the left cue should reflect increased activity of that hemisphere, hence, increased negativity. Therefore, as suggested by Harter and Anllo-Vento (1991), the differences are interpreted as contralateral negativities.

target ERPs and to improved discrimination of target features. Thus, a discrimination task was utilized in which target stimuli were presented following a centrally located arrow cue.

Experimental Design

Normal young adult volunteers ($n = 8$) viewed a computer display that flashed attention-directing cues which were followed by target stimuli requiring a discrimination and perhaps response. The centrally located cues were left or right arrows, or a plus sign (neutral cue) briefly flashed (280 msec) to fixation in random sequence. The cue was followed (after 680–1080 msec) by a stimulus located 10 degrees lateral to fixation (duration of 70 msec). The lateral stimulus was a square mask pattern (3.4 × 3.4 degrees) of colored pixels that had a "target" symbol embedded in it 50% of the time (Fig. 1).

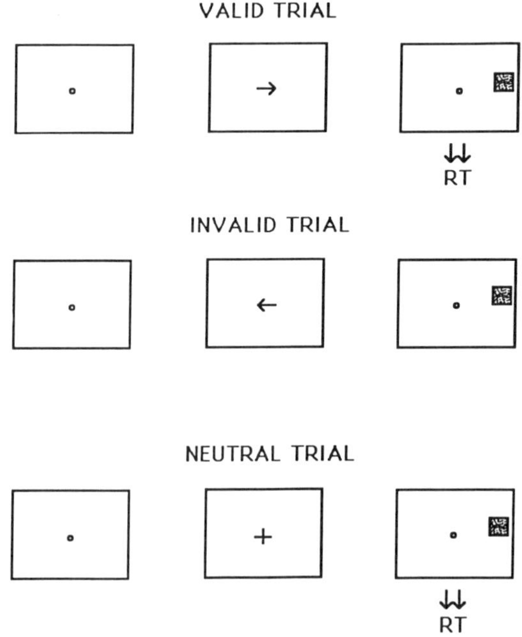

Figure 1.   Schematic representation of the stimulus displays. Only right field stimuli are represented, but stimuli were presented to right and left fields in random order. The three types of cue trials are indicated in the top, middle, and bottom rows. Each trial began with the subject fixating a central point on the video monitor, on which a cue (left, right, or neutral) was flashed. After 880–1080 msec, a lateralized stimulus was briefly flashed. If the trial was valid (top) or neutral (bottom) and the lateral stimulus contained a target symbol (not shown), a rapid manual response was required.

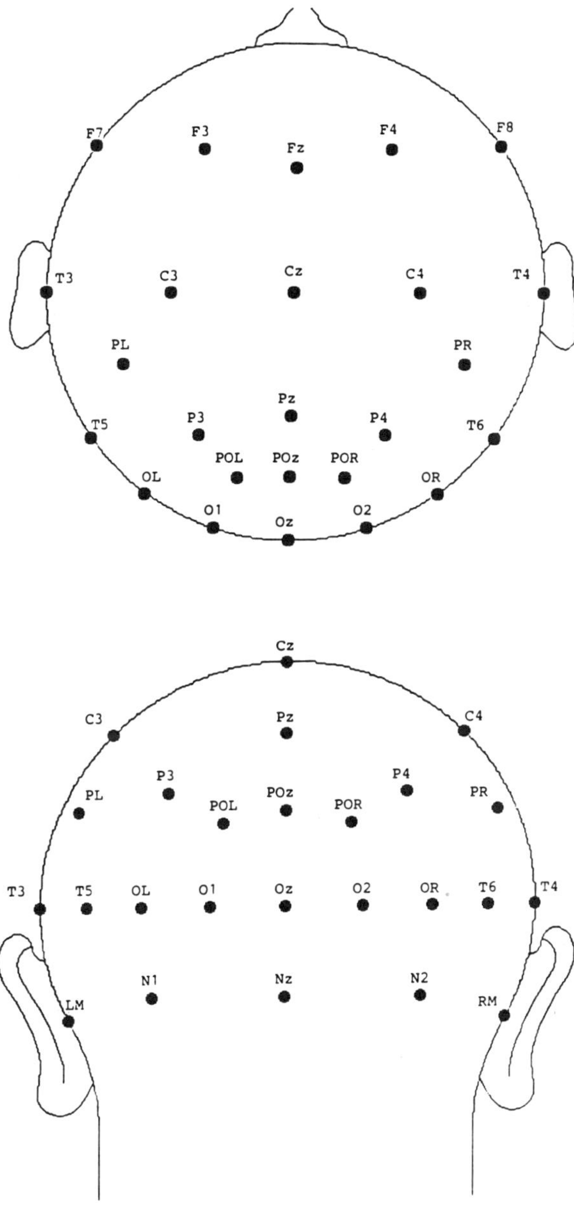

Figure 2.   Approximate locations of the electrodes used to record ERPs. At top, the view is from above the head; at bottom is a rear view. The projection of three-dimensional head onto the two-dimensional surface of the figure is radial for the top views and orthographic for the rear view.

Six percent of the trials were catch trials in which no stimulus followed the cue. Trials were separated by 2000 msec. The task was to covertly attend only to the cued visual half-field location and to respond with a button press (hand-counterbalanced) only if the target symbol was present in the mask pattern. If the cue was the neutral plus sign, the subjects performed the discrimination regardless of the side of the stimulus. Twenty runs of approximately 2.5 min duration (50 trials per run) were presented.

The electroencephalogram was recorded with a bandpass of 0.1–100 Hz and sampling rate of 256 samples/sec from 29 scalp sites (Fig. 2). Blinks were recorded with an electrode located below the right eye. All the preceding electrodes were referenced to the right mastoid process for recording and then rereferenced algebraically to an average of the left and right mastoids offline. Electrodes were placed at the outer ocular canthi to record lateral eye movements; the bandpass for this channel was 0.01–100 Hz. Following artifact rejection (i.e., for blinks, eye movements, and muscle activity), ERPs were computed offline beginning 700 msec before stimulus onset and continuing for 1000 msec post stimulus. One subject was excluded from the final analysis because of systematic eye movements to the cues, leaving seven subjects (four female). Repeated-measures ANOVA was used to evaluate effects in behavioral measures and ERPs.

Results and Discussion

PERFORMANCE AND SIGNAL DETECTION. Subjects were significantly faster in responding to validly cued as opposed to neutrally cued targets (688 vs. 726 msec). Discrimination performance was slightly higher for validly as compared to neutrally cued targets for percent correct detection, but did not reach significance (86% vs. 80%, $p < .053$); sensitivity (d′) was not significantly different (2.8 vs. 2.6). The failure to find significant increases in sensitivity while obtaining reasonably large RT effects (38 msec) may indicate that although subjects allocated attention based on the cue information, they were able to partially compensate for reductions in sensitivity in the divided attention task by reorienting attention to the iconic store of the target image to perform the task. Such a reorientation takes times, as reflected in the RT slowing, but results in high discrimination performance. However, another interpretation is that the divided (neutral cue) condition may also have resulted in subjects allocating additional attentional resources during those trials to compensate for the added requirement of detecting targets at either location.

ERPs TO THE CUES. After 250 msec latency, past the region principally associated with sensory processing of the physical aspects of the cue stimuli, the waveforms have two prominent patterns. Over the anterior scalp, there is a slow sustained negative shift that follows the P2 and lasts for several

ERPS TO CUES

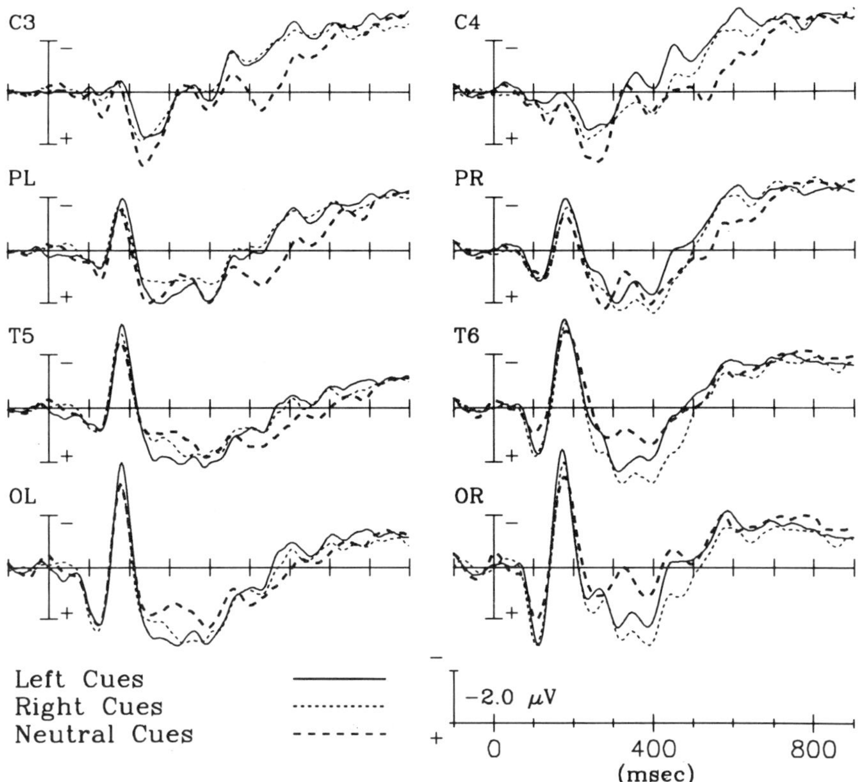

Figure 3.   Grand average ERPs over the 7 subjects to the left (*thin solid line*), right (*dotted line*), and neutral cues (*dashed line*). Traces are recordings from left and right central (C3 and C4), parietal (PL and PR), temporal (T5 and T6), and occipital (OL and OR) scalp sites.

hundred milliseconds. Posteriorly, the N1 peak is followed first by a late positive deflection (LPD) and then a slow negative shift lasting for several hundred milliseconds. In general, the slow negative shifts can be described as CNV-like (contingent negative variation) potentials that appear in tasks that have warning stimuli (i.e., the cues) and that are followed by stimuli requiring decisions or actions (e.g., McCallum, 1988; Walter et al., 1964). These shifts are thought to represent changes in cortical excitability (Caspars, Speckmann, and Lehmenkuhler, 1980; Deeke et al., 1990; Uhl et al., 1990).

In the latency range from 200 msec onward there are visible differences in the ERPs evoked to left, right, and neutral cues (Fig. 3). To better visualize the differences in these ERPs as a function of cue type, difference waves were formed by subtracting the ERPs in the neutral trials from those in either the right- or left-arrow trials. These subtractions remove ERP effects common to each type of cue trial and reveal the scalp potentials elicited as a function of leftward or rightward attentional orienting: It is the differences in these effects that are of principal interest (Fig. 4).

In response to the left-pointing and right-pointing cues (in comparison to the neutral cue trials), there were positive shifts over both hemispheres between 250 and 500 msec latency. These positive shifts resolved as negative-going potentials between 400 and 500 msec that were maximal over central scalp regions. The first significant differential effect of cue direction was similar to that reported by Harter et al. (1989). This effect was localized over parietal and temporal scalp regions in the latency range of 250–350 msec, and was manifest as increased negativity over the *right* hemisphere for *leftward* cues. In contrast, there was increased negativity over *left* hemisphere sites for *rightward* cues ($p < .002$ for interaction of cue direction × hemisphere of recording). Separate ANOVAs for the left and right hemisphere

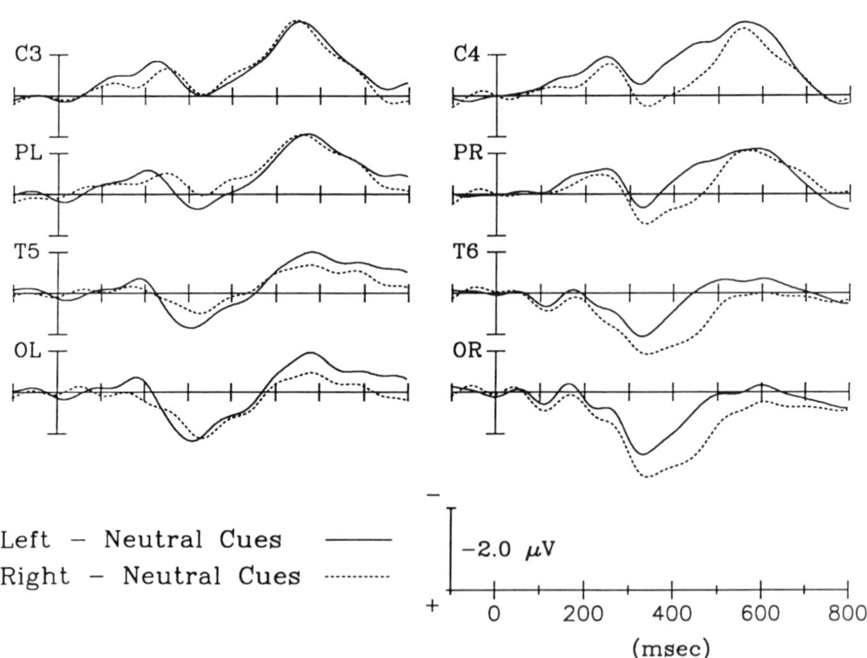

Figure 4.   Grand average difference waves to the cues obtained by subtracting the neutral cue traces from the left (*solid line*) and right (*dashed line*) ERPs.

Difference Maps To Cue
Left Minus Right Arrow

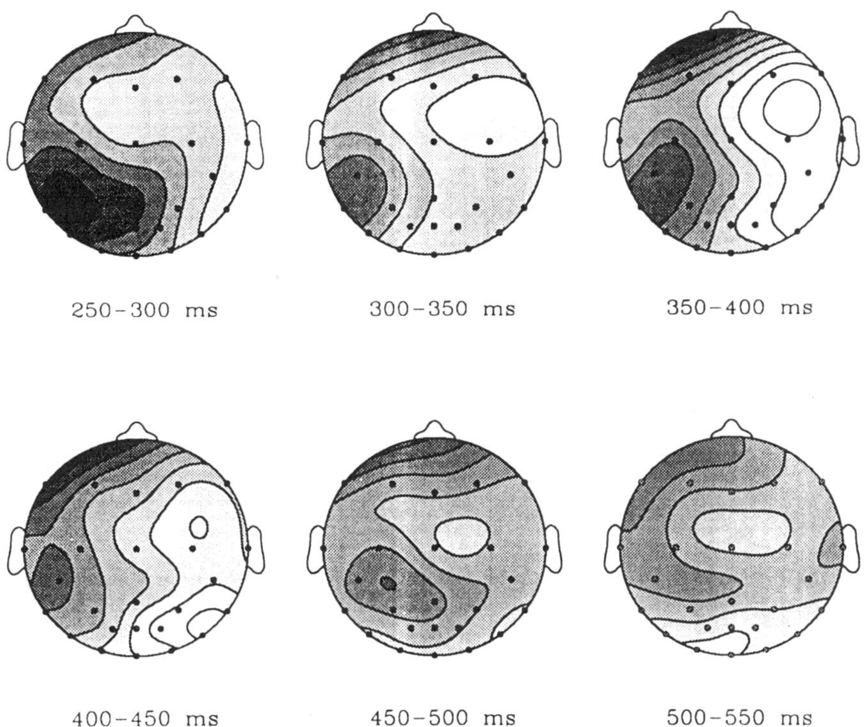

Figure 5.   Scalp topographic voltage maps at the successive time points. Lighter shades indicate negative voltages; darker shades indicate positive voltages; minimum voltage, $-0.96\,\mu$V; maximum voltage, $1.6\,\mu$V. All maps are scaled to the same maximum and minimum values. The view is from above and is presented in a radial projection. Electrodes are indicated as black dots. Topographical mapping performed using the spline interpolation method described by Perrin et al. (1989).

showed that the effect only reached significance for left hemisphere sites ($p < .04$).

Later in the waveform (300–500 msec latency), the effect of cuing was different. Direction of the cue resulted in differences in the ERPs only over the right hemisphere, such that leftward-pointing cues produced greater negativity than did rightward-pointing cues. Accordingly, separate ANOVAs for left and right hemisphere sites showed a significant difference between left and right cues for the right hemisphere sites ($p < .02$), but no such effect for the left hemisphere ($p < .25$). This right hemisphere effect appears to be different from that reported by Harter and colleagues, who observed the principal effects to be the early contralaterally negative EDAN (200–400 msec) and the later contralaterally positive LDAP (600–700). An interesting aspect of the 300- to 500-msec difference in the present data was a tendency for the early portion (300–400 msec) of the effect to have a more anterior scalp distribution than the later portion (400–500 msec), which was more occipital-temporal in scalp distribution (Fig. 5).

ERPs to the Lateral Stimuli. ERPs to the lateralized left-field and right-field nontarget stimuli are shown in Figure 6. Overlaid are the ERPs when the lateral stimulus was validly precued versus neutrally or invalidly precued. The stimuli elicited P1 (100–140 msec) and N1 (160–200 msec) components over posterior scalp sites; this was followed by a late positive deflection (LPD) from 250 to 500 msec latency. The early P1 component was significantly greater in amplitude for the validly cued stimuli over medial (O1 and O2) and lateral (OL and OR) occipital, and posterior temporal (T5 and T6) scalp sites ($p < .03$, $p < .01$, and $p < .04$, respectively). There were no statistically significant differences in the P1 validity effect between nontarget and target stimuli. Neither the sensory-evoked P1 nor the validity effect on P1 was different as a function of scalp sites contralateral versus ipsilateral to the stimulus. The scalp distribution of the sensory-evoked P1 component and the validity effect on P1 (valid minus invalid ERP) showed amplitude maxima over lateral occipital scalp sites (OL and OR) (Fig. 7). These effects on P1 are consistent with previous studies of spatial attention in both cuing paradigms (Mangun and Hillyard, 1991), and sustained attention paradigms (Mangun et al., 1993).

In addition to differences between valid and invalid cue conditions, it is also of interest to consider the ERPs to the lateral stimuli when preceded by a neutral cue. Figure 6 also shows the ERPs to lateral stimuli in each of the three cue conditions. In general, the ERPs to the left and right visual field stimuli (collapsed across targets and nontargets in the figure) in the neutral cue trials are of amplitude intermediate to that obtained in valid or invalid conditions. ANOVAs comparing valid versus neutral ERPs, and neutral versus invalid ERPs for the occipital P1 component, found no differences between conditions.

The posterior N1 component elicited by the validly cued nontarget stimuli also showed significant enhancements as compared to invalidly cued stimuli over lateral occipital ($p < .05$) and posterior temporal sites ($p < .02$). The posterior N1 component was also significantly affected by cue validity at lateral parietal sites (PL and PR; $p < .01$). Thus, the cue validity effects on the posterior N1 component were somewhat more broadly distributed over the scalp than was the P1 peak. However, unlike previous studies (e.g., Mangun and Hillyard, 1991), there were no significant validity

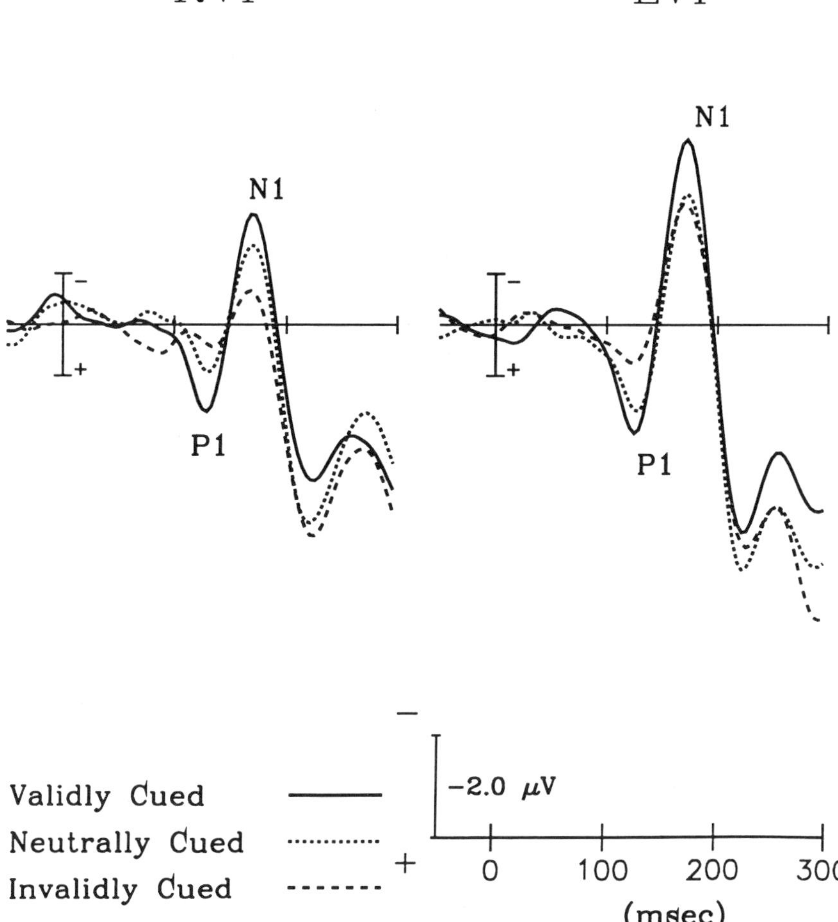

Figure 6.   Grand average ERPs to the lateralized nontarget stimuli. ERPs to left visual field (LVF) stimuli recorded from right hemisphere sites are shown at right. ERPs to right visual field (RVF) stimuli recorded from left hemisphere sites are shown at left. Overlaid are the ERPs when the stimulus had been validly cued (*solid line*), invalidly cued (*dashed line*), and neutrally cued (*dotted line*).

effects on the frontal or central N1 peaks. The significant posterior N1 validity effects were not different as a function of visual field of stimulus or hemisphere of recording. In contrast to the findings for the P1 component, there was a significant difference between the N1 components elicited by lateral stimuli that were validly versus neutrally cued [F(1,6) = 7.8, $p < .04$]; but there was no significant difference between neutral and invalid trials for the N1 component. Given the moderate number of subjects ($n = 7$) and the

SCALP TOPOGRAPHY OF
P1 VALIDITY EFFECT
(100–128 msec)

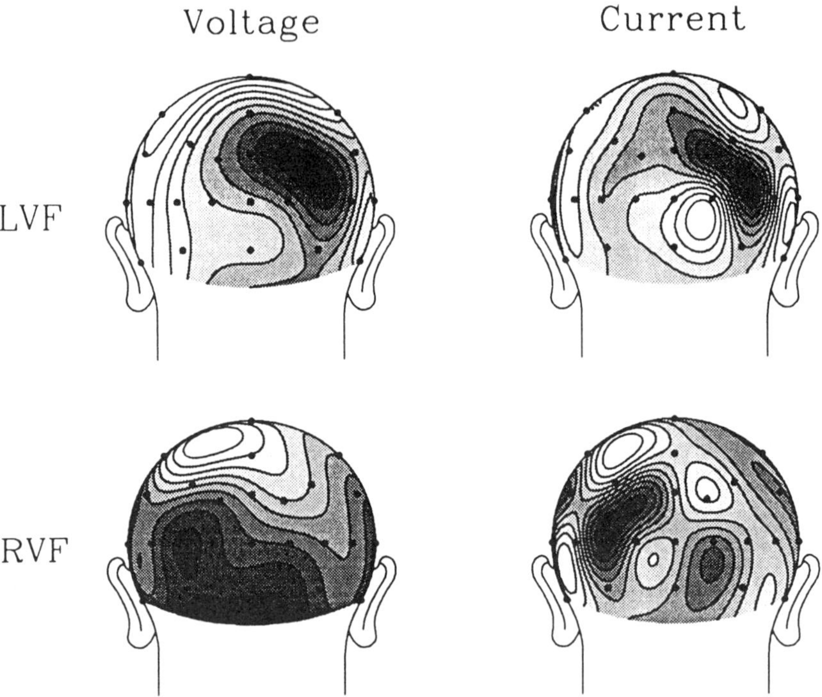

Figure 7.   Scalp voltage (left) and current density (right) topographic maps for the P1 attention effect (valid–invalid ERPs). Left visual field (LVF) is shown at top, and right (RVF) at bottom. Each map is individually scaled with positive voltage or current density in darker shades. The maximum and minimum values for each map are as follows: upper left map, max = 1.43 μV, min = -0.91 μV; upper right map, max = 22.65 μV/m^2, min = -13.9 μV/m^2; lower left map, max = 0.4 μV, min = -1.61 μV; lower right map, max = 19.6 μV/m^2, min = -16.7 μV/m^2.

small amplitude and statistical significance of the differences between valid, invalid, and neutral cues trials (or lack thereof), it is difficult to interpret the obtained patterns for either the N1 or P1 components with respect to the neutral cue condition. Nonetheless, the N1 pattern of significant differences between valid and neutral conditions (benefits) is of interest and should be investigated in future studies.

The late positive deflection (LPD) observed between 250 and 500 msec latency was significantly more positive for the invalidly cued stimulus ($p < .01$) and this effect had a broad scalp distribution. There was, however, a significant interaction of cue validity and scalp region that arose because the difference in the 250 to 500 msec latency range was larger over more anterior scalp regions. The LPD probably resulted in part from the positive polarity termination of a contingent negative variation (CNV) (McCallum, 1988) and a P300 (P3) component (Donchin, 1981). The contributions of motor-related potentials (e.g., Deeke, Scheid, and Kornhuber, 1969) to this LPD appears unlikely because no responses were made to these nontarget stimuli, and the LPD was actually more positive for the invalidly cued stimuli that never elicited a motor response.

Summary and Interpretations

The purpose of this chapter was to review cortical processes that are involved in the orienting of attention to spatial locations; in particular, those brain processes that result in attention-related changes in sensory-perceptual processing. These "executive" or control mechanisms can be measured directly using the ERP method, which permits an investigation of those brain processes that follow the delivery of cue instructions to the subjects. The resultant effect of these executive attentional processes on the processing of visual inputs can be evaluated separately by examining the ERPs to the subsequent task-relevant stimuli. Both executive and sensory-perceptual mechanisms represent information processing that precedes any task-relevant motor responses by tens or hundreds of milliseconds (Coles, 1988). Effects obtained during either the cue–target interval or post-target interval can be considered with respect to the elementary mental operations involved in spatial orienting and selection.

In the series of papers by Harter and his colleagues, ERPs were used in cuing paradigms in precisely the manner outlined above (e.g., Harter and Anllo-Vento, 1991). The present study is a replication and extension of those pioneering investigations, and provides additional information by combining multichannel mapping of such effects with signal detection methods to better understand the physiological and psychological mechanisms represented by ERPs in the cue–target interval. The present findings can be considered a replication and partial extension of the results of Harter and Anllo-Vento (1991) for their findings in adult subjects. However, a number of differences were also obtained.

Two main effects were obtained in the present study as a function of cue direction. The first was an early difference between left and right cues in the latency range from 250 to 350 msec. This effect appears to be the "early directing attention negativity" (EDAN; 200– 400 msec) of Harter and Anllo-Vento (1991). Thus, the EDAN appears to be replicated in the present study, although of rather small amplitude. The reduced size of the EDAN here may result from differences in the "neutral cue" condition between the two studies. Harter and Anllo-Vento used a middle (upward) cue that indicated attention was to be directed to a target which might appear above fixation, while here the third (neutral) cue was a plus sign that indicated divided attention. Thus, the present subjects may not have oriented attention to any particular location in the visual field in response to the neutral cue. One may consider, therefore, that any differences between neutral cue condition and arrow conditions in the ERPs reflect the process of orienting to a location, perhaps any location. The lack of a difference between the middle cue and arrow cues of Harter and Anllo-Vento may reflect that similar orienting "effort" may have been required in each of those three conditions, and as a consequence they obtained no differences in the waveforms.

Hemispheric Asymmetries

A much more significant difference between the previous results of Harter and colleagues and the current study is the difference between left- and right-pointing cues in the latency range from 300 to 500 msec. In the present study, there is a large difference in this latency range that is only present over the right hemisphere (see Fig. 4). The pattern of this effect is that the leftward-pointing cues lead to a more negative waveform in this latency range. This effect appears not to have been obtained in the studies of Harter and colleagues in either adults or children.

This 300- to 500-msec latency difference may represent a right hemisphere specialization for attentional orienting. Such an interpretation would be consistent with the neuropsychological literature on patients with lesions of the right or left hemisphere and the resulting deficits in attention that are commonly observed (DeRenzi, 1982; Heilman and Van Den Abell, 1979; Kinsbourne, 1987; Ogden, 1987). In contrast to the clinical neuropsychological view of right hemisphere function, Posner et al. (1984) reported no significant differences as a function of hemisphere of lesion in patient's ability to orient to lateral targets, nor did Holtzman and Gazzaniga (reviewed in Gazzaniga, 1987) obtain differences in the two hemispheres of split-brain patients in attentional orienting tasks. However, Mangun et al. (submitted) have obtained reaction time evidence in split-brain patients that shows a significant difference in the ability of the two hemispheres to orient attention in response to peripheral cues. Such findings of a right hemisphere specialization for visual attentional orienting in space may fit well with the present results of a right hemisphere difference between leftward and rightward attention-directing cues.

Modulations of Sensory Signals

Significant amplitude modulations of the sensory-evoked ERP components (P1 and N1) elicited by the lateralized stimuli were obtained here that appear to be similar or identical to those obtained in the expectancy-based, trial-by-trial, spatial cuing paradigm (Mangun and Hillyard, 1991), and the studies in children of Harter, Anllo-Vento and Wood (1989) and Harter and Anllo-Vento (1991). These early amplitude modulations with spatial attention also appear to be the same mechanisms invoked during sustained spatial attention paradigms that are so common in the ERP literature (Eason, 1981; Eason, Harter, and White, 1969; Harter, Aine, and Schroeder, 1982; Heinze et al., 1990; Hillyard and Münte, 1984; Luck et al., 1990; Mangun and Hillyard, 1987, 1988, 1990, 1991; Neville and Lawson, 1987; Rugg et al., 1987; Van Voorhis and Hillyard, 1977).

The attention-sensitive P1 component is maximal over occipital scalp regions that overlie the lateral occipital gyri of the extrastriate cortex (Homan, Herman, and Purdy, 1987) and appears to be generated by visual processes in this region as opposed to the earlier striate cortex (Mangun et al., 1993). The amplitude modulation of the P1 suggests a model of sensory gain control in the visual cortex as a function of attention (Hillyard and Mangun, 1987). Such a process can lead to relative differences in the perceptual representations of stimuli at attended versus unattended locations in the visual field. Thus, one may think of this process as a mechanism to improve the signal-to-noise ratio across the visual field, but probably not within the attended locus itself (Heinze et al., 1990). The early P1 component may well represent a human correlate of the attentional modulation of single neuron firing in extrastriate area V4 of the monkey (Moran and Desimone, 1985).

Conclusions

Considered together, these data suggest a picture of spatial attention and orienting very closely resembling that suggested by Harter and Anllo-Vento (1991) and Posner et al. (1984). When a human is confronted with information that requires attention to be oriented from its present location to another point in the visual field, several neural systems cooperate to accomplish this action. First, following a decoding of cue information, executive processes in anterior cortical regions, as well as lateral parietal and posterior temporal regions are active. These may reflect initial commands that subsequently prepare specific regions of the topographically mapped sensory pathways for an upcoming stimulus; the right hemisphere appears to be especially involved, exhibiting differential activity as a function of direction of attentional orienting. On arrival of sensory information in the primed cortical visual areas, the attended-location stimulus is processed with greater efficiency, thereby leading to a better perceptual representation of that event in addition to reducing interference from events at other spatial locations (cf. Hillyard, Luck, and Mangun, this volume). The final outcome is the enhanced perception of the stimulus and an increased ability to discriminate and respond to the essential information the stimulus contains.

Acknowledgments. Work supported by grant K21 MH00930 from NIMH, the Hitchcock Foundation and the McDonnell-Pew Foundation. Thanks to Todd Handy, Deborah Gartner, Patrick Brown, Flo Batt, and Clif Kussmaul for their assistance. Thanks also to Mado Proverbio and Michael Gazzaniga for helpful comments and criticisms.

References

Caspars H, Speckmann EJ, Lemenkuhler A (1980): Electrogenesis of cortical DC potentials. In: *Motivation, Motor and Sensory Processes of the Brain: Electrical Potentials, Behavior and Clinical Use*, Kornhuber HH, Deeke L, eds., pp. 3–16. Amsterdam: Elsevier.

Coles, MGH (1988): Modern mind-brain reading: Psychophysiology, physiology and cognition. *Psychophysiology* 26:251–269.

Deeke L, Scheid P, Kornhuber HH (1969): Distribution of readiness potential, pre-motion positivity and motor potential of the human cerebral cortex preceding voluntary finger movement. *Exp Brain Res* 7:158–168.

Deeke L, Lang W, Uhl F, Podreka I (1990): Looking where the action is: Negative DC shifts as indicators of cortical activity. In: *From Neuron to Action: An Appraisal of Fundamental and Clinical Research*, Deeke L, Eccles JC, Mountcastle VB, eds., pp. 25–41. Berlin: Springer-Verlag.

De Renzi E (1982): *Disorders of Space Exploration.* New York: Wiley.

Donchin E (1981): Surprise! . . . Surprise! *Psychophysiology* 18:493–513.

Downing CJ (1988): Expectancy and visual-spatial attention: Effects on perceptual quality. *J Exp Psychol Hum Percept Perform* 14:188–202.

Eason RG (1981): Visual evoked potential correlates of early neural filtering during selective attention. *Bull Psychon Soc* 18:203–206.

Eason RG, Harter M, White C (1969): Effects of attention and arousal on visually evoked cortical potentials. *Physiol Behav* 4:283–289.

Eriksen CW, Yeh YY (1985): Allocation of attention in the visual field. *J Exp Psychol Hum Percept Perform* 11:583–597.

Fischer B, Boch R (1985): Peripheral attention versus central fixation: Modulation of visual activity of prelunate cortical cells of the rhesus monkey. *Brain Res* 345:111–123.

Gazzaniga MS (1987): Perceptual and attentional processes following callosal section in humans. *Neuropsychologia* 25:119–133.

Harter MR, Anllo-Vento L (1991): Visual-spatial attention: Preparation and selection in children and adults. In: *Event-Related Brain Research*, Brunia CHM, Mulder G, Verbaten MN, eds., pp. 183–194. New York: Elsevier.

Harter MR, Aine C, Schroeder C (1982): Hemispheric differences in the neural processing of stimulus location and type: Effects of selective attention on visual evoked potentials. *Neuropsychologia* 20:421–438.

Harter MR, Anllo-Vento L, Wood FB (1989): Event-related potentials, spatial orienting, and reading disabilities. *Psychophysiology* 26:404–421.

Harter MR, Miller SL, Price NJ, LaLonde ME, Keyes AL (1989): Neural processes involved in directing attention. *J Cogn Neurosci* 1:223–237.

Hawkins HL, Hillyard SA, Luck SJ, Mouloua M, Downing CJ, Woodward DP (1990): Visual attention modulates signal detectability. *J Exp Psychol Hum Percept Perform* 16:802–811.

Heilman KM, Van Den Abell T (1979): Right hemisphere dominance for mediating cerebral activation. *Neuropsychologica* 17:315–321.

Heinze HJ, Luck SJ, Mangun GR, Hillyard SA (1990): Lateralized visual ERPs index focussed attention to bilateral stimulus arrays: I. Evidence for early selection. *Electroencephalogr Clin Neurophysiol* 75:511–527.

Hillyard SA, Mangun GR (1987): Sensory gating as a physiological mechanism for visual selective attention. In: *Current Trends in Event-Related Potential Research*, Johnson R, Parasuraman R, Rohrbaugh JW, eds., pp. 61–67. Amsterdam: Elsevier.

Hillyard SA, Münte TF (1984): Selective attention to color and locational cues: An analysis with event-related brain potentials. *Percept Psychophys* 36:185–198.

Hoffman JE, Nelson B (1981): Spatial selectivity in visual search. *Percept Psychophys* 30:283–290.

Homan RW, Herman J, Purdy P (1987): Cerebral location of international 10-20 system electrode placement. *Electroencephalogr Clin Neurophysiol* 66:376–382.

Jonides J (1981): Voluntary versus automatic control over the mind's eye's movement. In: *Attention and Performance IX*, Long J, Baddeley A, eds., pp. 187–203. Hillsdale, NJ: Erlbaum.

Kinsbourne M (1987): Mechanisms of unilateral neglect. In *Neurophysiological and Neuropsychological Aspects of Spatial Neglect*, Jeannerod M, ed., pp. 69–86. Amsterdam: Elsevier North-Holland.

Luck SJ, Heinze HJ, Mangun GR, Hillyard SA (1990): Visual event-related potentials index focused attention within bilateral stimulus arrays: II. Functional dissociation of P1 and N1 components. *Electroencephalogr Clin Neurophysiol* 75:528–542.

Mangun GR, Hillyard SA (1987): The spatial allocation of visual attention as indexed by event-related brain potentials. *Hum Factors* 29:195–212.

Mangun GR, Hillyard SA (1988): Spatial gradients of visual attention: Behavioral and electrophysiological evidence. *Electroencephalogr Clin Neurophysiol* 70:417–428.

Mangun GR, Hillyard SA (1990): Allocation of visual attention to spatial locations: Tradeoff functions for event-related brain potentials and detection performance. *Percept Psychophys* 47:532–550.

Mangun GR, Hillyard SA (1991): Modulations of sensory-evoked brain potentials indicate changes in perceptual processing during visual-spatial priming. *J Exp Psychol. Hum Percep Perf* 17:1057–1074.

Mangun GR, Hillyard SA, Luck SL (1993): Electrocortical substrates of visual selective attention. In: *Attention and Performance XIV*, Meyer D, Kornblum S, eds., pp. 219–243. Cambridge: MIT Press.

Mangun GR, Plager R, Loftus W, Hillyard SA, Luck SJ, Handy T, Clark V, Gazzaniga MS: Monitoring the visual world: Hemispheric asymmetries and subcortical processes in attention (submitted).

McCallum WC (1988): Potentials related to expectancy, preparation and motor activity. In: *Handbook of Electroencephalography and Clinical Neurophysiology: Human Event-Related Potentials*, Vol. 3, Picton TW, ed., pp. 427–534. New York: Elsevier.

Moran J, Desimone R (1985): Selective attention gates visual processing in the extrastriate cortex. *Science* 229:782–784.

Mountcastle VB, Motter BC, Steinmetz MA, Sestokas AK (1987): Common and differential effects of attentive fixation on the excitability of parietal and prestriate (V4) cortical visual neurons in the macaque monkey. *J Neurosc* 7:2239–2255.

Müller HJ, Findlay JM (1987): Sensitivity and criterion effects in the spatial cuing of visual attention. *Percept Psychophys* 42:383–399.

Müller HJ, Rabbitt PMA (1989): Reflexive and voluntary orienting of visual attention: Time course and activation and resistance to interruption. *J Exp Psychol Hum Percept Perform* 15:315–330.

Neville HJ, Lawson D (1987): Attention to central and peripheral visual space in a movement detection task: An event-related potential and behavioral study. I. Normal hearing adults. *Brain Res* 405:253–267.

Ogden JA (1987): The neglected left hemisphere and its contribution to visuospatial neglect. In: *Neurophysiological and Neuropsychological Aspects of Spatial Neglect*, Jeannerod M, ed. Amsterdam: Elsevier North-Holland.

Perrin F, Pernier J, Bertrand O, Echallier FJ (1989): Spherical splines for scalp potential and current density mapping. *Electroencephalogr Clin Neurophysiol* 72:184–187.

Petersen SE, Robinson DL, Currie JN (1989): Influences of lesions of parietal cortex on visual spatial attention in humans. *Exp Brain Res* 76:267–280.

Posner MI (1980): Orienting of attention. *Q J Exp Psychol* 32:3–25.

Posner MI, Nissen MJ, Ogden WC (1978): Attended and unattended processing modes: The role of set for spatial location. In: *Modes of Perceiving and Processing Information*, Pick HL, Saltzman IJ, eds., pp. 137–157. Hillsdale, NJ: Erlbaum.

Posner MI, Synder CRR, Davidson BJ (1980): Attention and the detection of signals. *J Exp Psychol Gen* 109:160–174.

Posner MI, Walker JA, Friedrich FJ, Rafal RD (1984): Effects of parietal injury on covert orienting of attention. *J Neurosci* 4:1863–1874.

Posner MI, Walker JA, Friedrich FJ, Rafal RD (1987): How do the parietal lobes direct covert attention? *Neuropsychologia* 25:135–145.

Rafal RD, Posner MI (1987): Deficits in human visual spatial attention following thalamic lesions. *Proc Natl Acad Sci* 84:7349–7353.

Rafal RD, Posner MI, Friedman JH, Inhoff AW, Bernstein E (1988): Orienting of visual attention in progressive supranuclear palsy. *Brain* 111:267–280.

Reinitz MT (1990): The effects of spatially directed attention on visual encoding. *Percept Psychophys* 47:497–505.

Rugg MD, Milner AD, Lines CR, Phalp R (1987): Modulation of visual event-related potentials by spatial and non-spatial visual selective attention. *Neuropsychologia* 25:85–96.

Shaw ML (1984): Division of attention among spatial locations: A fundamental difference between detection of letters and detection of luminance increments. In: *Attention and Performance X: Control of Language Processes*, Bouma H, Bouwhuis DG, eds., pp. 109–121. Hillsdale, NJ: Erlbaum.

Sperling G (1984): A unified theory of attention and signal detection. In: *Varieties of Attention*, Parasuraman R, Davies DR, eds., pp. 103–181. Orlando: Academic Press.

Sperling G, Dosher BA (1986): Strategy and optimization in human information processing. In: *Handbook of Perception and Performance*, Vol. 1, Boff K, Kaufman L, Thomas J, eds., pp. 2.1–2.65. New York: Wiley.

Spitzer H, Desimone R, Moran J (1988): Increased attention enhances both behavioral and neuronal performance. *Science* 240:338–340.

Uhl F, Franzen P, Serles W, Lang W, Lindinger G, Deeke L (1990): Anterior frontal cortex and the effects of proactive interference in paired associate learning: A DC potential study. *J Cogn Neurosci* 2:373–382.

Van der Heijeden AHC, Wolters G, Groep JC, Hagenaar R (1987): Single-letter recognition accuracy benefits from advance cuing of location. *Percept Psychophys* 42:503–509.

Van Voorhis ST, Hillyard SA (1977): Visual evoked potentials and selective attention to points in space. *Percept Psychophys* 22:54–62.

Walter WG, Cooper R, Aldridge VJ, McCallum WC, Winter CV (1964): Contingent negative variation: An electric sign of sensorimotor association and expectancy in the human brain. *Nature* (London) 203:380–384.

Chapter 4

The Order of Global- and Local-Level Information Processing: Electrophysiological Evidence for Parallel Perceptual Processes

H.-J. Heinze, Sönke Johannes, T. F. Münte, and George R. Mangun

The question of whether different levels of hierarchically structured visual stimuli are processed in parallel by separate systems or sequentially by a single processing resource was investigated by recording behavioral and event-related brain potential (ERP) measures in two divided-attention experiments. In the first experiment, letter stimuli were presented in which a large (global) letter was built up by the spatial arrangement of small (local) letters. Subjects were asked to respond to target letters that could occur with equal probability at the global or local level of the centrally presented stimuli. Both behavioral and electrophysiological results indicated a dissociation between the relative speed of responding to global or local target letters and the direction in which one level of information interfered with the processing of the other, suggesting the existence of separate systems involved in global and local pattern processing. Furthermore, ERP correlates of early target perception pointed to hemispheric differences of global and local pattern perception. Experiment 2 further tested this hemispheric asymmetry by using the same letter stimuli as in Experiment 1, but presenting them to positions at the left and right of fixation. ERP analysis strongly supported the assumption that separate perceptual systems process global- and local-level information in parallel with a left/right hemispheric asymmetry of local/global processing.

Cognitive Electrophysiology
H-J. Heinze, T.F. Münte, and G.R. Mangun, editors
© 1994 Birkhäuser Boston

Introduction

There has been a significant debate concerning the mechanisms by which hierarchically structured visual stimuli are processed. In particular, the question is whether there is a single resource processing global- and local-level information in a sequential "top-down" or "bottom-up" order, or whether different systems exist that process global- and local-level information independently and in parallel. In 1977, Navon published a series of experiments suggesting that global-level information is processed before local-level information (Navon, 1977). This "global precedence" hypothesis was based on the joint occurrence of two effects: a global response time (RT) advantage, and global interference. Navon found that subjects responded faster to global than to local targets (e.g., the large and small letters in Fig. 1), and that global-level information interfered with the processing of local-level information, but not vice versa. Because the global RT advantage occurred only when global and local patterns appeared within hierarchically organized structures, and not when the same-sized patterns were presented in isolation, Navon concluded that this effect indicated higher order cognitive mechanisms rather than retinal or sensory functions.

Although this joint occurrence of a global RT advantage and a global direction of interference has been observed in a number of subsequent studies (Boer and Keuss, 1982; Grice, Canham, and Boroughs, 1983; Lamb and Robertson, 1988; Martin, 1979; Navon and Norman, 1983; Pomerantz, 1983), it is by no means universal. Several investigators have demonstrated that RT advantage and the direction of interference may vary as a function of perceptual or attentional factors (Grice, Canham, and Boroughs, 1983; Kinchla and Wolfe, 1979; Lamb and Robertson, 1988, 1990; Miller, 1981; Pomerantz, 1983; Ward, 1982). For example, the relative discriminability, the size and the retinal location of global and local forms, as well as the size or distribution of the attended area, have been shown to have profound effects on RT advantage and interference. Such findings suggest that the

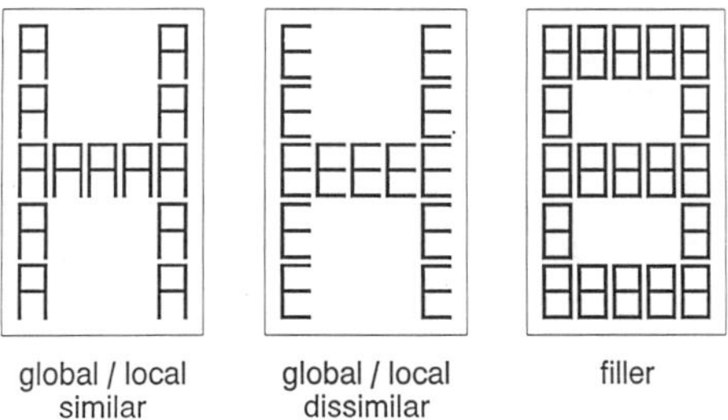

Figure 1.   Examples of two (of eight) hierarchically structured letter stimuli used in experiment. Left: global letter 'H' constructed from local letters 'A'; global and local letters are similar. Middle: global letter 'H' constructed from local letters 'E'; global and local letters are dissimilar. Right: neutral stimulus.

order of processing is not always from global to local, but can, under the right experimental conditions, be from local to global. Therefore, some authors have suggested that there is not a strictly sequential order, but that global- and local-level information might be processed in parallel or at least with a similar time course (Boer and Keuss, 1982; Hughes et al., 1984; Miller, 1981; Navon, 1981).

Recent neuropsychological studies in patients with unilateral brain lesions support the parallel-processing hypothesis demonstrating a left/right hemispheric asymmetry of global- and local-level information processing. Although patients with unilateral left or right frontal lesions exhibited the same pattern of global and local analysis as normal controls (Robertson, Lamb, and Knight, 1991), patients with left temporal-parietal lesions had difficulties with local pattern processing and patients with right temporal-parietal lesions with global pattern processing (Delis, Robertson, and Efron, 1986; Doyon and Milner, 1991; Lamb, Robertson, and Knight, 1989, 1990; Robertson and Delis, 1986). Robertson, Lamb, and Knight (1988) differentiated between lesions in the superior temporal gyrus and lesions in the rostral inferior parietal lobe, and proposed that the hemispheric global/local asymmetry is caused by a perceptual mechanism with a critical anatomic locus in the superior temporal gyrus.

A study by Lamb and Robertson (1989) in normal controls provided further evidence that different systems are involved in global- and local-level information processing. In a divided-attention paradigm, subjects were presented hierarchically structured letter stimuli to those in Figure 1. The task was to respond to target letters that could occur with equal probability either at the global level or at the local level. The authors found that the relative speed of responding to global and local targets varied with the visual angle subtended by the stimulus patterns: there was a local RT advantage when the larger letters subtended a visual angle of 6, 9, or 12 degrees, but global and local RTs did not differ when the global letters subtended only 3 degrees. Interference effects, on the other hand, exhibited a different pattern of results: the processing or local target letters was significantly influenced by global distractor letters, whereas local distractor letters did not interfere with global target processing, independent of the visual angle. From this dissociation between local RT advantage and global direction of interference, the authors concluded that RT advantage and interference do not reflect a sequential top-down or bottom-up processing of a single resource, but might indicate the activity of separate systems involved in global- and local-level information processing.

It is evident that a central question in this discussion is how behavioral measures such as RT advantage and interference are related to different processing stages and structures of global and local target recognition. To answer this question, methods are required that allow for a more direct measure of global and local processing. Event-related brain potentials (ERPs) are a particularly suitable method in that these provide a continuous record of processes occurring between stimulus and response. These scalp-recorded potentials reflect synchronous neural activity associated with sensory, motor, or cognitive processes and therefore help to identify different processing stages and structures (Coles et al., 1985; Hillyard and Picton, 1987).

The ERPs elicited by briefly presented visual stimuli consist of a series of positive and negative components that begin as early as 35–40 msec post stimulus and continue for hundreds of milliseconds thereafter. ERP components that are sensitive to visual selection processes include the P1 (peaking between 90 and 140 msec), N1 (160–200 msec), P2 (200–250 msec), and the N2 (200–400 msec) component (Mangun, Hillyard, and Luck, 1993). Whereas selection based on stimulus location is indexed by changes of the P1 and N1 components (Hillyard, Luck, and Mangun, this volume), selection of elementary stimulus features such as color, spatial frequency, spatial orientation, or contour leads to an enhancement of negative components over the posterior scalp sites in the latency range of the N2 component (Harter and Aine, 1984; Hillyard and Münte, 1984). Hillyard and Picton (1987) suggested that these negativities reflect further processing after the stimulus has been identified as belonging to the relevant category. Wijers et al. (1989) differentiated between two processes of feature selection, an early perceptual selection and a covert orienting of attention. They proposed that the early effect is indexed by a negativity with an onset of about 150 msec over the modality-specific areas that reflects the modulation of the feature-specific processing units by selective attention, whereas the controlled selection is associated with a later central negativity. In a number of recent studies (Heinze et al., 1990; Luck et al., 1990) a negative component occurred between 200 and 400 msec and was largest over the occipital-temporal scalp sites contralateral to the location of the target. This component was interpreted as the earliest sign of target recognition in the attended channel, probably similar to the N2 component that was observed in the study by Ritter, Simpson, and Vaughan (1983).

In light of these findings, Heinze and Münte (in press) conducted an ERP experiment to assess ERP correlates of global/local target processing and to determine how the relative speed of responding to global and local targets is related to these ERP measurements. In particular, the goal was to assess whether the time course and scalp distribution of the ERPs to global versus local targets provided evidence for a single processing resource, or instead supported the hypothesis that separate mechanisms for global- and local-level information processing exist. In a divided-attention paradigm similar to the one used by Lamb and Robertson (1989), hierarchical letter stimuli were presented and subjects asked to respond to target letters that could occur unpredictably either at the global or at the local level. ERP analysis revealed a posterior negative component with a left temporal maximum peaking at 250 msec (denoted as N250) that was the earliest sign of differential global/local target processing. This negativity was followed by a broad positive component (P3) with a central-parietal maximum. The N250 component resembled in important aspects the negative components in the N2 range related to target recognition in the attended channel. Therefore,

Heinze and Münte (in press) proposed that the N250 was an index of early global/local target perception, whereas the subsequent P3 component reflected later stages of target classification. Because the N250 to global versus local targets had widely overlapping but different time courses and topographical distributions, these findings were interpreted as electrophysiological evidence that initial global and local analyses are mediated by at least partially different processing structures.

The two experiments described in the current study addressed the following questions. First, in the Heinze and Münte (in press) experiment, electrophysiological correlates of global/local interference had not been investigated. As noted, an important argument for the existence of separate global/local processing structures is the finding that under certain conditions a local RT advantage or lack of global RT advantage is associated with a global direction of interference, this dissociation possibly reflecting the activity of different processing systems (Lamb and Robertson, 1989). Therefore, Experiment 1 aimed to determine whether this dissociation is also reflected in ERP correlates of global/local target processing, and whether there is electrophysiological evidence that global- and local-level patterns are processed in parallel at an early perceptual stage. Second, if the left/right hemisphere asymmetry of global/local processing is related to separate cortical perceptual structures, as proposed by a number of studies (e.g., Doyon and Milner, 1991; Lamb, Robertson, and Knight, 1989; Robertson, Lamb, and Knight, 1988), one would expect ERP correlates of global/local target perception to vary as a function of the hemisphere that primarily receives the stimulus. Experiment 2 investigated this question by presenting the same linguistic stimuli as in Experiment 1 at positions to the left and right of the fixation point.

Experiment 1

This experiment assessed behavioral and electrophysiological interference effects using the same divided-attention paradigm as in the Heinze and Münte (in press) study. Hierarchically structured linguistic stimuli were presented at the center of a video screen, and subjects had to respond to target letters that could occur with equal probability either at the global or at the local level. Interference effects were obtained by varying the similarity relations between target and distractor stimuli. According to Lamb and Robertson (1989), the letter 'H' is similar to the letter 'A' and dissimilar to the letters 'S' and 'E', whereas the letter 'S' is similar to the letter 'E' and dissimilar to the letters 'H' and 'A' (see Fig. 1). Interference, therefore, was measured as the interaction between similar and dissimilar target and distractor letters. Electrophysiological measurements were obtained by recording ERPs to letter stimuli that contained a target letter either at the global or at the local level or that did not contain a target letter.

Methods

Subjects. Eight right-handed male volunteers (age, 22–26 years) from the local student population served as subjects. They had normal or corrected-to-normal vision.

Stimuli. Stimuli were large capital letters composed of small capital letters: H, S, A, and E were used. Global letters were constructed from local letters in a 5×5 matrix (see Fig. 1). The global letters subtended 7 degrees vertically at a viewing distance of 70 cm. Global letters were 7.4 times as tall as local letters, and both global and local letters were 1.5 times as tall as they were wide. In separate runs, all four letters (H, S, A, E) could serve as targets. The letters H and S were combined with the letters E and A. Thus there were eight stimuli in all: H/A (global H and local A), H/E, S/A, S/E, A/H, A/S, E/H, and E/S. In addition, a "filler" stimulus was presented that was made up of large and small numerals, 8 (Fig. 1). Stimuli were white on a dark background and presented in the center of the screen. A dot in the center of the screen served as fixation point.

Procedure. In each run, the eight letter stimuli were each presented 10 times. A target letter occurred randomly in 50% of the letter stimuli (25% at the global level and 25% at the local level). The stimulus duration was 100 msec, and the interstimulus interval varied randomly between 800 and 1200 msec. The filler stimulus was alternated with the letter stimuli to prevent any carryover effects to the next stimulus. The subject's task was to indicate after each letter stimulus whether or not a target letter was present by pressing one of two buttons with the right index finger. The filler stimulus did not require a response. Instructions stressed both accuracy and speed. After eight practice runs in which each letter served twice as the target, subjects were presented with 40 runs.

Recording. The electroencephalogram (EEG) was recorded from standard left- and right-hemisphere scalp sites at frontal, central, parietal, occipital, and temporal electrode positions (International 10/20 System sites: F3, F4, C3, C4, P3, P4, O1, O2, T5, T6), referenced to the right mastoid. The horizontal electrooculogram (EOG) was recorded between the left and right external ocular canthi to monitor lateral eye position, and the vertical EOG was recorded from beneath the right eye and references to the right mastoid, to monitor blink activity. The recordings were made using tin electrodes mounted in a cap (Electro-Cap International). The potentials were amplified with a bandpass filter of 0.01–70 Hz. The voltages were digitized at 256 Hz and stored with stimulus and response codes for offline analysis.

Data Analysis. The EEG was averaged offline using an automatic artefact rejection system that rejected blinks, large eye movements, and excessive muscle activity. To assess early differential effects of global and local target processing on ERP waveforms, ERPs to letter stimuli without a target were subtracted from ERPs to letter stimuli that contained a target at the local or global level. These difference waves were quantified by calculating the mean amplitude (A) within successive time windows between 150 and 250, 200 and 300, 250 and 350 msec, etc. [denoted as A(150–250), A(200–300), A(250–350), etc.] with respect to the mean voltage during a 100-msec prestimulus baseline. To assess the onset of the difference waves, an onset measure was used that estimated the beginning of the first negative or positive deflection by linear fitting (Heinze and Münte, in press). For the original waveforms, the amplitude of the late positive component (P3) was measured as the mean voltage within the time window 350–750 msec after stimulus onset at central, parietal, and temporal leads. Because the P3 waves were often very broad and their peaks sometimes difficult to identify, P3 latency was assessed as the time required for the component to reach 50% of its mean amplitude over the measurement window of 350–750 msec. This area latency measure has the advantage of being more stable than the peak latency measure (Luck and Hillyard, 1990).

All behavioral and ERP measures were analyzed with repeated measures analyses of variance (ANOVAs), adjusting for nonsphericity with the Greenhouse–Geiser epsilon coefficient. RTs and hit rates to target stimuli were analyzed with two factors: Target Level (target global, target local) and Similarity (target/distractor, similar; target/distractor, dissimilar). The RTs and hit rates to nontarget stimuli were analyzed with the factor Similarity (global letter similar to target; local letter similar to target; both global and local letter dissimilar to target). ERP effects were assessed with three factors: Target Level (target, global; target, local). Similarity (target/distractor, similar; target/distractor, dissimilar), and Hemisphere (left, right), separately for the frontal, central, parietal, occipital, and temporal scalp site pairs.

Results

BEHAVIORAL MEASURES. RTs and hit rates are summarized in Table 1. RTs were faster and hit rates were higher for local than for global targets [main effect of Target Level, RT: $F(1,7) = 10.3$, $p < .02$; hit rate: $F(1,7) = 6.8$, $p < .04$]. RT interference was reflected as a main effect of Similarity [$F(1,7) = 14.7$, $p < .006$], indicating that targets with similar distractors were identified faster than targets with dissimilar distractors. As can be seen from Table 1, global distractors interfered more strongly with local targets than local distractors did with global targets. Nontarget stimuli with both global and local letters dissimilar to the target letter were identified faster and more accurately than nontargets with either global or local letters similar to the target letter [RT: $F(2,14) = 13.6$, $p < .001$; hit rate: $F(2,14) = 6.7$, $p < .01$].

Table 1. Mean RTs and % hits ($\pm$ SD) in the 7° condition of part 2 ($n = 8$).

	7°			
	Targets			
	Global		Local	
Distractor	Similar	Disimilar	Similar	Dissimilar
RT	487 ± 68	492 ± 70	455 ± 51	472 ± 51
% hits	92.8 ± 6	92.9 ± 6	96.2 ± 3	95.3 ± 4
	Nontargets			
	Global letter Similar to target	Local letter similar to target	Global/local dissimilar	
RT	551 ± 63	553 ± 61	529 ± 52	
% hits	94.8 ± 3	94.8 ± 3	97.1 ± 2	

ERP MEASURES. Figure 2 shows the grand average ERPs to global target, local target, and nontarget letter stimuli. Compared to the nontarget stimuli, target stimuli elicited a posterior negativity peaking at about 250 msec over the posterior scalp sites that was considerably smaller over the right hemisphere and exhibited a maximum at the temporal leads. This negativity was followed by a broad positivity (P3) with the largest amplitude at the parietal sites. P3 latency reflected the local RT advantage, being longer for global than for local targets [parietal: $F(1,7) = 5.7$, $p < .05$). P3 amplitude was not significantly different for global and local targets.

Figure 3 displays the grand average difference waves: ERPs to targets minus ERPs to nontargets at the temporal leads, where the N250 component was maximal. As can be seen, the earliest sign of global and local target processing consisted of a posterior negative component with a maximum over the left hemisphere peaking at about 250 msec, denoted as the N250 component. The N250 to local targets had an earlier onset [onset measure: $F(1,7) = 10.3$, $p < .02$] and a shorter duration than the N250 to global targets as reflected in the mean amplitude measurement in the 350–450 msec time range [main effect of Target Level: $F(1,7) = 5.8$, $p < .05$]. The N250 amplitude in the 200- to 300-msec time range was significantly larger over the left hemisphere [main effect of Hemisphere: $F(1,7) = 30.1$, $p < .001$] and displayed a left/right hemisphere asymmetry: over the left hemisphere, the N250 amplitude was higher to local than to global targets, whereas over the right hemisphere the N250 amplitude was about the same for global and local targets [A(200–300): Target level × Hemisphere interaction, $F(1,7) = 7.5$, $p < .03$).

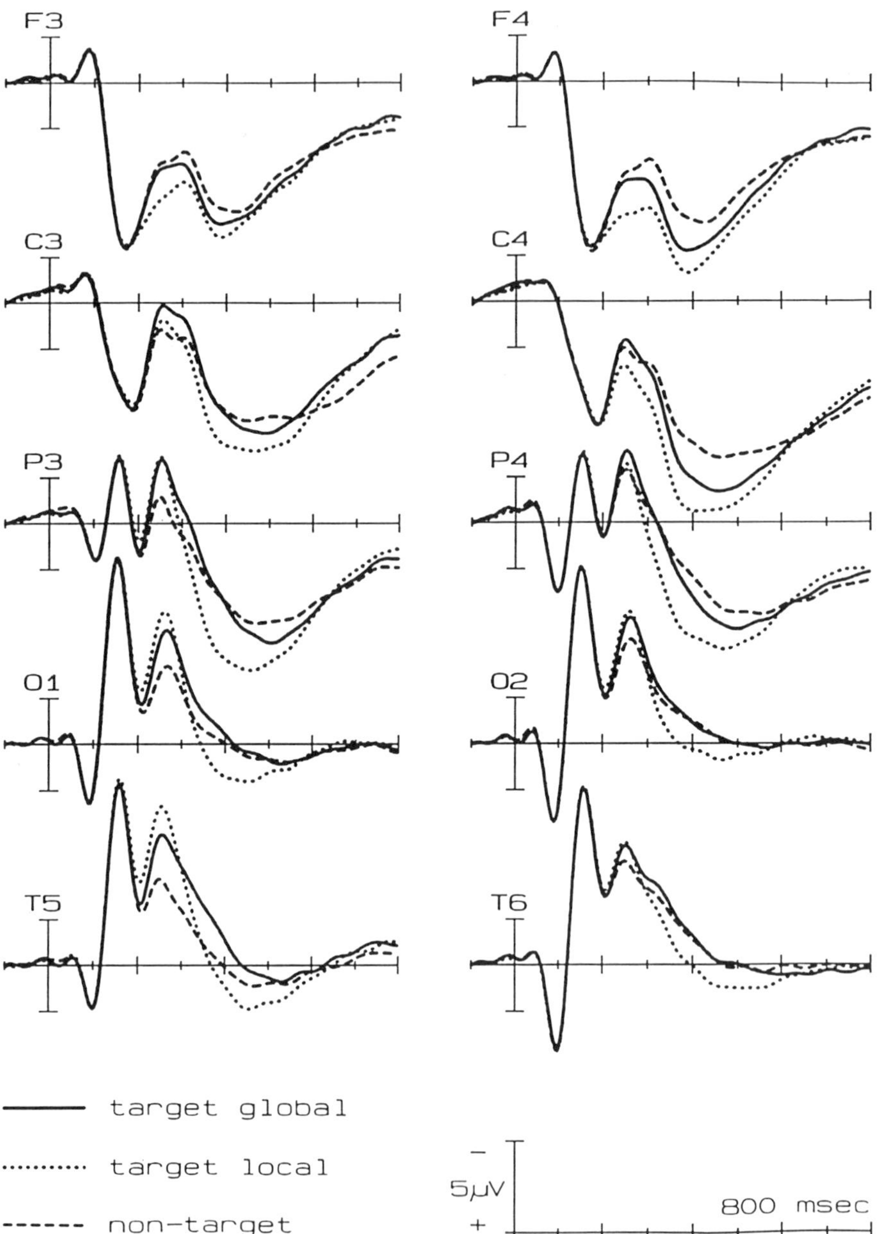

Figure 2.   Experiment 1. Grand average ERPs to stimuli containing target letters at global level (*solid line*), local level (*dotted line*), or without target letter (*dashed line*) at left and right frontal, central, parietal, occipital, and temporal electrode sites.

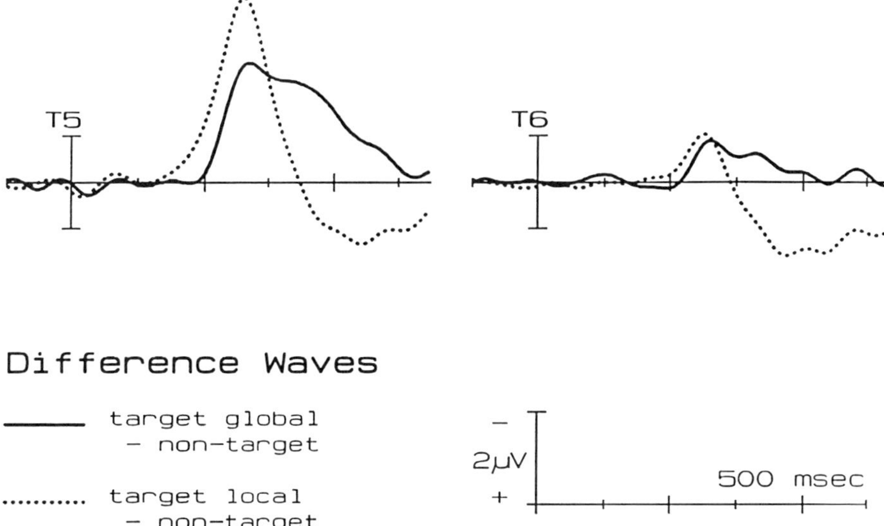

Figure 3.   Experiment 1. Grand average ERP difference waves at the left and right temporal electrode sites. *Solid line*: global targets minus nontargets; *dotted line*: local targets minus nontargets.

ERP effects of target–distractor similarity relations at the temporal scalp sites are displayed in Figure 4. Difference waves were calculated by subtracting nontarget stimuli with both local and global letters being dissimilar to the target letter from target stimuli, separately for global and local targets and for distractors similar or dissimilar to the target. Global and local distractor letters displayed the same direction of early ERP interference effects, the resonance to targets with similar distractors being more positive in the 300- to 400-msec time range [A(300–400): main effect of Similarity, $F(1,7) = 8.8$, $p < .02$]. This interference effect was stronger for local than for global distractors [Target Level × Similarity interaction: $F(1,7) = 8.0$, $p < .03$] and started earlier for local than for global distractors [A(250–350): significant Target Level × Similarity interaction, $F(1,7) = 42.5$, $p < .001$, but no significant main effect of Similarity].

The P3 component also displayed interference effects, but in a different way (Fig. 5). P3 latency was longer for global targets with similar than with dissimilar local distractors, but shorter for local targets with similar than with dissimilar global distractors [Target Level × Similarity interaction, $F(1,7) = 5.7$, $p < .05$].

Discussion

The results agree with previous findings that advantage of global/local target identification and direction of interference do not necessarily covary (Lamb

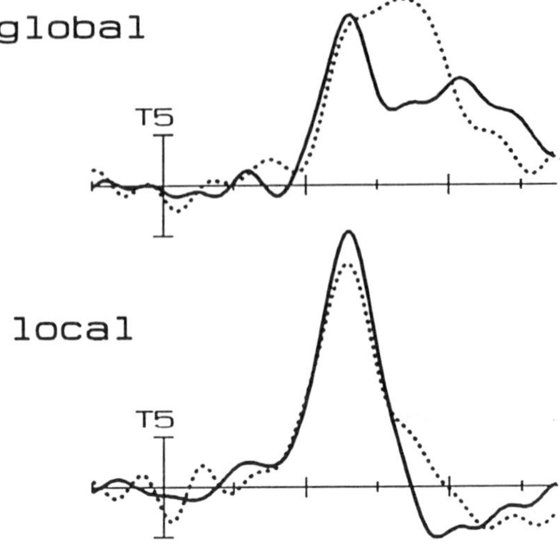

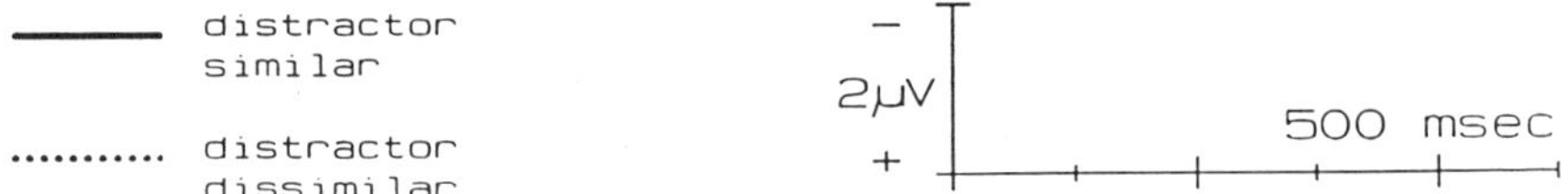

Figure 4.   Experiment 1. Grand average ERP difference waves at left temporal electrode site. Top: global targets with similar local distractors minus nontargets (*solid line*) and global targets with dissimilar local distractors minus nontargets (*dotted line*). Bottom: local targets with similar global distractors minus nontargets (*solid line*), and local targets with dissimilar global distractors minus nontargets (*dotted line*).

and Robertson, 1989; Heinze and Münte, in press). The behavioral data showed that RT advantage and interference are dissociable: There was a local RT advantage, but global distractors appeared to interfere even more strongly with local target processing than local distractors with global target processing. ERP analysis replicated the findings of the Heinze and Münte (in press) study, revealing a posterior negative component (N250) that was the earliest sign of differential global/local target processing. As argued in this study (Heinze and Münte, in press), the N250 component resembled in important aspects the negative components in the N2 range that have been observed in previous studies on visual selective attention: that is, the N250 occurred between 200 and 400 msec, was largest over the posterior

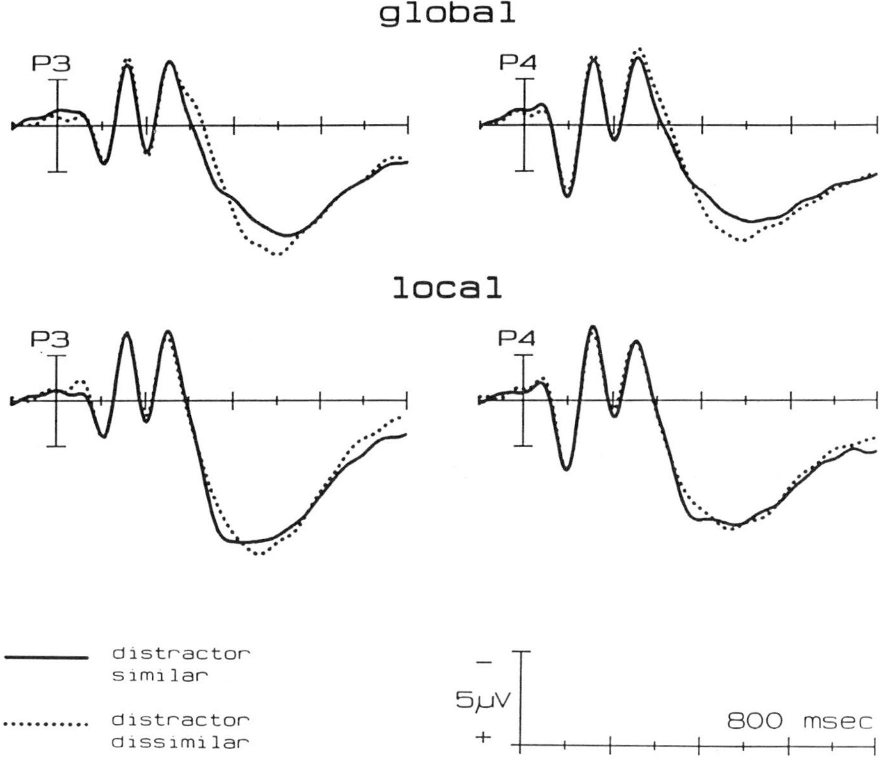

Figure 5.   Experiment 1. Grand average ERPs to global targets (top) and local targets (bottom) with similar (*solid line*) and dissimilar (*dotted line*) distractors at left (P3) and right (P4) parietal electrode sites.

scalp sites, and was the earliest sign of target recognition. Therefore, this component can be interpreted as a sign of early global/local target perception, whereas the positivity (P3) subsequent to the N250 indicates a later stage of target evaluation. Target–distractor similarity reactions were reflected in both N250 and P3 changes, but in different ways: Global and local distractors similarly affected the N250 waveform across a widely overlapping time range. This interference effects were stronger and started earlier for local than for global distractors. P3 latency, on the other hand, showed an interaction between target level and target–distractor similarity reactions.

These results provide additional arguments for the hypothesis that separate perceptual systems exist that process global- and local-level information in parallel. Both RT and P3 latency showed a local advantage of target evaluation, but also a global as well as a local direction of interference. Following previous arguments (Lamb and Robertson, 1989; Heinze and Münte, in press), this dissociation could be explained by the assumption that at an early stage both global- and local-level patterns are processed independently and in parallel, thereby leading to global as well as local interference effects. This view is supported by the early ERP effects. The overlapping but different time course and topographical distribution of the N250 to global and local targets suggests the existence of least partially separate structures involved in early global and local target perception. Further, the fact that ERP correlates of early target perception exhibited a similar pattern of global and local interference effects within a widely overlapping time range provides strong evidence that at an early perceptual stage global- and local-level information is processed in parallel and independently.

As in the Heinze and Münte study, the N250 amplitude showed a left hemisphere preponderance and a left/right hemisphere asymmetry, the local N250 amplitude being relatively larger over the left hemisphere. This distribution might partially reflect a left hemispheric specialization for linguistic stimuli and left/right hemisphere differences of global and local target processing. If this is true, one would expect that hierarchically structured linguistic stimuli, presented at positions left and right from the central fixation point, would elicit an interaction between the visual field of the stimulus and behavioral/electrophysiological measures of global and local target processing. This hypothesis was investigated in the second experiment.

Experiment 2

Methods

SUBJECTS. Nine subjects (two men) from the local student population (age, 20–27 years) who did not participate in Experiment 1 served as subjects. All were right handed and had normal or corrected-to-normal vision.

STIMULI AND PROCEDURE. The eight letter stimuli were the same as in Experiment 1; filler stimuli were not used. Stimuli were presented at positions 7.5 degrees left and right from the central fixation point as measured to the inner margin. At a viewing distance of 110 cm, the stimuli subtended an visual angle of 8 degrees vertically. Thus, the stimuli were slightly larger than in Experiment 1 to compensate for changes in the cortical magnification factor in moving from fixation to lateralized stimuli. In each run, each stimulus was presented four times at the left and right position in a random order. Stimulus duration was 100 msec; the interstimulus interval varied between 1200 and 1600 msec. Subjects were asked to maintain fixation, to attend to both sides, and to indicate after each stimulus whether or not a target letter was present by pressing one of two buttons with the right index finger. Instructions stressed both accuracy and speed. On each of two days, subjects were given 54 runs after 12 practice runs.

RECORDING AND DATA ANALYSIS. Recording and data analysis were the same as in Experiment 1, except that the ANOVA included the Visual Field of stimulus (left, right) as an additional factor. The analysis was restricted to the early (N250) components because the major question addressed in this study was the processing of global/local level information at an early perceptual stage.

Results

BEHAVIORAL RESULTS. Mean RTs and hit rates are summarized in Table 2. Subjects responded faster to global than to local targets [main effect of Target Level: $F(1,8) = 41.9$, $p < .001$]. Furthermore, stimuli in the right visual field were identified faster than stimuli in the left visual field [main effect of Visual Field: $F(1,8) = 16.7$, $p < .004$]. A significant Visual Field $\times$ Target Level interaction [$F(1,8) = 7.3$, $p < .03$) indicated that this left hemisphere RT advantage was relatively stronger for local than for global targets. Analysis of interference effects yielded a dissociation between the global RT advantage and the direction of interference: global targets were identified faster with similar than with dissimilar local distractors, while local targets were identified faster with dissimilar than with similar global distractors [Target Level $\times$ Similarity interaction: $F(1,8) = 25.33$, $p < .002$]. This effect was stronger for right than for left field stimuli [Target

Table 2.

	Global			
	Left targets		Right targets	
Distractor	Similar	Dissimilar	Similar	Dissimilar
RT	476 ± 36	479 ± 28	454 ± 37	476 ± 34
% hits	93.0 ± 3	92.9 ± 3	92.0 ± 2	92.4 ± 2

	Local			
	Left targets		Right targets	
Distractor	Similar	Dissimilar	Similar	Dissimilar
RT	538 ± 48	526 ± 28	512 ± 17	477 ± 13
% hits	81.3 ± 4	90.0 ± 3	86.3 ± 3	92.4 ± 2

Level × Visual Field × Similarity interaction: $F(1,8) = 35.65$, $p < .001$]. Hit rate analysis yielded similar results, the hit rate being higher for global than for local targets [$F(1,8) = 39.8$, $p < .001$] and for right than for left visual field targets [$F(1,8) = 12.3$, $p < .01$] with a relative left hemisphere advantage for local targets [Visual Field × Target Level interaction: $F(1,8) = 10.8$, $p < .01$]. Furthermore, the hit rate was higher for local targets with dissimilar than with similar global distractors, while local similar and dissimilar distractors had no differential impact on global targets [Target Level × Similarity interaction: $F(1,8) = 18.2$, $p < .003$].

ERP ANALYSIS. Figures 6A and 6B display the grand average ERPs to global target, local target, and nontarget stimuli in the left and right visual field. As in Experiment 1, both global and local targets in the left and right visual field elicited a posterior negativity with a left temporal maximum as the earliest sign of global/local target processing. The target minus nontarget difference waves (Fig. 7) showed that this posterior negativity peaked at about 250 msec (denoted as N250) and had about the same onset, but a longer time course, for local than for global targets.

ERP interference effects on the N250 component at the temporal leads are presented in Figure 8. As in Experiment 1, these difference waves were calculated by subtracting EPRs to nontarget stimuli, with both global and local letters dissimilar to the target letter, from ERPs to target stimuli separately for global/local targets and similar/dissimilar distractors. Right visual field stimuli (Fig. 8B) displayed strong interference effects in the 200- to 300-msec time range: for both global and local distractors the amplitude of the N250 to local and global targets was higher for dissimilar than for similar distractor letters. Left visual field stimuli (Fig. 8A) showed a similar pattern of early interference; these effects, however, were stronger for global than for local distractors. Statistical analysis confirmed a left/right hemisphere asymmetry of the N250 mean amplitude in the 200- to 300-msec time range at temporal scalp sites [A(200–300): main effect of Hemisphere ($F(1,8) = 14.86$, $p < .005$], but did not show a significant main effect of target level. The N250 onset was not significantly different between global and local targets. The main effect of similarity was significant [$F(1,8) = 12.25$, $p < .009$], indicating that both global and local target stimuli exhibited a higher mean amplitude in the 200- to 300-msec time range for dissimilar than for similar distractor letters. Further, there was a significant Target Level × Similarity × Visual Field interaction [A(200–300): $F(1,8) = 14.22$, $p < .006$]. Post hoc comparisons showed that for right visual field stimuli both global and local distractors produced a significant effect of similarity, while for left visual field stimuli this effect was significant for global but not for local distractors.

Discussion

The results of Experiment 2 provide electrophysiological as well as behavioral evidence for hemispheric differences of global and local pattern process-

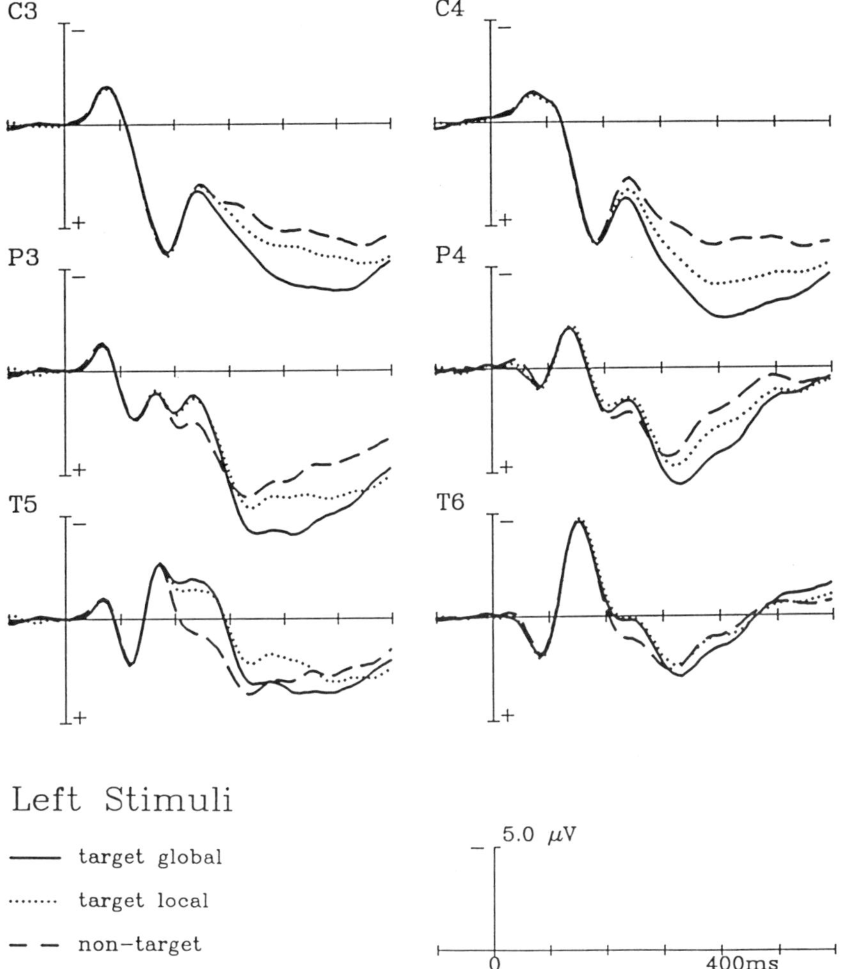

Figure 6.   Experiment 2: Grand average ERPs to left (A) and right visual field (B) global targets (*solid line*), local targets (*dotted line*), and nontargets (*dashed line*) at left and right central, parietal, and temporal electrode sites.

ing. As in Experiment 1, the behavioral data revealed a dissociation between RT advantage and the direction of interference, thus arguing for partially independent, parallel processing systems for global and local levels of processing (Lamb and Robertson, 1989). Further, the relative local RT advantage for stimuli in the right visual field is compatible with the assumption of a left hemispheric preponderance for local information processing.

ERP analysis showed that target letters in both the left and right visual field elicited a posterior negative component (N250) that was the earliest sign

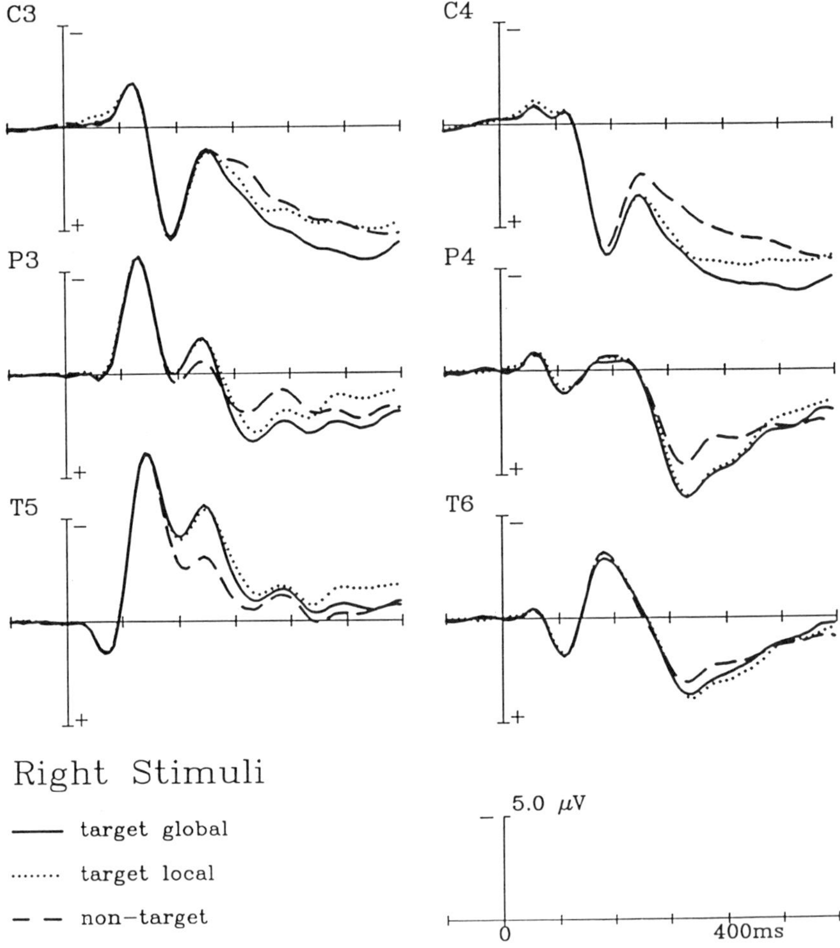

Figure 6.   (*continued*)

of global/local target perception. Similar to Experiment 1, this component had a left temporal maximum, possibly partially reflecting a left hemispheric preponderance for linguistic stimuli, as also suggested by the faster RTs and higher hit rates for right than for left visual field stimuli. In accordance with previous findings (Heinze and Münte, in press), there was a dissociation between early and later stages of processing: the global RT advantage was not paralleled by an earlier onset of the global N250. Therefore, a global RT advantage does not necessarily imply a global-to-local order of target perception. Further, whereas the RT measures showed a Target Level ×

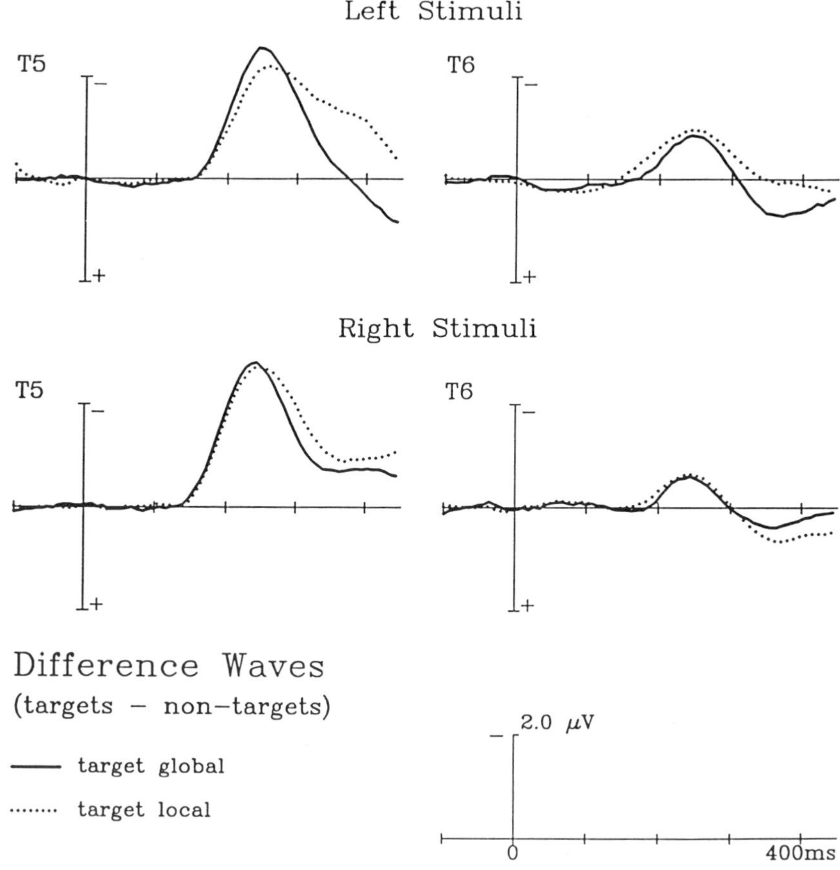

Figure 7.   Experiment 2. Grand average ERP difference waves at left (T5) and right (T6) temporal scalp sites for left (top) and right visual field (bottom) stimuli. *Solid line*: global targets minus nontargets; *dotted line*: local targets minus nontargets.

Similarity interaction, global and local distractors produced a similar pattern of early ERP interference effects.

The fact that the nature of the global (local) distractor letters significantly influenced ERP correlates of early local (global) target letter perception strongly suggest that at this early perceptual stage both global- and local-level information is available and processed independently, whereas the global advantage of target identification occurs at a later stage of processing. The differences between left and right field stimuli possibly reflect a left hemispheric preponderance for local-level information: local distractors seemed to have less impact on global target processing for left than for right field stimuli.

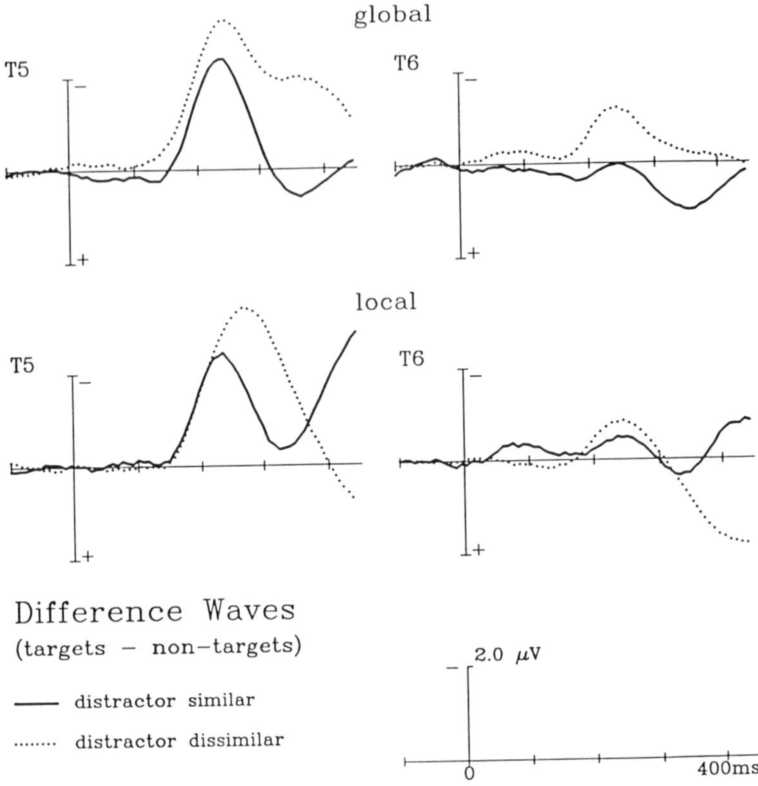

Figure 8.   Experiment 2. Grand average ERP difference waves at left and right temporal electrode sites for left (A) and right visual field (B) stimuli. Top: global targets with similar local distractors minus nontargets (*solid line*), and global targets with dissimilar local distractors minus nontargets (*dotted line*). Bottom: local targets with similar global distractors minus nontargets (*solid line*) and local targets with dissimilar global distractors minus nontargets (*dotted line*).

General Discussion

The current study investigated the mechanisms of global and local processing within hierarchically structured visual stimuli. ERPs and behavioral measures were utilized to examine whether global and local analyses were performed serially in top-down or bottom-up fashion (e.g., Navon, 1977), or proceeded in partially or wholly separate systems (e.g., Lamb and Robertson, 1989). The results in both experiments of this study suggest that, in part, global- and local-level analyses proceed in separate, parallel processing systems. Thus, the current findings are in line with prior evidence for separate mechanisms for global and local processing that was derived from behavioral

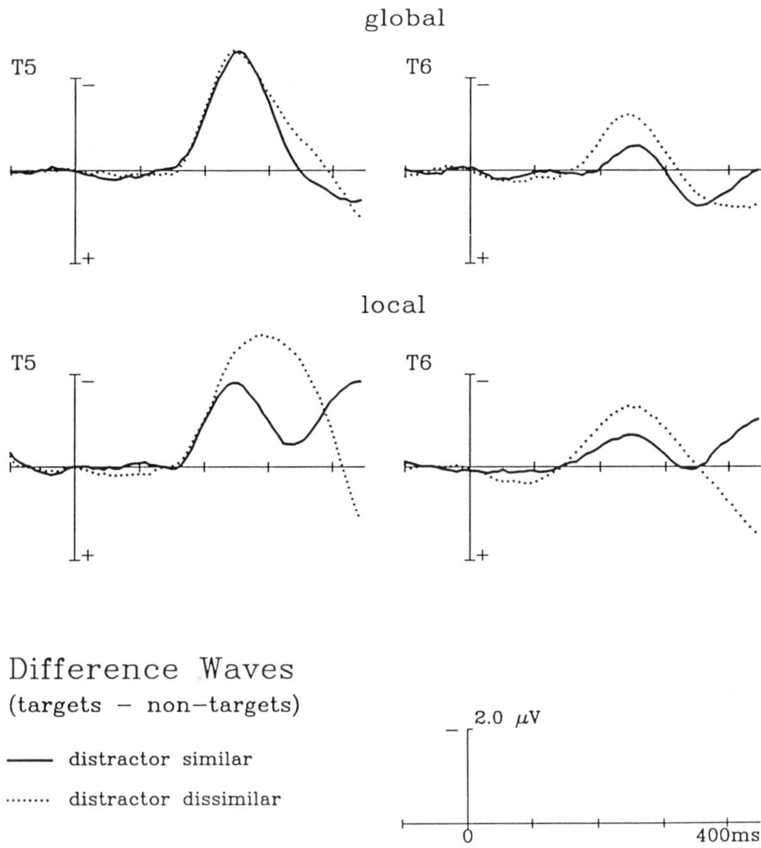

Figure 8.   (*continued*)

(Lamb and Robertson, 1989) and neuropsychological (Robertson, Lamb, and Knight, 1988) studies. The current data also replicate and extend previous electrophysiological findings that suggested separate systems for global and local processing within complex stimuli (Heinze and Münte, in press).

The brain systems that mediate global versus local processing have been suggested by studies that examined differences in such processing in the two cerebral hemispheres. As reviewed earlier, a number of behavioral studies demonstrated left/right hemispheric differences in the effects of parietal-temporal lesions on global/local processing. The present data provide electrophysiological support for a hemispheric asymmetry. Together, the evidence from focal lesion work and electrophysiological recordings suggest that the left hemisphere processes local-level information with greater facility, while the right hemisphere preferentially processes global-level information. This distinction is probably not a hemispheric "specialization" per se, but rather a lateralized bias in the level of information most efficiently processed by each cerebral hemisphere.

In conclusion, these combined behavioral and ERP experiments support the view that, in part, separate systems mediate the analysis of global and local information. Moreover, the ERP data presented here provide increased information about the time course and relative contributions of early versus later processing stages in the perception of hierarchical stimuli. Such analyses will lead us to clearer understanding of the perception of the parts and wholes of complex visual objects.

References

Boer LC, Keuss PJG (1982): Global precedence as a postperceptual effect: An analysis of speed-accuracy tradeoff functions. *Percept Psychophys* 31:358–366.

Coles, MGH, Gratton G, Bashore TR, Eriksen CW, Donchin E (1985): A psychophysiological investigation of the continuous flow model of human information processing. *J Exp Psychol Hum Percept Perform* 11:529–553.

Delis D, Robertson LC, Efron R (1986): Hemispheric specalization of memory for visual hierarchical stimuli. *Neuropsychologia* 24:205–214.

Doyon J, Milner B (1991): Right temporal-lobe contribution to global visual processing. *Neuropsychologia* 29:343–360.

Grice GC, Canham L, Boroughs JM (1983): Forest before trees? It depends where you look. *Percept Psychophys* 33:121–128.

Harter MR, Aine CJ (1984): Brain mechanisms of visual selective attention. In: *Varieties of Attention*, Parasuraman R, Davies R, eds., pp. 293–321. London: Academic Press.

Heinze HJ, Münte TF: Electrophysiological correlates of hierarchical stimulus processing: Dissociation between onset and later stages of global and local target processing. *Neuropsychologia* (in press).

Heinze HJ, Luck SJ, Mangun GR, Hillyard SA (1990): Visual event-related potentials index focused attention within bilateral stimulus arrays. I. Evidence for early selection. *Electroencephalogr Clin Neurophysiol* 75:511–527.

Hillyard SA, Münte TF (1984): Selective attention to color and location: An analysis with event-related brain potentials. *Percept Psychophys* 36:185–198.

Hillyard SA, Picton TW (1987): Electrophysiology of cognition. In: *Handbook of Physiology: Higher Functions of the Nervous System, Section 1: The Nervous System: Vol. V, Higher Functions of the Brain, Part 2*, Plum F, ed., pp. 519–584. Bethesda, MD: American Physiological Society.

Hughes HC, Layton WM, Baird JC, Lester LS (1984): Global precedence in visual pattern recognition, *Percept Psychophys* 35:361–370.

Kinchla RA, Wolfe JM (1979): The order of visual processing: "Top down", "bottom up", or "middle out." *Percept Psychophys* 25:225–231.

Lamb MR, Robertson LC (1988): The processing of hierarchical stimuli: Effects of retinal locus, locational uncertainty, and stimulus identity. *Percept Psychophys* 44:172–181.

Lamb MR, Robertson LC (1989): Do response time advantage and interference reflect the order of processing of global- and local-level information? *Percept Psychophys* 46:254–258.

Lamb MR, Robertson LC (1990): Component mechanisms underlying the processing of hierarchically organized patterns: Interferences from patients with unlateral cortical lesions. *J Exp Psychol Learn Mem Cognit* 16:471– 483.

Lamb MR, Robertson LC, Knight RT (1989): Attention and interference in the processing of hierarchical patterns: Interferences from patients with right and left temporal-parietal lesions. *Neuropsychologia* 27:471–483.

Lamb MR, Robertson LC, Knight RT (1990): Component mechanisms underlying the processing of hierarchically organized patterns: Inferences from patients with unilateral cortical lesions. *J Exp Psychol Learn Mem Cognit* 16:471–483.

Luck SJ, Heinze HJ, Mangun GR, Hillyard SA (1990): Visual event-related potentials index focused attention within bilateral stimulus arrays. II. Functional dissociation of P1 and N1 components. *Electroencephalogr Clin Neurophysiol* 75:528–542.

Luck SJ, Hillyard SA (1990): Electrophysiological evidence for parallel and serial processing during visual search. *Percept Psychophys* 48:603–607.

Mangun GR, Hillyard SA, Luck SJ: Electrocortical substrates of visual selective attention. In: *Attention and Performance XIV*, Meyer D, Kornblum S, eds. Cambridge: MIT Press 1993, pp. 219–243.

Martin M (1979): Local and global processing: The role of sparsity. *Mem Cognit* 7:476–484.

Miller J (1981): Global precedence in attention and decision. *J Exp Psychol Hum Percept Perform* 7:1161–1174.

Navon D (1977): Forest before trees: The precedence of global features in visual perception. *Cognit Psychol* 9:353–383.

Navon D (1981): Do attention and decision follow perception? Comment on Miller. *J Exp Psychol Hum Percept Perform* 7:1175–1182.

Navon D, Norman J (1983): Does global precedence really depend on visual angle? *J Exp Psychol Hum Percept Perform* 9:955–965.

Pomerantz JR (1983): Global and local precedence: Selective attention in form and motion perception. *J Exp Psychol Gen* 112:516–540.

Ritter W, Simpson R, Vaughan HG (1983): Event-related potential correlates of two stages of information processing in physical and semantic discrimination tasks. *Psychophysiology* 20:168–179.

Robertson LC, Delis DC (1986): "Part-whole" processing in unilateral brain damaged patients: Dysfunction of hierarchical organization. *Neuropsychologia* 24:363–370.

Robertson LC, Lamb MR, Knight RT (1988): Effects of lesions of temporal-parietal junction on perceptual and attentional processing in humans. *J Neurosci* 8:3757–3769.

Robertson LC, Lamb MR, Knight RT (1991): Normal global-local analysis in patients with dorsolateral frontal lobe lesions. *Neuropsychologia* 29:959–967.

Ward LM (1982): Determinants of attention to local and global features of visual forms. *J Exp Psychol Hum Percept Perform* 8:562–581.

Wijers AA, Mulder G, Okita T, Mulder LJM, Scheffers MK (1989): Attention to colour: An analysis of selection, controlled search, and motor activation, using event-related potentials. *Psychophysiology* 26:89–109.

Chapter 5

Event-Related Potentials and Stimulus Repetition in Direct and Indirect Tests of Memory

MICHAEL D. RUGG AND MICHAEL C. DOYLE

In this chapter we give an overview of some of the studies in our laboratory in which event-related brain potentials (ERPs) have been recorded in tasks involving the repetition of stimuli such as words, nonwords, and pictures, and outline our current interpretation of the findings that have emerged. A major aim of this work has been to determine the conditions under which ERPs are sensitive to stimulus repetition when it is incidental to the experimental task, and hence when no discriminative response is required on the basis of whether an item is repeated or unrepeated.

When ERPs are sensitive to repetition in such circumstances, there are grounds for thinking that they reflect the differential processing accorded unrepeated and repeated stimuli, and are therefore in some way sensitive to memory formation or retrieval. Such ERP effects can be exploited to investigate, among other issues, which attributes of a stimulus are incidentally encoded in memory during the course of its processing, and the extent to which encoding and retrieval operations are under strategic or attentional control. After discussing the effects of stimulus repetition on ERPs in indirect memory tasks (i.e., tasks in which repetition is incidental to the experimental task), we shall briefly turn to studies in which memory for repeated items has been assessed directly, in the form of tests of recognition memory. Throughout we concentrate on functional issues, paying little regard to the possible neural bases of these effects. Work relevant to this latter issue can

Cognitive Electrophysiology
H-J. Heinze, T.F. Münte, and G.R. Mangun, editors
© 1994 Birkhäuser Boston

be found in Halgren and Smith (1987), Heit, Smith, and Halgren (1990), Potter et al. (1992), Rugg et al. (1991), Smith and Halgren (1989), and Stapleton and Halgren (1986).

Studies with Indirect Tasks

The basic form of the indirect task we have used in several studies of the effects of stimulus repetition on ERPs is illustrated in Figure 1. As can be seen from that figure, the task requires that subjects detect and respond to "target" items interspersed among nontargets, a proportion of which are repetitions of previously presented items.

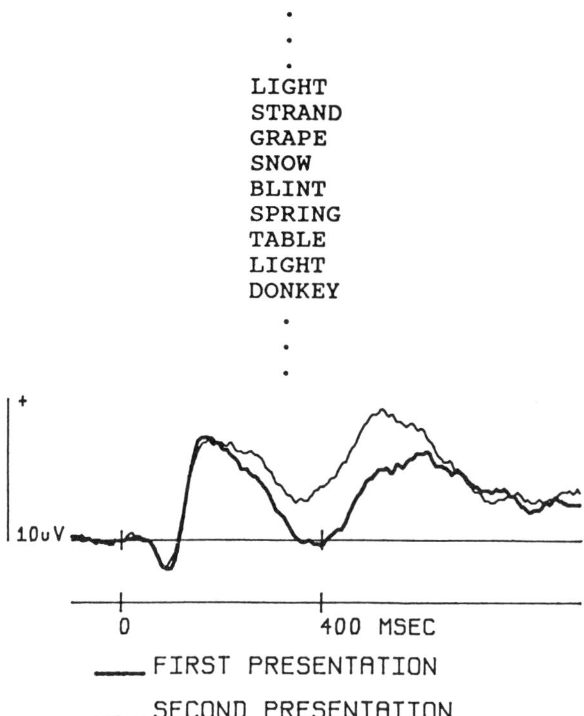

Figure 1.   Top: Example of part of stimulus sequence in indirect memory task referred to in text. In this example, the task is to respond promptly to nonwords (e.g., BLINT), and to withhold responses to all other stimuli. Effects of repetition on ERPs are assessed by comparing the waveforms evoked by first and second presentations of repeated words. Stimuli are typically presented for 300 msec, with intertrial interval of about 3 sec. Bottom: "ERP repetition effect." Grand average waveforms ($n = 16$), from Cz, evoked by first and second presentations of words in lexical decision task similar to that illustrated. Interitem lag, 6.

The task has the following merits: (i) The delay (interitem lag) between first and second presentations is easily manipulated, while holding all other aspects of the task constant; (ii) control is effected over stimulus processing. To perform the task accurately, subjects must process the items to the level required to discriminate 'targets' from 'nontargets' (e.g., nonwords from words), and performance can be assessed by standard measures of speed and accuracy. (iii) Although a target/nontarget decision is required for each item, no overt response is made to the nontargets, minimizing the influence of response-related components to the ERPs evoked by the items of experimental interest. (iv) The same (implicit) response is required to both first and second presentations of repeating items; any effects of repetition are therefore unconfounded with response type. And, (v) less relevant to present concerns, the indirect nature of the task makes it suitable for use with memory-disordered subjects, many of whom show normal or near-normal performance on tasks like lexical decision.

Against the positive features of this task must be set the disadvantage that there is no means, independent of the ERP data, of ascertaining whether repeated and unrepeated items receive differential processing. Thus the finding that ERPs do not differ as a function of repetition is potentially ambiguous. Without further evidence, it is impossible to know whether this finding indicates that repeated and unrepeated items received equivalent processing, or whether the two classes of item were processed differentially, but in a manner not reflected in ERPs.

Figure 1 shows the typical effects of word repetition on ERPs when the target/nontarget discrimination requires a lexical decision. In the example shown, first and second presentations of the words were separated by about six intervening items. In comparison to the ERPs to first presentations, those to repeats exhibit a relative positivity, which onsets around 250 msec post stimulus and persists for some 300 msec or more. We shall refer to this difference between first and second presentations as the *ERP repetition effect*. This term is purely descriptive; it should not be taken to imply that we believe the effect to involve only a single underlying component, or to reflect only a single cognitive process. Indeed, as will become evident, we believe the data point in quite the opposite direction.

Repetition Effects with Nonlexical Stimuli

Other than words, what kinds of stimuli give rise to the ERP repetition effect? The answer to this question should provide important clues about the nature of the codes or representations that must in some sense overlap if the effect is to occur. For example, were the effect to be evoked by words but not by wordlike, pronounceable nonwords (e.g., BLINT), it could be argued that it was mediated by processing at the semantic level (a level of processing available only to words), but not at orthographic or phonological levels (levels of analysis to which both words and nonwords can be subjected).

In fact, the effect is evoked by pronounceable nonwords (Rugg, 1987; Rugg and Nagy, 1987), suggesting that it is unlikely to be mediated solely by the overlap of semantic processing.

On the other hand, it is not the case that the repetition of *any* orthographic string is sufficient to produce an effect on ERPs. In their second experiment, Rugg and Nagy (1987) employed both orthographically legal nonwords (i.e., "wordlike," pronounceable letter strings such as BLINT or FICT), and orthographically illegal, unpronounceable nonwords (e.g., TJRKDE). These two types of nonword were presented in a task in which "targets" consisted of equal proportions of legal and illegal nonwords in which one of the inner letters had been replaced by an "@" character (e.g., FI@T, TJR@DE). These targets were chosen so as to equalize the difficulty of discriminating each type of critical nonword from the target set. Immediate repetition of the legal nonwords yielded an ERP repetition effect, whereas no effect was observed when illegal items repeated. Rugg and Nagy (1987) argued that the wordlike orthographic structure of legal nonwords allows them to gain access to the visual lexical processing system, leading to partial activation of the representations of words sharing similar orthographic properties (e.g., BLINT might partially activate the representation of BLINK). By contrast, the deviant orthographic structure of illegal nonwords precludes their being processed in this manner. Rugg and Nagy (1987) therefore concluded that the ERP repetition effect depended on the employment of letter strings that activate preexisting representations in lexical memory. This, they claimed, was inconsistent with the view that the effects of repetition on ERPs arise solely as a result of the influence of "episodic" memory. If overlap between a current and a recent processing episode is sufficient for the effect to occur, they argued, then there is no reason why illegal items should not elicit it.

The findings of Rugg and Nagy (1987) clearly showed that mere repetition is insufficient to evoke the ERP repetition effect. In their account of these findings, however, Rugg and Nagy neglected the fact that their legal and illegal nonwords differed not only in orthographic structure, and thus how wordlike they were visually, but also in their phonological properties; while all the legal items could easily be assigned a pronunciation (and thus a unique name), this was not true for the illegal items. Therefore, until an experiment is conducted in which orthographic and phonological properties of nonwords are varied independently, the exact reason for the dissociation between legal and illegal nonwords shown by Rugg and Nagy (1987) will remain uncertain.

In collaboration with S. Thomas, D. Perrett, and M. Harries, we have since investigated the effects of repeating two kinds of pictorial stimuli. Examples of the stimuli are shown in Figure 2. The pictures were drawn at random from a large corpus on a video disc. They depicted a wide variety of scenes and objects, and had not been seen previously by the experimental subjects. Other similar stimuli were subjected to distortion by the addition

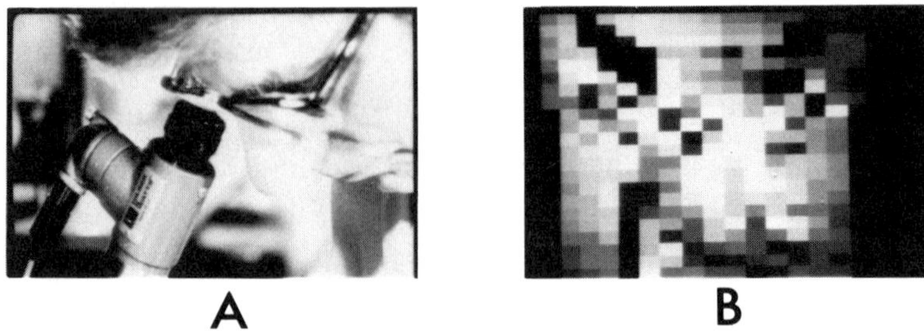

Figure 2.   Examples of normal (**A**) and distorted (**B**) stimuli employed to investigate effects of picture meaningfulness on ERP repetition effect.

of spurious high spatial frequency components, yielding the kind of picture seen in Figure 2B. Thus, although physical features (e.g., size, color, and brightness, although not, of course, spatial frequency) of the two classes of stimuli illustrated in Figure 2 were on average roughly similar, they differed in their representational content. The task used with these stimuli was somewhat different from the one illustrated in Figure 1. The requirement was to respond whenever a picture was repeated immediately ($p = .44$), but otherwise to withhold a response. Of interest were the ERPs to "non-target" repetitions, stimuli that repeated between five and nine items after their first presentation ($p = .26$).

Twelve subjects participated in the main experiment. For a further 8 subjects, the task was changed to one of responding to *all* repeats, and behavioral data only were collected. This subsidiary experiment showed that although subjects were better at detecting lagged repetitions of the normal than the distorted pictures, they were nonetheless highly accurate at detecting these latter items (phit-pfalse alarm = .69). We therefore consider it unlikely that differences in the main experiment between the ERPs to lagged repetitions of the two kinds of stimuli can be attributed simply to the inability to recognize distorted pictures during the interitem lag.

The principal findings of this experiment are shown in Figure 3. A robust effect of repetition is apparent in the ERPs to the normal pictures, taking the characteristic form of a relative increase in positivity to repeated items. In contrast, no such effect is evident in the waveforms evoked by the distorted pictures. Thus the manipulation of picture "meaningfulness" appears to have a similar effect on the ERP repetition effect to that of orthographic legality.

What do the results of these experiments say about the stimulus attributes necessary to engender an ERP repetition effect? On the one hand, the effect can be evoked by a range of stimuli, including items (legal nonwords) that have no semantic or lexical representation. On the other hand, *some* minimal informational content appears necessary, in that mere letter strings, or abstract visual patterns, are almost wholly ineffective in evoking the effect. We believe that a promising hypothesis is that the effect

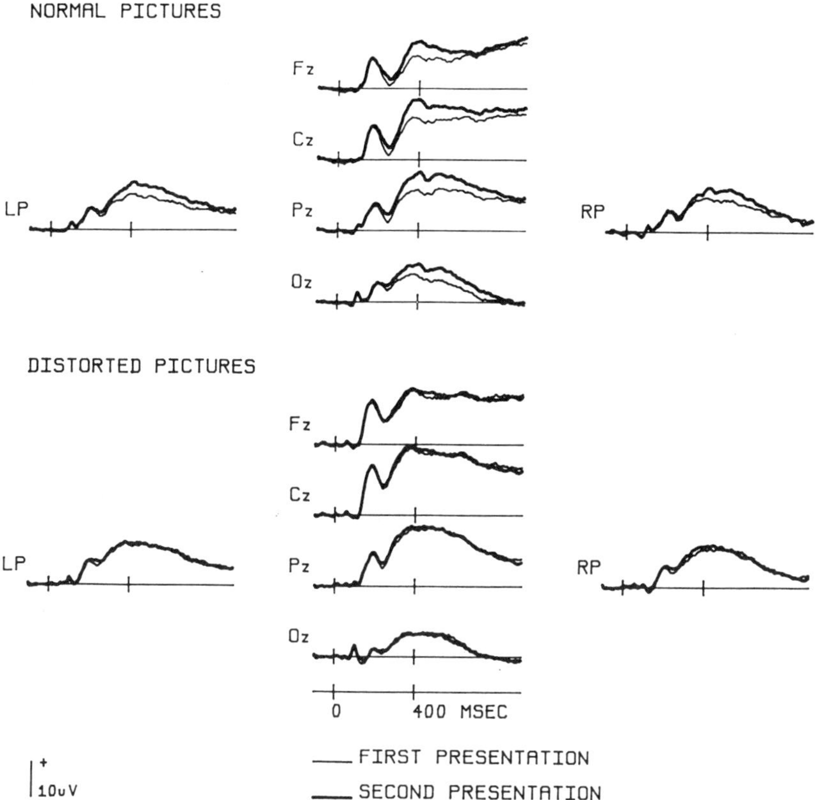

Figure 3.   Grand average ERPs ($n = 12$) evoked by first and second presentations of normal and distorted pictures. Fz, Cz, Pz, Oz are an anterior-posterior chain of midline electrodes; LP and RP were placed over left and right inferior parietal sites.

is evoked only by the repetition of stimuli possessing attributes that lead to the generation of one or more *unitized* codes. For instance, although a legal nonword can be represented by a unitary phonological code (corresponding to its pronunciation), this is not the case for illegal nonwords of the kind employed by Rugg and Nagy (1987), the encoding of which cannot proceed much beyond the representation of their constituent letters in the correct sequential order. Similarly, although normal pictures can be encoded in terms of their phonological and semantic attributes, there is little scope for unitizing the representation of distorted pictures.

Intermodal ERP Repetition Effects

The hypothesis that the ERP repetition effect is dependent on the use of stimuli capable of generating unitized codes accounts for the effects of orthographic legality and picture distortion described previously, but it says little about the nature of the codes themselves. For instance, is "code overlap" at any representational level (orthographic, semantic, phonological, etc.) equally potent at causing repetition effects, or are some codes more important than others?

One way to address this issue is to compare the effects of item repetition within and across sensory modalities (see Brown et al., 1991, and Kirsner, Dunn, and Standon, 1989, for reviews and examples of relevant behavioral work). If repetition effects depend on the overlap solely of "central" codes, such as those generated by semantic processing, a change of modality between first and second presentations will have no effect. On the other hand, if the effects of repetition reflect overlap of more peripheral, modality-specific processing the effects will be smaller when items are repeated in a different modality than when modality is held constant.

In collaboration with C. Melan of the University of Strasbourg, we have investigated the effects of modality change by using a multimodal version of the lexical decision task described above (Rugg, Doyle and Melan, in press). In this experiment, subjects were exposed to a random sequence of stimuli, half auditory and half visual; 17% of the stimuli were nonwords demanding a response, while the remainder were words. Critical items were split among four randomly interleaved conditions: in the *intramodal* conditions, items were presented twice in the same modality (i.e., either visual-visual or auditory-auditory), with an interitem lag of six; in the *intermodal* conditions, items changed modality between their first and second presentations (i.e., visual-auditory, auditory-visual), and were again separated by a lag of six.

This design addresses the question of the dependence of the ERP repetition effect on modality-specific processing. If the effect reflects only modality-specific operations, there should be little in the way of an inter-modal effect. If, on the other hand, the effect depends solely on the overlap of semantic processing, then on the assumption that visually and auditorily presented words activate a common semantic memory system intermodal repetition effects should be as large as intramodal effects.

Subjects were slower and slightly less accurate to detect auditory than visual nonwords [reaction time (RT): 1132 msec vs. 813 msec, respectively; error rate: 3.8% vs. 1.8%]. The ERP data are shown in Figures 4 and 5.

Auditory-auditory repetition gave rise to a sizeable, widespread, and sustained repetition effect, which onset approximately 100 msec later than that observed with visual repetition. As is also evident from Figures 4 and 5, the effects of visual-auditory repetition were essentially identical to those of intramodal repetition. Intra- and intermodal repetition effects were, however, not equivalent for visually presented repeats. As can be seen, visual-visual repetition yielded a repetition effect onsetting around 300 msec. Although auditory-visual repetition also gave rise to a large and robust effect, its onset was some 100 msec later than in the intramodal condition, leading to a significant difference in the size of the two effects before 500 msec post stimulus.

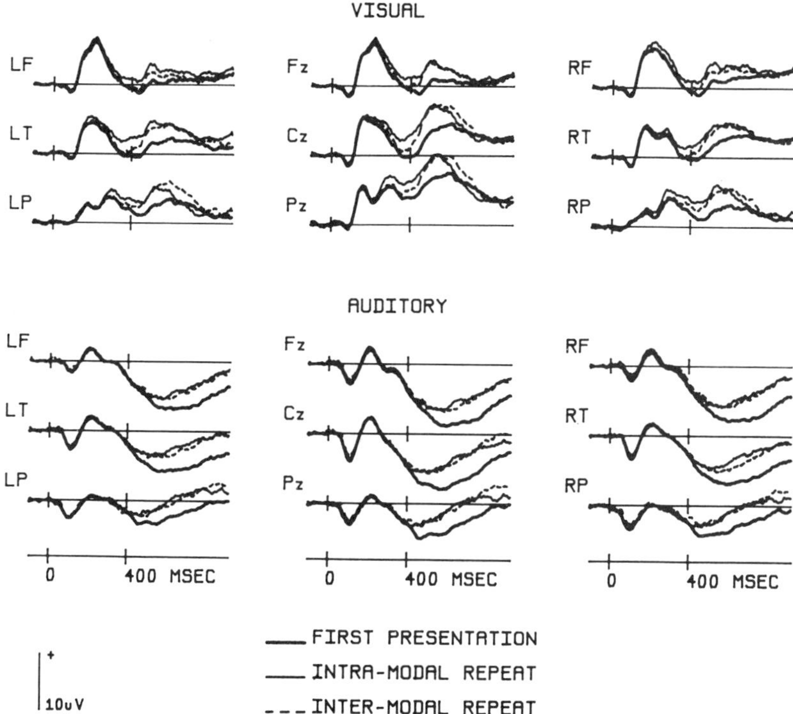

Figure 4.   Grand average ERPs ($n = 16$) from experiment investigating intramodal vs. intermodal word repetition. Visual: ERPs to visually presented words first presented visually (*thin solid line*) or auditorily (*dashed line*); Auditory: ERPs to auditorily presented words first presented auditorily (*thin solid line*) or visually (*dashed line*). LF, RF, LT, RT, LP, and RP refer to electrodes over left and right frontal, left and right temporal, and left and right inferior parietal sites, respectively.

These findings suggest that the ERP repetition effect exhibits an element of modality specificity, albeit in a rather subtle fashion. The dissimilarity of the visual-visual and auditory-visual effects is arguably sufficient to indicate that the intramodal effect cannot be attributed entirely to the overlap of modality nonspecific (e.g., semantic) processing.

The different effects of visual-visual and auditory-visual repetition seen in this experiment clearly need further study. In particular, it is not clear from these findings whether these differing effects reflect the dissociation of the ERP repetition effect into multiple components, differentially sensitive to intra-versus intermodal repetition or merely the delayed onset of the effect in the intermodal condition. Favoring the second of these alternatives, we are currently testing the hypothesis (derived from contemporary views of the relationship between auditory and visual word recognition (e.g., Ellis and

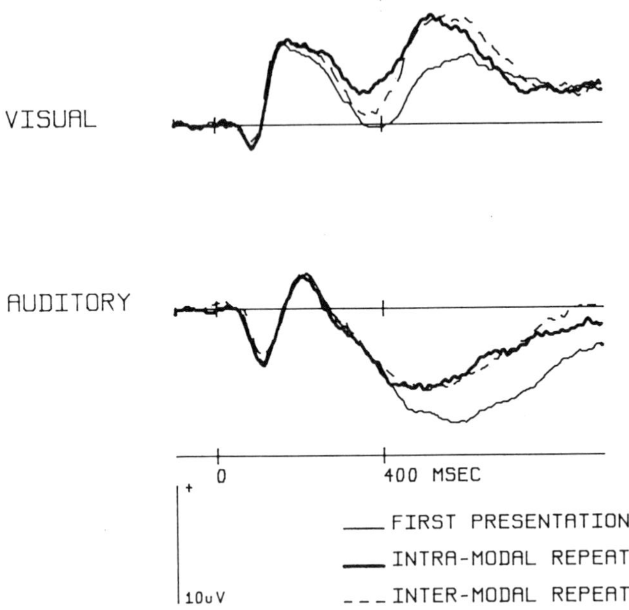

Figure 5.   Overlay of ERPs illustrated in Figure 4. Cz electrode site.

Young, 1988, p. 222) that these findings reflect differences in the codes formed from auditorily and visually presented words. By this hypothesis, subjects generated phonological codes from both auditorily *and* visually presented words. The auditory-auditory and visual-auditory conditions therefore gave rise to equivalent amounts of overlap at the phonological level and therefore to equivalent ERP effects. By contrast, orthographic codes were formed only when words were presented visually. Thus, while visual-visual repetition involves overlap of orthographic processing, the auditory-visual condition does not. For this condition, overlap is only possible at phonological or semantic levels. As it takes longer for visually-presented words to generate representations at these levels than at the orthographic level, visual-visual repetition effects onset earlier than auditory-visual effects.

Effects of Interitem Lag

In two separate experiments, Nagy and Rugg (1989) compared immediate word repetition with repetition after 6 intervening items, and repetition after 6 versus 19 items. In each experiment the ERP repetition effect tended to be smaller for the longer lag, but these differences did not approach significance. Similarly, Bentin and Peled (1990) found equivalent ERP repetition effects in a lexical decision task when repetition was immediate compared to when it occurred after a lag of 15 intervening items.[1] Interestingly, Bentin and Peled did find a lag effect in a second task, in which subjects were required simply to observe and try to remember each word; now, no effect was apparent at lag 15, whereas at lag 0 an effect equivalent to that occurring during lexical decision was found.

The results of Bentin and Peled (1990) suggested that the durability of the ERP repetition effect is task dependent. Also, the effects of lag with stimuli other than words (e.g., nonwords) have yet to be investigated. Nonetheless it is clear that the influence of lag can be slight up to about 20 intervening items (i.e., about 1 min). As discussed by Nagy and Rugg (1989), it is therefore unlikely that the ERP repetition effect depends on the maintenance of words in short-term or "working" memory. Were this to be the case, the effect should vary as a function of whether interitem lag exceeds short-term memory capacity, which, for unrelated words, is no more than seven or eight items (Baddeley, 1990).

Fractionation of the Effect

The studies discussed so far provide no strong evidence to suggest that the ERP repetition effect can be dissociated into more than one component. Two manipulations, word frequency and the interval between first and second presentations, do however appear to dissociate the ERP repetition effect into at least two underlying components. Both these variables were investigated by Rugg (1990a).

Word frequency, the normative frequency with which a printed word occurs in its native language, is of interest to workers in the fields of word recognition and memory for several reasons. First, it is an important determinant of the efficiency with which words can be processed in a variety of tasks (e.g., Monsell, 1991). Second, frequency interacts with intraexperimental repetition such that the beneficial effects of repetition on behavior are greater for low- than for high-frequency words. This finding has led to suggestions that word frequency and intraexperimental repetition influence a common mechanism (Monsell, 1985). Third, low-frequency words are better

---

[1] In the studies of Rugg (1987) and Nagy and Rugg (1989), the positive-going shift associated with the immediate repetition of a word (or nonword; Rugg, 1987, experiment 2) was preceded by a transient negative shift in the region of the P200 component. A similar effect is evident in the lag 0 data of Bentin and Peled (1990), although it did not attain significance in that study. Rugg (1987) and Nagy and Rugg (1989) speculated that, unlike the later effects of repetition, this very early effect was specific to immediate repetition. Unfortunately, this transient negativity has proved highly elusive; we have been unable to find it in a variety of subsequent studies employing immediate repetition [both published (Rugg and Nagy, 1987; Rugg, Furda, and Lorist, 1988, and unpublished data], and have nothing further to say about it here.

recognized than high-frequency words in tests of recognition memory (Glanzer and Bowles, 1976).

Rugg (1990a) investigated the repetition of high- and low-frequency words over two different delays. In phase 1 of the study, subjects responded to nonword targets interspersed in a series of high- and low-frequency words, some of which repeated with an interitem lag of six. There then followed a break of 15 min, after which subjects performed an identical task, the critical items in which were either entirely new words or words that had been presented only once during phase 1.

The major findings from phase 1 of this study are illustrated in Figure 6. Three aspects of the data are of interest. First, the amplitude of a prominent negative-going deflection peaking around 400 msec is larger for first presentations of low- than of high-frequency words. Second, the effect of repetition on this region of the waveform interacted with word frequency, such that the negative deflection was attenuated more by the repetition of low-frequency than of high-frequency words. Third, a later region of the waveform, encompassing what looks like a late positive wave, was enhanced only by the repetition of low-frequency items. The findings from phase 2 of the study were somewhat different. Repetition effects were found only for low-frequency items, and only in the region of the late positive wave. The preceding negative deflection was larger for low- than high-frequency words irrespective of whether these were new or had been presented previously in phase 1.

The relevance of these findings to an understanding of the functional

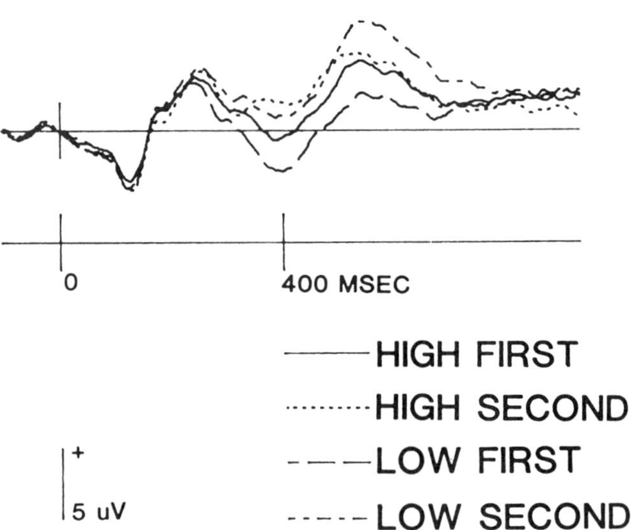

Figure 6.   Grand average ($n = 16$) ERPs from Cz electrode of phase 1 of Rugg (1990a). ERPs were evoked by first and second presentations of high-frequency and low-frequency words.

significance of the ERP repetition effect is discussed next. Whatever the validity of that discussion, these data suggest that the effect can be fractionated into at least two components [see Young and Rugg (1992) for a replication and extension of these results]. One component, centered on the negative-going deflection peaking around 400 msec, appears sensitive to both word frequency and repetition, but is only responsive to repetition over relatively short intervals. The second component, which encompasses a late positive deflection and subsequent slow wave, appears to be enhanced solely by the repetition of low-frequency words and, in contrast to the preceding negativity, remains sensitive to the repetition of these items when second presentations occur as long as 15–20 min after first presentations. On the basis of their differing onset latencies, these two aspects of the ERP repetition effect will be referred to as the *early* and *late* effects, respectively.

Components Underlying the ERP Repetition Effect

Which previously studied ERP components might contribute to the ERP repetition effect? One obvious candidate is the P3 (also known as P3b, P300, and the late-positive component). This component is sensitive to the subjective probability and task relevance of the stimulus evoking it, has a peak latency proportional to the time required to evaluate or categorize the evoking stimulus, and possesses a centroparietal scalp distribution (for reviews, see Donchin and Fabiani, 1991; Johnson, 1988; Verleger, 1988). In several respects, therefore, it is a good candidate as a major contributor to the positive-going modulation of ERPs evoked by repeated items. One might well expect a relatively rare event such as stimulus repetition to evoke an enhanced P3, and hence for this component to contribute to either (or both) the early and late effects described here.

A second component whose modulation might underlie the effects of repetition on ERPs is the N400 (for reviews, see Fischler and Raney (1991); Kutas and Van Petten (1988); Kutas and Kluender, this volume). Modulation of N400 has been reported in a wide variety of paradigms, most notably as a consequence of variations in sentential context ("contextual priming": *it was his first day at WORK* vs. *he takes his coffee with cream and WORK*), and in tasks varying the semantic relationship between word pairs ("semantic priming": doctor– NURSE vs. curry–NURSE). In such tasks, the N400 is larger when evoked by unprimed than primed words; that is, ERPs evoked by primed items are more positive going than are those evoked by unprimed ones. This priming effect typically onsets around 250 msec, about the same latency as the divergence of ERPs evoked by repeated as opposed to unrepeated items. If any effects of priming a stimulus by its prior presentation (repetition priming) are functionally equivalent to contextual and semantic priming, then it is plausible that the modulation of N400 may play a role in the ERP repetition effect.

Rugg (1987; see also Rugg, 1985) attempted to address this issue by comparing immediate word repetition effects with the effects of following a word by a strong semantic associate. Repetition gave rise to a large and widespread effect, whereas semantic priming produced a much smaller, more anteriorly distributed effect. This study therefore suggested that repetition and semantic priming do not give rise to qualitatively similar ERP modulations when compared under the same conditions, raising the possibility that the two manipulations do not reflect the modulation of the same component or components. However, the semantic priming effects in this study were so small as to be of only marginal significance, and a replication in which these effects are more similar in size to those of repetition is badly needed. By contrast, the recent reports by Besson, Kutas, and Van Petten (1990, 1992) that contextual priming and word repetition appear to have interactive effects on the N400 component provide strong evidence implicating this component in the ERP repetition effect.

On the assumption that the ERP repetition effect can largely be understood in terms of the modulation of N400 and P3, with no need to invoke the existence of previously unidentified components, what is the relative contribution of the two components to the early and late effects?

In the case of the early effect, there are two reasons for thinking that the modulation of P3 plays little or no part. First, only some kinds of repeated stimuli evoke any repetition effect at all. For example, in the legal/illegal nonword experiment of Rugg and Nagy (1987), the repetition of each class of item was equally rare and "task relevant," yet only legal nonwords gave rise to a repetition effect, a finding hard to reconcile with the functional properties of the P3 described earlier. A similar argument can be made in the case of the picture-repetition experiment described earlier and was made also by Bentin and Peled (1990).

Second, if the early effect is to be attributed to the enhancement of P3, it is unclear why some manipulations that cause the effect to diminish in size should do so by increasing the positivity of ERPs to *unrepeated* items, rather than by reducing the positivity of ERPs to repeats. For example, Rugg, Furda, and Lorist (1988) investigated the effects of immediate word repetition in three different tasks, which differed by virtue of the targets to which a response was required. Two of the tasks (detection of nonwords and animal names) required relatively "deep" processing of the stimuli, but the remaining task (detection of words written in lowercase letters) did not. A large and robust ERP repetition effect was found for the first two tasks, but not for case detection, when the effects of repetition were small and temporally limited. The small size of the repetition effect in the case task arose not because the ERPs to repeated words were less positive going than those in the other tasks, but because ERPs to unrepeated items were *more* positive going.

The data from the picture repetition experiment described earlier show the same pattern. As can be seen in Figure 7, from about 350 msec onward the amplitude of the ERPs to repeated normal pictures is roughly equivalent

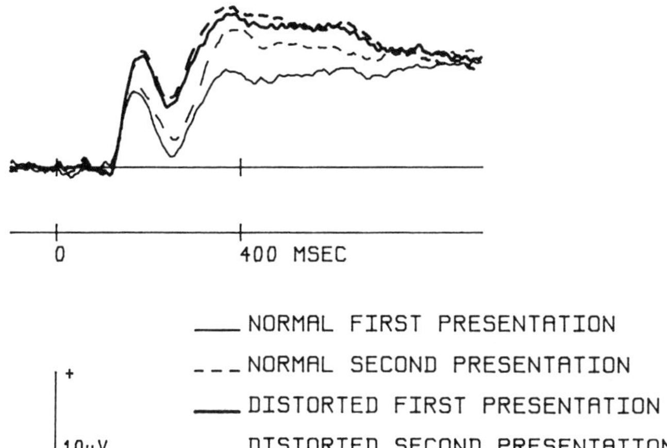

Figure 7.   Grand average ($n = 16$) ERPs overlaid as function of picture type (normal vs. distorted) and first vs. second presentation.

to that of the ERPs to both the first and second presentations of distorted stimuli, which do not differ; the "odd waveform out" is the one evoked by normal pictures on their first presentation. Thus as in the study of Rugg, Furda, and Lorist (1988), changes in the size of the ERP repetition effect in this experiment were mediated more by changes in ERPs to first than to second presentations.

As argued originally by Rugg, Furda, and Lorist (1988), such findings seem incompatible with the idea that the basis of the ERP repetition effect is the enhancement of P3 in ERPs to repeated items in most studies. For this idea to be valid, it would be necessary to suppose that in the Rugg, Furda, and Lorist case-detection task unrepeated items evoked larger P3s than in the other two tasks, a supposition that seems implausible in light of the apparent functional properties of the P3. The findings illustrated in Figure 7 are equally difficult to account for in terms of changes in P3. There seems no reason why unrepeated distorted pictures should evoke P3s larger than those evoked by unrepeated normal stimuli.

Rejecting the idea that the ERP repetition effect reflected changes in amplitude of P3, Rugg, Furda, and Lorist (1988) proposed instead that it reflected the attenuation of a negative-going wave evoked by first presentations. They argued that in their case-detection task this wave was not present, hence ERPs to both first and second presentations in this task resembled those to second presentations in the other two tasks. An analogous account would appear possible for the picture-repetition experiment described here. For the reasons already discussed, the obvious candidate component corresponding to this wave is the N400.

Although changes in the amplitude of P3 appear to contribute little to the early ERP repetition effect, this may be less true of the late effect, that is, that part of the ERP repetition effect that appears attributable to the frequency-sensitive late positive wave (hereafter called the "P600") evident in Figure 6. Repeated low-frequency words elicit a P600 that is not only greater in amplitude than that evoked by the same words on their first presentations, but which is also larger than the P600 evoked by repeated high-frequency items. Thus the P600 may indeed be a specific response to the repetition of certain kinds of stimuli, in which case it is possible that its generators overlap at least partially with those of the classical P3 component.

Functional Significance of ERP Repetition Effect

The Early Effect

If, as argued here, the early ERP repetition effect reflects changes in the amplitude of the N400 component, what does this imply about its likely functional significance? A good starting point is to ask what processes might be shared by the different experimental manipulations that appear to modulate N400. Rugg (1990a, see also Halgren and Smith, 1987, for a similar view) suggested that these various manipulations have in common the fact that each changes the ease with which an item can be integrated with its context.

The primary motivation for this suggestion comes from an important class of accounts of semantic and contextual priming (e.g., Forster, 1981; Kroll, 1990; Seidenberg et al., 1984). These authors proposed that much of the facilitation in behavioral responding caused by such priming stems not from differences in the speed with which primed and unprimed items are identified (i.e., from changes in "lexical access'), but from differences in the ease with which, once identified, they can be integrated or assimilated with the priming context. From this perspective, differences in the amplitude of N400 evoked by semantically or contextually primed and unprimed words covary with differences in the ease with which such items can be integrated with their contexts.

The starting point for the contextual integration hypothesis of N400 is the idea that this process is an obligatory by-product of the processing of any stimulus event. It is part of what allows an event to be enduringly encoded as a distinctive episode and hence subsequently to be retrieved and discriminated from other, otherwise similar events. The hypothesis proposes that the amplitude of N400 is inversely proportional to the ease with which a stimulus is integrated with its context and is modulated by any variable that impinges on the integration process.

In the case of priming by a sentence context, ease of integration is a function of the constraints created by preceding items: DOG is easier to integrate in the sentence *he picked up the leash and called the DOG,* than in the equally meaningful sentence *he was walking along the street when he saw the DOG*. Similar integrative processes are held to operate in the typical semantic priming situation (cf. leash-dog vs. leaf-dog, e.g., de Groot, 1985). In the case of repetition, ease of integration is modulated not by the direct manipulation of context, but by the reoccurrence of an item within the *same* context.

How does this hypothesis account for the finding that the ERP repetition effect is not evoked by all stimuli or in all tasks? We suppose that codes are generated or activated at several levels of analysis in the course of stimulus processing. These codes form the internal representation of the stimulus, and it is these that are integrated with the context in which the stimulus has been presented. This integrative process is somehow reflected by N400. When no code or only an impoverished code is generated (e.g., in the cases of illegal nonwords, abstract pictures, or words processed only to the level of their global physical features), little or no integrative activity can take place, no N400 is generated, and there is no scope for the attenuation of N400 by stimulus repetition.

To a first approximation, this hypothesis appears able to account for much of what is currently known about the ERP repetition effect and to provide a framework for linking the effect with the modulation of ERPs in other kinds of priming task. The hypothesis also appears able to account for the finding that isolated low-frequency words evoke larger N400s than more common items (Rugg, 1990a), in that it seems plausible to suppose that low-frequency words are, in general, more difficult to integrate with the context of their presentation (however, see Van Petten et al., 1991, for an opposing view). N400 has also been reported to be smaller when evoked by illegal rather than legal nonwords in both the visual and auditory modalities (Holcomb and Neville, 1990). Within the present framework, this finding is of course interpreted as reflecting the fact that unitized codes cannot be generated by illegal items; hence, little contextual integration of these stimuli is possible and N400 is small.

Finally, the contextual integration hypothesis would also seem able to account for the finding that differences in N400 evoked by high- and low-frequency words diminish as a function of increasing serial position within a sentence (Van Petten and Kutas, 1990); according to the present hypothesis, this too represents an example of a manipulation of context. As the context provided by the preceding text becomes increasingly rich and constraining, it increases the ease with which each newly presented word can be integrated.

Although parsimonious, the context integration hypothesis is presently deficient in a number of respects. The term context is badly underspecified. It is employed to describe entities as diverse as the linguistic and thematic schemas generated by sentence fragments and the general environment in which an experiment is conducted. This looseness of usage provides the hypothesis with explanatory power that it probably does not really possess, and makes it difficult to test empirically. A further poorly specified aspect of the hypothesis, although one which seems more amenable to investigation, concerns the mechanism by which different codes are selected for contextual integration. Once generated, are codes automatically integrated, or do only those codes that are the focus of attention become bound to their context? And one would of course like to know which contextual attributes are subject to obligatory processing (cf. Hasher and Zacks, 1979, 1984), and which are under attentional control. Finally, what predictions does this hypothesis make about the mnestic role of the processes reflected by N400?

This last question is worthy of further comment. In view of the importance of contextual factors in memory encoding and retrieval (Tulving, 1983), it would be surprising, if the context integration hypothesis has any validity at all, if there were not *some* relationship between performance on direct tests of memory and the processes reflected by N400 and its modulation. Attempts to demonstrate such a relationship have not yet been accepted, however. For example, Rugg (1990b) was unable to find any relationship between the size of the ERP repetition effect evoked by a word and the probability that the word would subsequently be retrieved in a test of free recall. And in two studies employing direct tests of recognition memory, one with normal subjects (Rugg and Nagy, 1989), and one studying patients who had undergone anterior temporal lobectomy (Rugg et al., 1991), we could find no evidence that the processes reflected by the modulation of N400 contribute in any significant way to the ability to discriminate repeated and unrepeated words.

The Late Effect

The foregoing discussion applies only to the early repetition effect, that is, the part of the ERP repetition effect that appears to result from the attenuation of a component akin to N400. As already noted, the study of Rugg (1990a) suggested that changes in N400 can be overlapped by the modulation of a subsequent late positive component, P600. Rugg (1990a) suggested that this component was sensitive to the relative familiarity of the word that was evoking it.

The concept of relative familiarity comes from "dual-process" models of recognition memory (Jacoby and Dallas, 1981; Jacoby and Kelley, 1991; Mandler, 1980), and is central to the accounts by such models of the advantage of low frequency words in recognition memory. Briefly, dual-process models propose that there are two independent bases for recognition memory judgments. One basis for such judgments is a consciously mediated memory search for an appropriate prior episode using the test item as a retrieval cue. Independently, judgments about prior occurrences can be based on the test item's relative familiarity, and the component of recognition memory that is thought to underlie the word frequency effect. Relative familiarity refers to the disparity between the familiarity engendered by an item in a recognition memory test, and the familiarity it would be expected to engender given its typical frequency of occurrence in everyday life. A high level of relative familiarity indicates that the item must have been experienced more recently than is typically the case and, in the context of a recognition memory task, is therefore likely to have been present at study. The word frequency effect comes about because the gain in familiarity resulting from intraexperimental repetition (that is, exposure both at study and test) is greater, relative to preexperimental levels, for low than for high-frequency words. That is, relative familiarity differentiates 'old' and 'new' low-frequency words to a greater extent than it does old and new high-frequency items.

Rugg's (1990a) suggestion that P600 amplitude is sensitive to relative familiarity was based on his finding that this component was enhanced when evoked by repeated low-frequency words, items which, according to the foregoing reasoning, should have engendered especially high levels of relative familiarity. As the effects of repetition on P600 appeared to outlast those in the N400 latency range, Rugg (1990a) further suggested that the processes responsible for the modulation of the two components are independent [see Van Petten et al. (1991) for a similar view regarding the independence of the processes underlying N400 and P600, but a very different interpretation of their functional significance].

ERPs in Direct Tests of Recognition Memory

In the last part of this chapter we briefly discuss ERPs recorded in direct tests of recognition memory, concentrating on the hypothesis that the P600 is sensitive to relative familiarity.

A number of studies have reported differences between ERPs evoked by correctly detected "old" and "new" words in recognition memory tests (Bentin and Moscovitch, 1989; Friedman, 1990; Johnson, Pfefferbaum, and Kopell, 1985; Karis, Fabiani, and Donchin, 1984; Neville et al., 1986; Rugg and Nagy, 1989; Smith and Halgren, 1989). When the interval between study and test is relatively long, these old/new ERP differences consist largely of the enhancement of a centroparietally distributed late positive component, identified by most investigators as the P3. Why P3 should be larger when evoked by old rather than new words has been a matter for debate. The effect seems unlikely to reflect differences in probability between the two classes of stimuli (Rugg and Nagy, 1989), and has variously been attributed to a "target" effect (Karis, Fabiani, and Donchin, 1984), to the outcome of "matching" operations in memory (Neville et al., 1986), and to the effects of decision confidence (Karis, Fabiani, and Donchin, 1984). According to Rugg (1990a), a further source of variance in the amplitude of late positive components recorded during tests of recognition memory should be the relative familiarity of the evoking item.

Word Frequency Effects in Recognition Memory and ERPs

If P600 is indeed enhanced as a function of relative familiarity, differences between P600s evoked by old and new low-frequency words should be greater than those between equivalent high-frequency items. Furthermore, this interaction between old/new status and frequency should be unattributable to differences in other variables known to be associated with the amplitude and latency of the "classical" P3, such as RT, response confidence and probability. This is what we found in the two studies described next.

In the first study, subjects initially made special lexical decisions to a random sequence of 100 nonwords, 100 low-frequency words (7 or fewer occurrences per million) and 100 high-frequency words (130 or more occurrences per million). After solving visual puzzles for about 15–20 min, subjects were then required to discriminate the words shown in the lexical decision task from equal numbers of new words. As is typically the case, recognition performance was reliably better for low- than for high-frequency items (phit-pfalse alarm of .51 and .37, respectively). Importantly, RTs for old and new decisions were almost the same for the two world classes (778 and 850 msec, respectively, for old and new judgments of low-frequency words, and 776 and 868 msec, respectively, for old and new judgments of high-frequency items).

The ERPs from this experiment are shown in Figures 8 and 9. Correctly categorized old items evoked larger P600 waves than correctly categorized new items did, and this old/new difference was larger for low than for high-frequency items. Thus, as predicted, P600 differed to a greater extent when evoked by old and new low-frequency words than when evoked by old and new words of high frequency. This interaction occurred despite the fact that the items forming the ERPs had all been correctly categorized as old or new.

The scalp distribution of the old/new effects in this study is worthy of comment. As suggested by Figure 8, these effects were significantly larger over the left than the right hemisphere, especially at the parietal sites. A very similar finding was reported by Neville et al. (1986). Such an asymmetry is consistent with neuropsychological evidence pointing to the lateralisation of verbal memory function to the left hemisphere (Smith, 1989), but its significance is otherwise unclear.

The findings described here demonstrate that the interactive effects of word frequency and repetition on P600 can be found in direct as well as indirect tests of memory. The fact that the interaction was found in ERPs evoked by correctly categorized items rules out a rather trivial interpretation of the earlier findings of Rugg (1990a), namely, that the P600 component was evoked whenever a repeat was detected, and that subjects were simply less likely to detect the repetition of high- than low-frequency words. The findings described so far do not, however, eliminate the possibility that the interactive

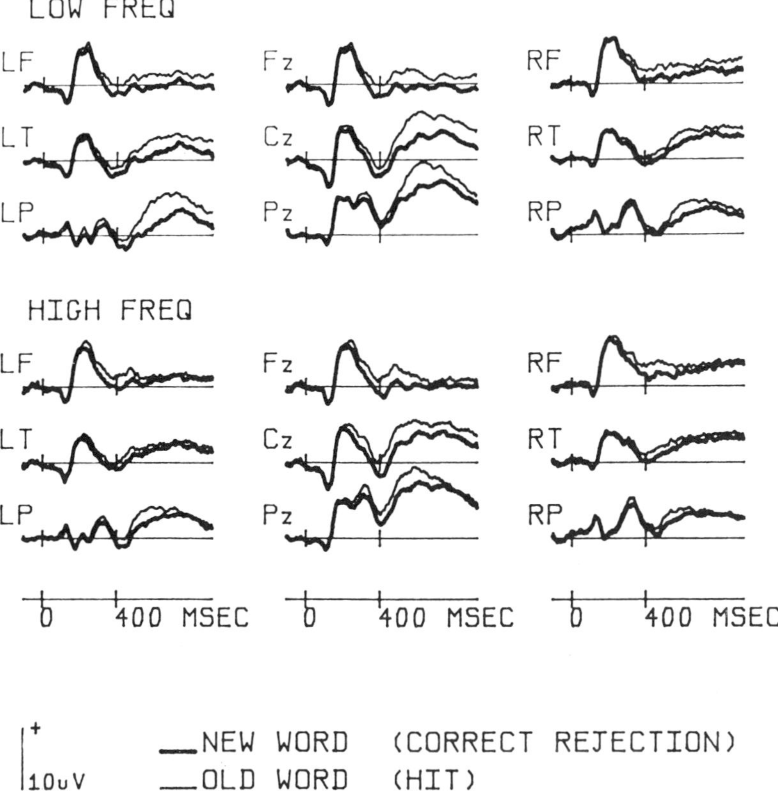

Figure 8.   Grand average ($n = 16$) ERPs evoked by correctly classified high-frequency and low-frequency words in recognition memory test. Electrode sites as in Figure 4.

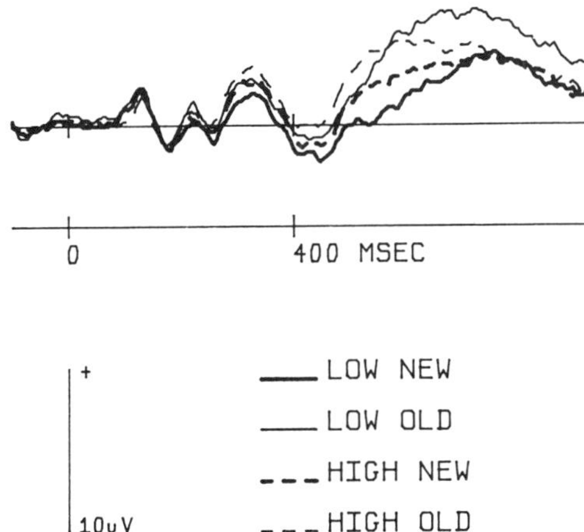

Figure 9.   Overlay of ERPs illustrated in Figure 8 (left parietal electrode).

effects of frequency and repetition on P600 are confounded with response confidence. Not only are low-frequency words recognized more accurately than those of high frequency, they are also recognized more confidently (Glanzer and Adams, 1990). Hence if the size of the old/new modulation of P600 is influenced by confidence, there may be no need to invoke an explanation in terms of relative familiarity to account for its modulation being larger with low-frequency words.

We (Rugg and Doyle, 1992) therefore replicated the experiment, adding the requirement to make a binary decision (high/low) about the confidence of each recognition judgment. As expected, correctly categorized low-frequency words were more likely to be associated with confident decisions than were high-frequency words (73% vs. 59%).

This finding indicates that, in our previous experiment, it is almost certain that the ERPs to high-frequency words were formed from a higher proportion of trials associated with low confidence decisions than were those to low-frequency words. However, when in the second study we formed ERPs only from trials associated with highly confident responses, the pattern of the results was qualitatively very similar to that of the first experiment.

As can be seen in Figure 10, old/new effects were once again larger for low-frequency than for high-frequency words and again, as before, predominated over the left hemisphere.

In summary, in both indirect and direct tests of memory for isolated low- and high-frequency words, P600 amplitude varies as if it is sensitive to the relative familiarity of the item evoking it. These findings therefore offer support for the idea that the disparity between the intra- and extraexperimental familiarity of an item is computed online, as proposed by dual-process theories of recognition. An important next step in the development of the hypothesis that P600 indexes relative familiarity will be to determine whether it varies when the contribution made by relative familiarity to recognition performance is manipulated by variables other than word frequency. Early indications are promising (Potter et al., 1992).

Summary and Concluding Comments

We have argued that the effects of stimulus repetition on the ERP, in both indirect and direct memory tasks, reflect the modulation of at least two underlying components: a negative wave related to or identical with the N400, and a late-positive component allied to the P3. These two components appear to be associated with independent processes. We have hypothesized that the N400 reflects the process(es) of integrating the attributes of an item with its context, while the late-positive component is sensitive to relative familiarity. We have yet to address the questions of how, and under what conditions, these two putative contributors to the ERP repetition effect interact.

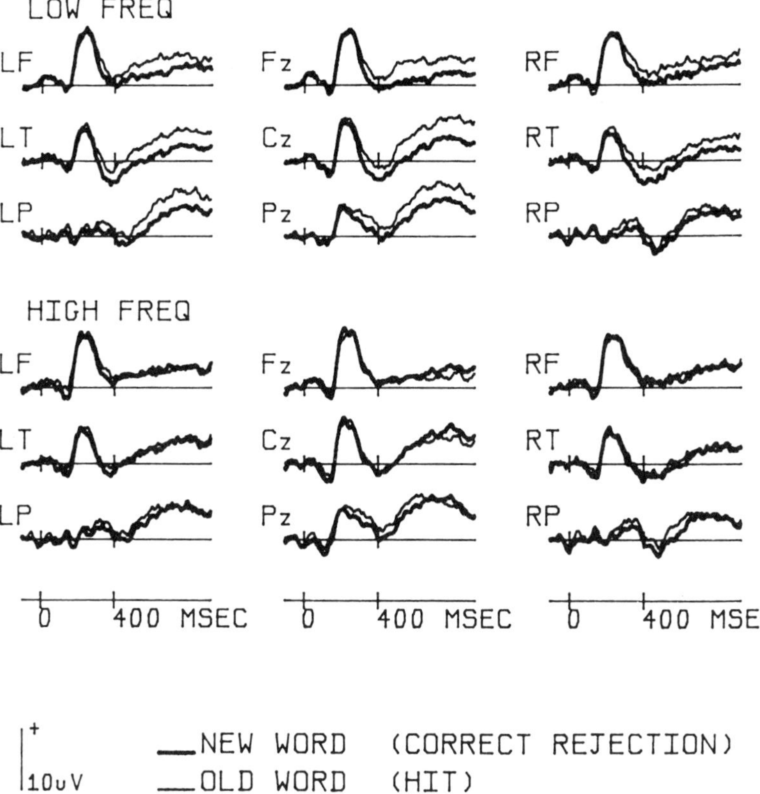

Figure 10.   Grand average ($n = 16$) ERPs evoked by correctly classified high-frequency and low-frequency words in the test phase of Rugg and Doyle (1992). ERPs formed only from trials associated with highly confident responses. Electrode sites as in Figure 4.

Whether or not any of the ideas outlined in this chapter are shown to be correct, we believe the findings we have described emphasize the utility of using both indirect and direct tests to investigate the effects of stimulus repetition on ERPs. While indirect tests permit such effects to be studied in relatively pure form, direct tests provide the information necessary to allow the effects to be placed in a behavioral context. When similar findings emerge in both kinds of test, we can be reasonably confident that we are observing manifestations of processes that are not tightly bound to a single experimental task or paradigm, and which are therefore likely to be of some general significance. Whether we have made any progress in discerning the nature of their significance, time will tell.

Acknowledgments. This chapter is based on a presentation by the first author at the conference *New Developments in Event-Related Potentials*, Hanover, Germany, May 1991. We thank Leun Otten for her comments on an earlier version. The work described here was supported by the Wellcome Trust and the Medical Research Council of the United Kingdom.

References

Baddeley A (1990): *Human Memory, Theory and Practice*, London: Erlbaum.

Bentin S, Moscovitch M (1989): Psychophysiological indices of implicit memory performance. *Bull Psychon Soc* 28:346–352.

Bentin S, Peled BS (1990): The contribution of stimulus encoding strategies and decision-related factors to the repetition effect for words: electrophysiological evidence. *Mem Cognit* 18:359–366.

Besson M, Kutas M, Van Petten C (1990): ERP signs of semantic congruity and word repetition in sentences. In: *Psychophysiological Brain Research*, Vol. 1, Brunia CHM, Gaillard AWK, Kok A, eds. Tilburg: Tilburg University Press.

Besson M, Kutas M, Van Petten C (1992): An event-related potential (ERP) analysis of semantic congruity and repetition effects in sentences. *J Cognit Neurosci* 4:132–149.

Brown, AS, Neblett DR, Jones TC, Mitchell DB (1991): Transfer of processing in repetition priming: some inappropriate findings. *J Exp Psychol Learn Mem Cognit* 17:514–525.

de Groot, AMB (1985): Word-context effects in word naming and lexical decision. *Q J Exp Psychol* 37A:281–298.

Donchin E, Fabiani M (1991): The use of event-related brain potentials in the study of memory: Is P300 a measure of event distinctiveness? In: *Psychophysiology: Central and Autonomic System Approaches*, Jennings JR, Coles MGH, eds. Chichester: Wiley.

Ellis AW, Young AW (1988): *Human Cognitive Neuropsychology*. London: Erlbaum.

Fischler I, Raney GE (1991): Language by eye: Behavioral and psychophysiological approaches to reading. In: *Handbook of Cognitive Psychophysiology: Central and Autonomic System Approaches*, Jennings JR, Coles MGH, eds. Chichester: Wiley.

Forster KL (1981): Priming and the effects of sentence and lexical contexts on naming time: Evidence for autonomous lexical processing. *Q J Exp Psychol* 33A:465–496.

Friedman D (1990): ERPs during continuous recognition memory for words. *Biol Psychol* 30:61–88.

Glanzer M, Adams JK (1990): The mirror effect in recognition memory: Data and theory. *J Exp Psychol Learn Mem Cognit* 16:5–16.

Glanzer M, Bowles N (1976): Analysis of the word frequency effect in recognition memory. *J Exp Psychol Learn Mem Cognit* 2:21–31.

Halgren E, Smith ME (1987): Cognitive evoked potentials as modulatory processes in human memory formation and retrieval. *Hum Neurobiol* 6:129–139.

Hasher L, Zacks RT (1979): Automatic and effortful processes in memory. *J Exp Psychol Gen* 108:356–388.

Hasher L, Zacks RT (1984): Automatic processing of fundamental information: The case of frequency of occurrence. *Am Psychol* 39:1372–1388.

Heit G, Smith ME, Halgren E (1990): Neuronal activity in the human medial temporal lobe during recognition memory. *Brain* 113:1093–1112.

Holcomb PJ, Neville HJ (1990): Auditory and semantic priming in lexical decision: A comparison using event-related brain potentials. *Lang Cognit Processes* 5:281–312.

Jacoby LL, Dallas M (1981): On the relationship between autobiographical memory and perceptual learning. *J Exp Psychol Gen* 3:306–340.

Jacoby LL, Kelley C (1991): Unconscious influences of memory: Dissociations and automaticity. In: *The Neuropsychology of Consciousness*, Milner AD, Rugg MD, eds. London: Academic Press.

Johnson R (1988): The amplitude of the P300 component of the event-related potential: Review and synthesis. In: *Advances in Psychophysiology*, Ackles PK, Jennings JR, Coles MGH, eds. Greenwich: JAI Press.

Johnson R, Pfefferbaum A, Kopell BS (1985): P300 and long-term memory: Latency predicts recognition performance. *Psychophysiology* 22:497–507.

Karis D, Fabiani M, Donchin E (1984): "P300" and memory: Individual differences in the Von Restorff effect. *Cognit Psychol* 16:177–216.

Kirsner K, Dunn JC, Standen P (1989): Domain-specific resources in word recognition. In: *Implicit Memory: Theoretical Issues*, Lewandowsky S, Dunn JC, Kirsner K, eds. Hillsdale, NJ: Erlbaum.

Kroll JF (1990): Recognizing words and pictures in sentence contexts: A test of lexical modularity. *J Exp Psychol Learn Mem Cognit* 16:747–759.

Kutas M, Van Petten C (1988): ERP studies of language. In: *Advances in Psychophysiology*, Ackles PK, Jennings JR, Coles MGH, eds., pp. 139–188. Greenwich: JAI Press.

Mandler G (1980): Recognizing: The judgment of previous occurrence. *Psychol Rev* 87:252–271.

Monsell S (1985): Repetition and the lexicon. In: *Progress in the Psychology of Language*, Vol. 2, Ellis AW, ed., pp. 147–195. London: Erlbaum.

Monsell S (1991): The nature and locus of word frequency effects in reading. In: *Basic Processes in Reading*, Besner D, Humphrey GW, eds. Hillsdale, NJ: Erlbaum.

Nagy ME, Rugg MD (1989): Modulation of event-related potentials by word repetition: The effects of inter-item lag. *Psychophysiology* 26:431–436.

Neville HJ, Kutas M, Chesney G, Schmidt AL (1986): Event-related brain potentials during initial encoding and recognition memory of congruous and incongruous words. *J Mem Lang* 25:75–92.

Potter DD, Pickles CD, Roberts RC, Rugg MD (1992): The effects of scopolamine on event-related potentials in a continuous recognition memory task. *Psychophysiology* 29:29–37.

Rugg MD (1985): The effects of word repetition and semantic priming on event-related potentials. *Psychophysiology* 22:642–647.

Rugg MD (1987): Dissociation of semantic priming, word and non-word repetition by event-related potentials. *Q J Exp Psychol* 39A:123–148.

Rugg MD (1990a): Event-related potentials dissociate repetition effects of high and low frequency words. *Mem Cognit* 18:367–379.

Rugg MD (1990b): The recall of repeated and unrepeated words: An ERP analysis. In: *Psychophysiological Brian Research*, Vol. 2, Brunia CHM, Gaillard AWK, Kok A, eds. Tilburg: Tilburg University Press.

Rugg MD, Doyle MC (1992) Event-related potentials and recognition memory for high and low frequency words. *J Cognit Neurosci* 1992, 4:69–79.

Rugg MD, Nagy ME (1987): Lexical contribution to non-word repetition effects: Evidence from event-related potentials. *Mem Cognit* 15:473–481.

Rugg MD, Nagy ME (1989): Event-related potentials and recognition memory for words. *Electroencephalogr Clin Neurophysiol* 72:395–406.

Rugg MD, Furda J, Lorist M (1988): The effects of task on the modulation of event-related potentials by word repetition. *Psychophysiology* 25:55–63.

Rugg MD, Roberts RC, Potter DD, Pickles CD, Nagy ME (1991) Event-related potentials related to recognition memory: Effects of unilateral temporal lobectomy and temporal lobe epilepsy. *Brain* 114:2313–2332.

Rull MD, Doyle MC, Melan C. An event-related potential study of written- and across-modality word repetition. *Lang Cognit Pro*, (in press).

Seidenberg MS, Waters GS, Sanders M, Langer P (1984): Pre- and postlexical loci of contextual effects on word recognition. *Mem Cognit* 12:315–328.

Smith ME, Halgren E (1989): Dissociation of recognition memory components following temporal lobe lesions. *J Exp Psychol Learn Mem Cognit* 15:50–60.

Smith ME, Stapleton JM, Halgren E (1986): Human medial temporal lobe potentials evoked in memory and language tasks. *Electroencephalogr Clin Neurophysiol* 63:145–159.

Smith ML (1989): Memory disorders associated with temporal-lobe lesions. In: *Handbook of Neuropsychology*, Vol. 3, Squire L, Gainott G, eds. Amsterdam: Elsevier.

Tulving E (1983): *Elements of Episodic Memory*. Oxford: Oxford University Press.

Van Petten C, Kutas M (1990): Interactions between sentence context and word frequency in event-related brain potentials. *Mem Cognit* 18:380–393.

Van Petten C, Kutas M, Kluender R, Mitchiner M, McIsaac H (1991): Fractionating the word repetition effect with event-related potentials. *J Cognit Neurosci* 3:129–150.

Verleger R (1988) Event-related potentials and memory: A critique of the context updating hypothesis and an alternative interpretation of P3. *Behav Brain Sci* 11:341–354.

Young MP, Rugg MD (1992): Word frequency and multiple repetition as determinants of the modulation of event-related potentials in a semantic classification task. *Psychophysiology* 6:664–676.

Chapter 6

Slow Potentials During Long-Term Memory Retrieval

FRANK RÖSLER, MARTIN HEIL AND ERWIN HENNIGHAUSEN

Almost half a century ago, Karl Lashley summarized his decade-long effort on the localization of the engram with the following, somewhat pessimistic statement: "it is not possible to demonstrate the isolated localization of a memory trace anywhere in the nervous system. Limited regions may be essential for learning or retention of a particular activity, but ... the engram is represented throughout the region" (Lashley, 1950). As far as neocortical structures are concerned, this position seems to be still valid. Neither experimentally induced lesions in animals nor those which occur naturally in human subjects support the idea that specific mnemonic contents can be narrowly localized anywhere in the cortex. A localization of memory functions could be established only insofar as that specific anatomic structures were found to be essential for the *process* of storage and retrieval but not for the engram itself. Moreover and surprisingly, these structures, which may be seen as relay stations within larger functional circuits, are for the most part not localized in the neocortex but in more ancient regions of the brain, for example, in the diencephalon, the basal forebrain, the hippocampus or the amygdalae (see, for example, the summarizing theories of (Mishkin and Appenzeller, 1987, or Markowitsch, 1985)). How these relay stations communicate with neocortical regions during storage and retrieval and where in the neocortex specific memory contents are held are still open questions.

In the light of these circumstances it may sound hubrical if someone

Cognitive Electrophysiology
H-J. Heinze, T.F. Münte, and G.R. Mangun, editors
© 1994 Birkhäuser Boston

claims that it is possible to monitor the contribution of neocortical structures to memory functions by means of the electroencephalogram. Nevertheless, the data that we summarize in this chapter do indeed suggest a close relationship between neocortical structures and memory retrieval processes. Moreover, these data also suggest that some neocortical structures are specifically related to the processing of particular mnemonic codes.

The methodological basis of our studies are the so-called slow event-related potentials (slow ERPs). ERPs can be classified along the continuum of their temporal extension: At one end there are the short-lived, transient shifts that have a clear peaklike appearance; at the other end there are slow deflections which may prevail for a couple of hundred milliseconds or more. The latter can show either a short rise and fall time and a more or less constant amplitude between on- and offset or a more gradual, ramplike increase and resolution. In any case these slow potentials have not a well-defined peak but a broad temporal extension. The best known slow wave phenomena are the processing negativity (Nd; Hansen and Hillyard, 1983), the readiness potential (RP; Kornhuber and Deecke, 1965) and the contingent negative variation (CNV; Rohrbaugh and Gaillard, 1983; Walter et al., 1964) and all three are manifestations of longer lasting functional states of the brain: attention to particular stimulus "channels," preparation for motor performance, and event anticipation, respectively.

These three, however, are not the only slow wave phenomena that can be related to particular cognitive states. Scanning through the literature reveals that slow negative waves have been observed in a wide variety of tasks, among others during short-term memory scanning and perceptual scanning of displays (Horst, Ruchkin, and Munson, 1987; Looren de Jong, Kok, and van Rooy, 1987; Wijers et al., 1989), during planning and execution of motor responses (Deecke et al., 1987; Lang, M. et al., 1988), during more conceptual tasks as mental arithmetic (Rösler and Heil, 1991; Rösler, Schumacher, and Sojka, 1990; Ruchkin et al., 1988), picture naming (Stuss et al., 1984; Stuss, Picton, and Cerri, 1986), mental rotation (Farah and Peronnet, 1989; Rösler, Schumacher, and Sojka, 1990), concept formation (Lang, M. et al., 1987; Uhl et al., 1990) or anticipation and processing of feedback stimuli (Brunia and Damen, 1988; Delisle, Stuss, and Picton, 1986; Grünewald et al., 1984; Rösler and Heil, 1991). For summaries, see Birbaumer et al. (1990); Haider, Groll-Knapp, and Ganglberger (1981); McCallum and Curry (1993).

A closer look at these data suggests that slow waves may be of particular interest for monitoring "higher" cognitive functions. First, slow waves seem to have a task-specific topography. Tasks with semantic material are accompanied by slow waves that have a maximum over more frontal regions of the brain, while tasks which require the processing of nonsemantic material, for example, mental rotation of images, evoke slow negative shifts with a more parietal maximum. Second, slow waves seem to have an amplitude that is related to task difficulty or processing effort. For example, the negativity that can be observed during short-term memory scanning shows a systematic increase of amplitude with increasing memory load (Wijers et al., 1989). Likewise the negativity that accompanies mental rotation tasks increases with the angular disparity between the two objects to be compared (Peronnet and Farah, 1989; Rösler, Schumacher, and Sojka, 1990). Finally, some studies provide evidence that the duration of these slow waves is related to the duration of particular processing stages. Nd is a convincing example; it persists as long as the evidence of a particular channel has to be analyzed and discriminated against the evidence of another, competing channel (Hillyard and Hansen, 1986). In summary, these psychophysiological findings suggest that slow waves, in particular negative slow waves, seem to be a manifestation of longer lasting cognitive states. Their topography may reveal differences in task quality; their amplitude, differences in the amount of effort that has to be invested to solve a task, and their duration may be related to the duration of particular processing stages.

This functional hypothesis is further corroborated and extended by neurophysiological findings about the generating mechanisms of slow waves. There is convincing evidence that slow waves originate predominantly from cortical structures (Creutzfeldt, 1983; Speckmann, Caspers, and Elger, 1984). According to the present knowledge, slow waves seem to reflect changes of the basic activity level of cortical tissue, that is, synchronized changes of the resting potentials of cortical neurons at their apical dendrites. Moreover, the polarity of slow waves provides clues about the type of change. In general, it seems to hold that slow negative potentials result from an increase of excitatory postsynaptic potentials (EPSPs) in the upper layers of the cortex; thus, they indicate an increase of activity or an increase of the responsiveness of the underlying cortical areas. Slow positive potentials, on the other hand, go together with a decrease of EPSPs or even an increase of inhibitory postsynaptic potentials (IPSPs). Therefore, they indicate a relative decrease of activity or a relative inhibition of particular cortical areas. Elbert and Rockstroh (1987) suggested that slow waves are manifestations of a threshold regulating mechanism that adjusts cortical areas according to expected actual processing demands.

Considering these presumed generating mechanisms of slow waves the hypothesis about their functional significance can be extended toward a neuropsychological dimension. A particular slow wave pattern being evoked during a particular cognitive task may reveal which cortical areas are in a relatively higher or lower state of activation. Thus, monitoring slow waves during cognition may provide an online method for a functional differentiation of cortical areas. This neuropsychological hypothesis is corroborated by studies in which regional cerebral blood flow and slow waves were monitored in the same task conditions. These studies revealed that negative potentials appear with substantial amplitude selectively above those cortical areas which also show a high metabolic activity (Lang, M. et al., 1988).

In this chapter, we report some recent studies in which slow waves were monitored during long-term memory (LTM) retrieval. To overcome some limitations of traditional LTM retrieval tasks, we made use of a special experimental paradigm, the so-called FAN-effect paradigm that was introduced into memory research by Anderson and colleagues (e.g., Anderson, 1974; Reder and Anderson, 1980; Reder and Ross, 1983). The empirical basis of this paradigm is the following observation: The time to decide whether two memory representations have an episodic link or not depends on the total number of links that branch out from each representation. If each representation has connections to many other representations, the decision will be very difficult and time consuming; if, on the other hand, there are only very few other connections, the decision will be easy and fast. An example may help to illustrate this fact: Consider you have to decide whether two researchers, authors A and B, ever published a paper together. Assume that both authors published only two papers in their total career. In that case you would be pretty fast with retrieving the necessary fact. But now consider that both authors were very active, that they published about 50 papers, and that you are familiar with all. In that case it would need a very thorough search through all the entries that exist in your memory about the publications of these two authors until you come up with the correct conclusion. And this more extensive search would result in a considerably longer decision time.

It is easy to conceive how the FAN-effect may be used to manipulate memory retrieval processes systematically. What is necessary is an experimentally established memory structure in which the FAN of different representations varies in a well-controlled manner. Retrieval tasks can then be constructed that activate different subsets of an associative structure and which therefore vary in difficulty. The paradigm can also be used to study differences in the quality of retrieval processes. For example, an associative network can be elaborated for different types of material, semantic entities as words or concepts, and nonsemantic entities as spatial positions or colors, respectively. Within one type of material the quality of associations can be manipulated as well, for example, associations can be elaborated between elements that belong either to the same or to a different level of an hierarchically ordered set.

Experiment 1: Associations Between Semantic Concepts

In our first experiment of this series (Rösler, Heil, and Glowalla, 1993), subjects had to learn associations between semantic concepts. The items to be learned were nouns that were presented as elements of distinct lists. Each list comprised seven elements: a list title, three specific concepts or tokens, and three general concepts or types. The general concepts were always superordinates of the specific concepts (see Table 1 for examples). One particular specific concept appeared either in only one, in only two, or in

Table 1. Design of Experiment 1.*

List #	Label	Token	Type
1	STATION	furniture	table
		island	Malta
		fruit	banana
2	LAWYER	sports	football
		car	Porsche
		dog	beagle
3	SCHOOL	furniture	table
		car	Porsche
		vegetables	*carrot*
4	RESTAURANT	furniture	table
		city	Berlin
		dog	*dachshund*
5	LECTURE	sports	football
		city	Berlin
		fruit	orange
6	APARTMENT	sports	football
		island	Malta
		vegetables	beans

* The table shows a subset of the memory structure that was established by means of an extensive learning procedure. Ss had to learn associations between a particular list label, category names (tokens) and category exemplars (types), respectively. In the example probes of exemplars with FAN 1 are given in italics, probes with FAN 3 are underlined. See text for further explanation.

three different lists. If it appeared in only one list (e.g., as carrot or dachshund in Table 1) it was episodically linked to the elements of only one list. Thus it had a FAN of one. In case a specific concept appeared in two different lists it was episodically linked to the other elements of both lists and it had a FAN of two (e.g., Porsche or Berlin in Table 1). Finally, a specific concept appeared in three lists it had a FAN of three. It was episodically linked to all other elements of the three lists.

If the subject is now asked to decide whether two concepts belong to the same list or not a different subset of the associative structure has to be activated and searched. This subset increases systematically from concepts having FAN 1 to those having FAN 3. The retrieval times for negative probes, that is, pairs of test words that are not elements of the same list and on which the subject has to decide with "no, there is not a common link," are given in Figure 1. It can be seen that there is an almost linear increase of the retrieval/decision time with increasing FAN for specific concepts. This supports the claim that the retrieval process varies in difficulty, if specific concepts with a different FAN have to be checked for their connectedness.

For general concepts, the story is different. If one considers how the material is constructed, one has to conclude that these general concepts always have an explicit FAN of three. Thus it is not too surprising that the

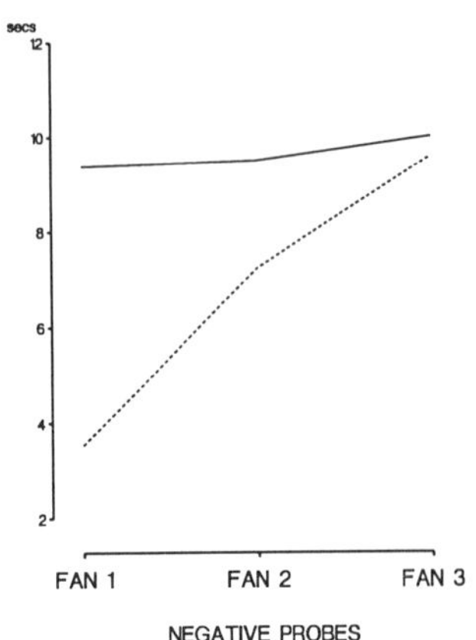

Figure 1.   Average retrieval times obtained in the semantic FAN experiment. Subjects had to decide whether two concepts shared a common episodic link (belonged to the same list) or not. Shown are the retrieval times for negative probes, i.e., pairs of concepts for which the correct decision was "no, they do not share a link."

retrieval times for general concepts do not vary in dependence of whether they are associated with a specific concept having a FAN of one, two or three respectively. A closer analysis of the retrieval process itself, however, suggests that retrieving associations between general concepts does not involve exactly the same processes as retrieving associations between specific concepts (Heil, Rösler, and Hennighausen, 1993). The difference concerns the route how the particular links are accessed. With specific concepts this access is straightforward. The system checks whether there is a common episodic link or not. With general concepts this access is indirect; it is accomplished via the specific links. The general concepts are in a certain sense redundant to the specific ones; they have not to be learned explicitly. If a subject has to decide whether an island and a car belongs to the same list, he or she can retrieve this fact by first activating the specific counterparts (e.g., Malta and Porsche). And this seems to be what subjects actually do in this situation. One can conclude, therefore, that the retrieval process varies in difficulty for specific concepts having a different FAN but that it varies also in quality, if one compares access to specific and to general concepts, respectively. With general concepts an additional process has to be assumed that first generates specific instances.

We expected that these two experimental manipulations should become manifest in distinct slow wave effects. In particular, we hypothesized: (1) a variation of the difficulty of retrieval processes should produce either different amplitude levels or different durations of retrieval-related slow waves, and (2) qualitatively different retrieval processes should be associated with topographically different slow wave patterns.

Methods

Nine healthy subjects (students of the Philips-University of Marburg) came to the lab on two consecutive days. On the first day they had to learn the material by heart. This was a set of 18 lists constructed as outlined in Table 1. Knowledge about these lists was 100% at the end of the training session (see Rösler, Heil, and Glowalla, 1993, for details). On the second day the retrieval test took place with the EEG recorded; 400 test trials were presented in five blocks with 80 trials each. Subjects initiated trials by briefly lifting one of two fingers. After a delay of 1 sec two test words appeared in the center of the screen and remained visible for a total of 15 sec, independently from whether or not the subject responded within the 15 sec. The subjects were instructed to respond as quickly and as accurately as possible. A bonus was paid depending on the percentage of correct responses. The two responses (*yes*: "the two concepts are elements of the same list", or *no*: "they are not") were given by lifting either the index or the middle finger of either the right or the left hand. The responding hand was varied across subjects.

The EEG was recorded monopolar from left, central, and right frontal cortex (F3, Fz, F4), from the vertex (Cz), and from left, central, and right parietal cortex (P3, Pz, P4) with linked mastoids as reference. Eye movement and blink artefacts were monitored with one electrode pair placed right to the upper canthus and left to the lower canthus of the left eye. Blinks were corrected according to Gratton, Coles, and Donchin (1983); epochs with other artefacts were rejected. Low-frequency cutoff of the recording system was set to $-3\,dB$ at $0.0013\,Hz$ (tc = 120 sec) and upper-frequency cutoff was set to $-3\,dB$ at $35\,Hz$. Electrode impedance was kept below $1\,k\Omega$ in all cases. The activity of all channels was AD converted with 12-bit resolution and a rate of 64 samples/sec.

Results and Discussion

The experiment revealed several systematic slow wave effects that seem to be related to search in long-term memory. First, the overall topography reveals a pronounced, DC-like negativity over the left frontal cortex (Fig. 2). This negativity appeared in all retrieval conditions irrespective of a particular type of probe (general or specific), a particular level of FAN, or a particular response (yes or no). As the following experiments reveal, this negativity seems to be specific to retrieval processes that have to be performed on semantic contents. A left frontal negative slow wave was also observed by Lang, M. et al. (1987), while subjects had to learn paired associates of words. Similarly, Stuss, Picton, and Cerri (1986) described a kind of slow wave that appeared during object naming and which was also

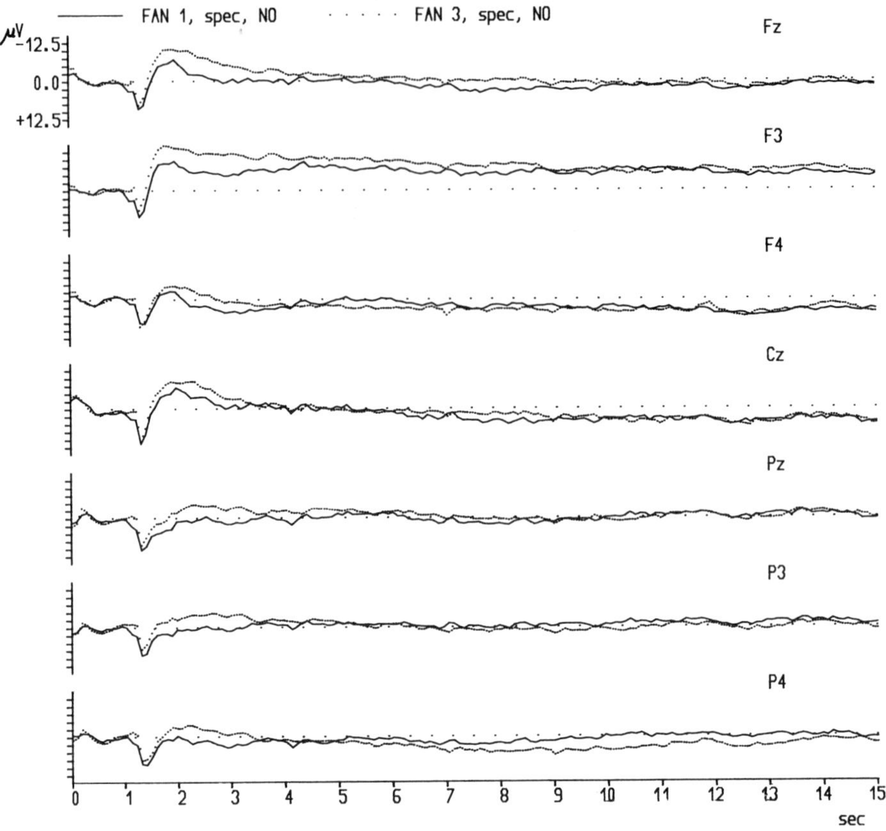

Figure 2.   Event-related activity during retrieval of associations between semantic concepts. Grand averages were obtained with negative probes having either a FAN of one (continuous line) or a FAN of three (dotted line). Probe onset was at 1 sec. Notice pronounced sustained negativity at F3.

substantially negative over frontocentral regions. Considering the functional interpretation of negative slow waves given earlier, all these findings agree on the hypothesis that the left frontal cortex is particularly involved when semantic material has either to be stored into or retrieved from LTM. This would also be in line with neuropsychological evidence provided by lesion studies (e.g., Risse, Rubens, and Jordan, 1984).

The second finding of this study concerns the experimental manipulation of retrieval difficulty. The amplitude of the negative potential that was observed over the frontal cortex varied systematically with different FAN; it was largest with FAN 3, intermediate with FAN 2, and less pronounced but still substantial with FAN 1 (see Fig. 2). This result suggested that the overall negativity over the frontal cortex is modulated by task difficulty.

Moreover, the fact that the main experimental manipulation became manifest exactly at those locations where the overall negativity had its maximum substantiates the assumption that this frontal negativity is specifically related to the process of retrieval or activation of permanently stored semantic associations.

Finally, the experimental factor type of concept general versus specific became also apparent as a slow wave effect. Decisions on the relatedness of general concepts were always accompanied by a more pronounced negativity than decisions on the relatedness of specific concepts. However, this amplitude difference had a more parietal distribution, and it emerged later in time than the amplitude difference observed between the different FAN conditions. The difference from factor FAN was already present 500 msec after probe onset, while the difference from type of concept emerged no earlier than 1000–1500 msec after probe onset.

In summary, the first experiment in which the FAN paradigm was employed revealed that slow waves seem to be systematically related to memory retrieval processes. The slow wave pattern had a specific topography, and two topographically distinct amplitude variations were found to be related to particular retrieval conditions. Both effects were associated with a modulation of a negative slow wave pattern in such a manner that the always more demanding condition evoked a more substantial negative-going amplitude. This is in line with our general working hypothesis, which assumes a systematic relationship between the amplitude of negative waves and the amount of effort that has to be invested to solve a task. The topographic difference between the two effects, that is, that factor FAN had a more frontal expression and factor type of concept a more parietal one, is in line with the psychological analysis of the task. It has to be assumed that both experimental manipulations tap different mechanisms of the involved retrieval processes. Factor FAN concerns a difficulty manipulation; it relates to the fact that a different number of associations has to be searched. The factor type of concept, on the other hand, concerns a manipulation of retrieval strategy. When subjects had to decide on the relatedness of general concepts, they activated specific instances before they searched these for common links. Subjects had reported during debriefing that they often accessed or generated these specific instances by means of imagery. This nonsemantic strategy would explain why the effect had its maximum expression over the parietal cortex.

Experiment 2: Associations Between Semantic Concepts Revisited

The first study was performed with limited equipment. There were only a few electrodes and moreover the EEG was not recorded with genuinely DC- but with AC-coupled amplifiers having a time constant of 120 sec. To test for the reliability of the effects just mentioned and to get a more complete picture about the topography, we repeated the experiment.

Methods

Another sample of 11 healthy students were trained and tested with the same semantic material as used in experiment 1. Training and testing were accomplished as previously outlined. However, apart from the recording equipment there were some other minor methodological changes. A test trial was no longer initiated by the subject but presented in a computer-controlled mode with a varying intertrial interval of 2 sec. A trial began with the presentation of a fixation point. After a baseline recording epoch of 3 sec, there was an acoustic warning stimulus, which was followed by the test words 1 sec later. The test words were visible for 14 sec. The EEG was recorded with proper DC characteristics from 17 locations: from F7, F3, Fz, F4, F8, T3, C3, Cz, C4, T4, T5, P3, Pz, P4, T6, O1 and O2. The EOG was monitored for artifact control by means of three channels, which recorded the vertical, lateral, and radial vector component, respectively, of eye movements and blinks. The DC recordings were corrected for drift artifacts by a method suggested by Hennighausen, Heil, and Rösler (1993). Trials with other artifacts, as substantial eye movements or blinks, were rejected by inspection of all single trials. Slow wave activity was measured by means of average amplitudes for consecutive intervals of 500 msec. All effects were tested by means of MANOVA, and only those with a reliability of at least $p(F) < .01$ are reported.

Results and Discussion

The overall topography observed with this replication was very similar to the one already described with the limited set of electrodes. There is again a clear left frontal preponderance of a negative slow shift that is switched on after probe onset and which gradually resolves with progressive time, that is, with increasing likelihood that a decision about the relatedness of the two probe words had been achieved (Fig. 5, top row). There is also some negative activity going on over the occipital cortex, a phenomenon that we were not able to observe in Experiment 1. This negativity is not too surprising, because the probe was presented throughout the total recording epoch and subjects could process the visual stimuli permanently.

The slow wave amplitude difference resulting from different levels of FAN could also be replicated. Again FAN 3 evoked the most pronounced negative amplitude and FAN 1 the least pronounced. This difference appeared immediately after probe onset and again had its maximum expression over frontocentral locations with a preponderance to the left hemisphere (see Fig. 3).

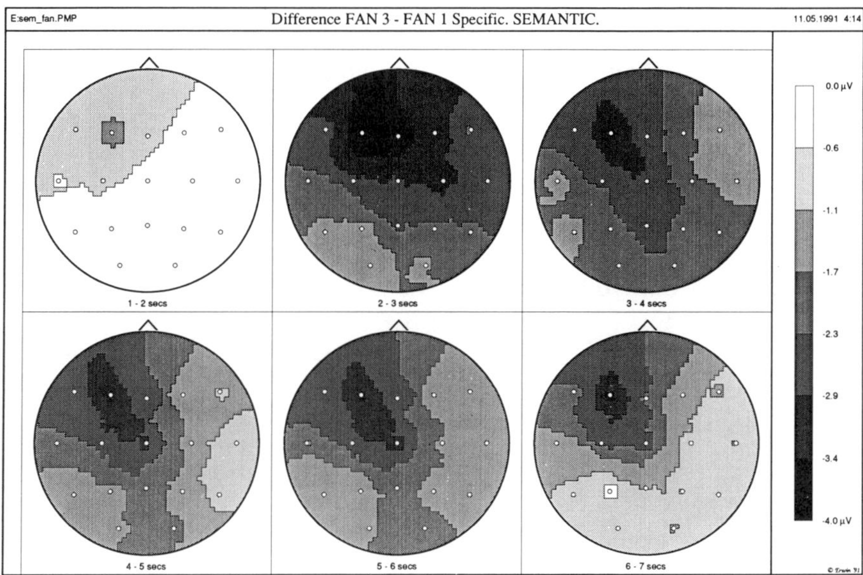

Figure 3.   Difference maps of event-related activity during retrieval of associations between semantic concepts show the difference between the grand averages obtained with negative probes having a FAN of three and a FAN of one. More negativity in condition FAN 3 results in a larger difference amplitude. The maps show the average amplitude difference of consecutive time intervals after probe onset beginning with interval 1, 2 sec (top left) and ending with interval 6, 7 sec (bottom right). Notice pronounced sustained negativity over left frontocentral cortex.

The slow wave difference from the factor type of concept was not as pronounced as in Experiment 1. The overall topography of this effect when expressed as a difference wave (general specific) revealed as before a maximum that was more posterior and later than the maximum observed in the difference wave obtained from subtracting the potentials of FAN 1 from those of FAN 3. However, the effect extended more toward frontal sites than in Experiment 1. One explanation for this different outcome of the two experiments may be found in interindividual differences. As outlined earlier, the effect from factor type of concept seems to be a correlate of a specific strategy that is employed when links of general concepts are retrieved. This strategy, which says that specific instances are activated on the presentation of general probes by means of imagery, is very likely not used by all subjects to the same extent. Debriefing revealed that there were also some subjects who had actually learned the general concepts explicitly. In addition, the ability to make use of the process of imagery may also vary across subjects. Unfortunately, we did not control these factors explicitly. Therefore, it could be that our second sample included fewer subjects who used this particular retrieval strategy.

Nevertheless, the replication revealed two very reliable effects. The clear left frontal maximum of the negative slow wave was found in both studies, and the variation of retrieval difficulty became apparent in both studies with the same topography and the same polarity: the left frontal negativity increased with increasing difficulty.

Experiment 3: Associations Between Pictures and Spatial Locations

The objective of this study (Heil, Rösler, and Hennighausen, 1990; Rösler, Heil, and Hennighausen, 1993) was to investigate whether and how the overall topography of the slow wave pattern changes, when the retrieval process concerns associations between completely nonsemantic memory representations. To this end, subjects had to learn links between line drawings of real objects and locations in a checkerboard-type grid (Fig. 4). The situation is similar to that created in the children's game "Memory."

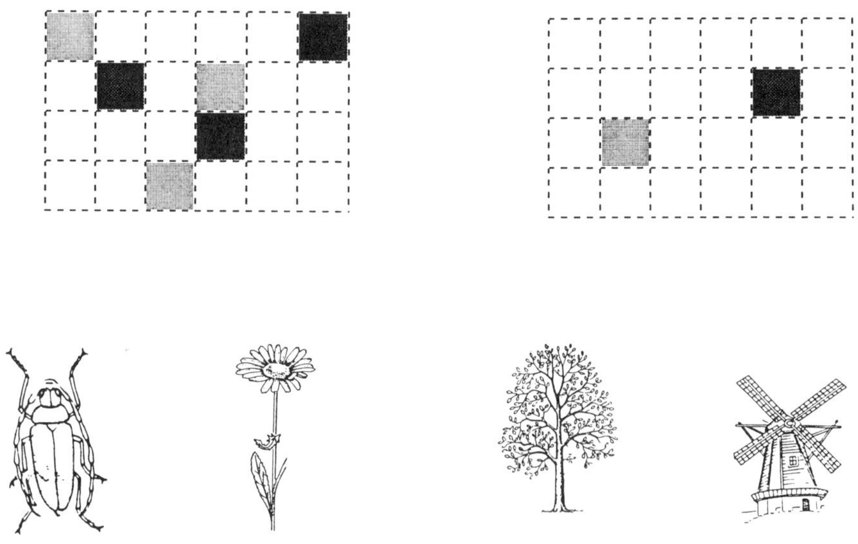

Figure 4.   Construction principle of material used in Experiment 3. Subjects learned associations between locations in a grid and linedrawings, the latter showing either animate or inanimate real objects. The left panel presents two objects that had a FAN of three, i.e., they were each associated with three different locations in the grid (e.g., the beatle was associated with the locations shown in light grey, the flower with the locations shown in dark grey). In the right panel, two objects are presented that both had a FAN of one. See text for further explanation.

The grid pattern had three rows and six columns of uniformly gray squares. During learning, a line drawing was presented below the grid and one square of the grid was specified by a brightness change. The subject had to memorize the associations between the line drawing and the particular location in the grid. The material was constructed according to the same principles as outlined in Experiment 1, that is, there was a systematic variation of FAN and a variation of the type of object. FAN was varied as follows. A particular line drawing could be associated with only one location, with two locations, or with three, respectively. The type of object manipulation concerned the concreteness of the object that was depicted by the line drawing. Specific objects were realized by pictures of actually existing and prototypical objects (e.g., a tower windmill, a post windmill, a modern windmill motor). General objects were more abstract pictures that represented no details but just the typical features of an object category.

Methods

A sample of 11 healthy students, who had not participated in Experiments 1 or 2, was recruited. Again, they were trained with the material extensively until they had a complete command of it. On the next day they were tested and the EEG was recorded. A test trial always included two line drawings that were presented side by side without the grid. The subject had to decide whether both objects shared a common grid position. The response was given by lifting one of two fingers. The test trials were constructed such that the two items were either both of type specific or of type general, and that both had either a FAN of one, two, or three, respectively. In addition, there were some filler trials. These showed two line drawings that had either a different FAN or which were of a different type. (These filler trials were not analyzed.) The EEG was recorded from the same 17 locations as in Experiment 2. Likewise, the EOG was recorded with three channels. Artifact handling and statistical analysis was as in Experiment 2.

Results and Discussion

For brevity only the overall topography obtained with this variant of the FAN paradigm is reported here (for a full account of this study see Rösler, Heil, and Hennighausen (1993)). There was again a very pronounced negative slow wave that emerged shortly after the presentation of the test trials and which gradually resolved toward the end of the recording epoch. The topography of this slow wave, however, was completely different from that observed with the semantic FAN (Fig. 5). The maximum was now found over the parietal cortex while there was hardly any negativity over frontal areas. The frontal sites were even somewhat positive. This supports our hypothesis that the slow wave pattern being observed during memory retrieval has a code-specific topography. Moreover, the topography evoked

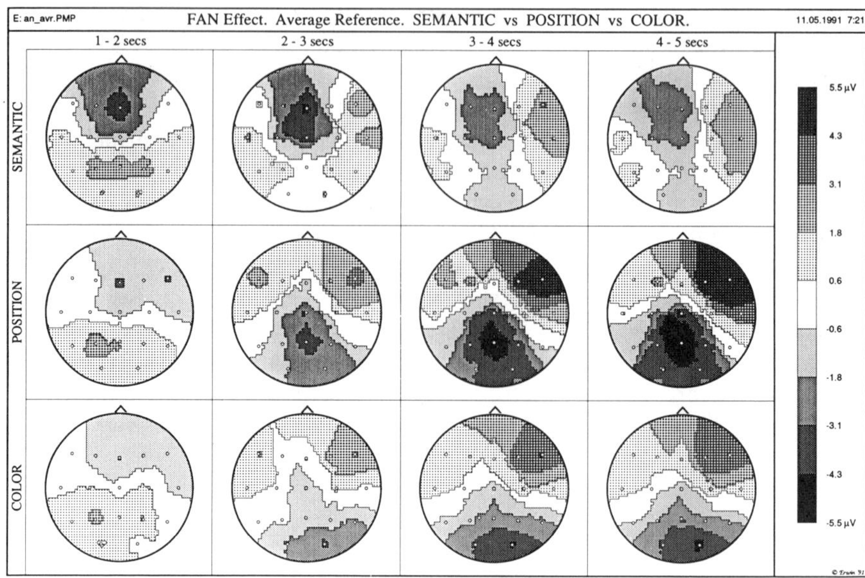

Figure 5.   Event-related brain activity observed during retrieval of different types of associations from long-term memory. Top row: data from experiment 2, retrieval of semantic associations; middle row: data from Experiment 3, retrieval of associations between pictures and spatial locations; bottom row: data from experiment 4, retrieval of associations between pictures and color patches. Presented are grand averages of the overall topography observed within consecutive intervals of 1 sec after probe onset. Maps were computed by means of spline interpolations. To permit a comparison across exeriments, data were rescaled with respect to an "average reference." Relative negativity is presented as a densely stippled pattern; relative positivity as a coarsly stippled pattern.

by this nonsemantic retrieval condition is in line with neuropsychological findings on the functional division of the cortex. The negativity was most pronounced exactly above those cortical areas that are functionally related to the processing of spatial representations (e.g., Kosslyn, 1987).

Experiment 4: Associations Between Pictures and Color Patches

With our final experiment (Heil, Rösler, and Hennighausen, 1990; Rösler, Heil, and Hennighausen, 1993) we wanted to test more thoroughly the material-specific topography of memory-related slow waves. An obvious question concerns the functional distinctiveness of the topography: is it only related to gross functional differences, as semantic versus nonsemantic, or is the functional resolution possibly much finer such that also more subtle differences can be detected? As far as the sensitivity of the method is concerned, the results of Experiment 1 were already promising. They had revealed a topographical difference within the domain of semantic material alone. This difference was found to be related to the strategy employed for retrieving specific versus general concepts. On the other hand, this effect was very likely also bound to a semantic versus nonsemantic distinction, because, as it was argued, the presentation of general concepts may have induced a nonsemantic strategy for generating specific instances. With the current experiment, we pursued a more direct approach to determine functional differences. We used another nonsemantic type of material, color-picture associations, and hypothesized that the retrieval processes should evoke a slow wave pattern topographically different from that observed in Experiment 3.

Methods

The same line drawings as in Experiment 3 were used, but instead of spatial locations they were paired with different color patches. During learning each line drawing was presented above a grid with three rows and six columns of color patches. A color patch to be associated with a particular picture was flashing throughout the presentation of a trial. To exclude any positional clue the color patches were rearranged in the grid on the presentation of each new learning trial. Thus the subject had indeed to learn associations between pictures and colors, not pictures and locations. All other details of the material were as in Experiment 3: there was the same variation of FAN—one, two, or three colors could be associated with one particular picture, and there were again the specific line drawings showing real objects and the general ones showing abstract category representations. All other methodological and procedural details were the same as in Experiment 3. Eleven healthy students who had not participated in any of the former experiments were recruited for this study.

Results and Discussion

Again we report only about the overall topography. (For more details see Rösler, Heil, and Hennighausen (1993)). As before, the retrieval process is associated with a well-pronounced negative slow wave. The time course is similar to Experiment 3; it has a short rise time and a gradual resolution. However, as expected, the topography is different. Compared to the topography observed with the spatial locations the maximum of the negativity has now moved more toward occipital sites (see Fig. 5). This result clearly demonstrates once more that the topography of the slow wave pattern observed during memory retrieval is code specific. However, it is not just the gross distinction between semantic and nonsemantic codes that becomes manifest in a distinct topography. Obviously, differences within the domain of nonsemantic codes can be detected by means of this approach as well. It is important to note that the overall topography observed in this study is also in agreement with neuropsychological findings on the cortical representation of color-processing mechanisms. As summarized by Meadows (1974), all types of color agnosias are caused either by lesions within areas 18 and 19 or by a disconnection of these areas from other functional units.

Summary and Conclusions

The set of experiments reported here proves that it is possible to monitor long-term memory functions by means of slow event-related brain potentials. All four experiments revealed a pronounced negative slow wave that is temporally related to the process of retrieval. In each case this negativity emerged immediately after the presentation of a memory probe, and it resolved at about the same time when an overt response indicated the end of the search process. The topography of the negative slow wave was found to be closely related to the quality of the retrieval process; it changed when different memory codes were accessed. The maximum of the plateau-like negativity was found over frontal areas with semantic material, over parietal areas with spatial material, and over occipital areas with color material. Moreover, the topography was also affected when distinct retrieval strategies were induced by different probes belonging to one type of material. Finally, the amplitude of the negative slow wave was found to be systematically related to the difficulty of the retrieval process. Probes that induced a more extensive search, as indicated by response times, were always associated by a larger negative amplitude. This amplitude modulation was most pronounced at those electrode locations where the negativity had its absolute maximum. Taken together, these findings suggest that the observed slow wave effects were indeed specific to the memory retrieval task.

The topography of the three slow wave patterns was congruent with the functional division of the cortex as it is suggested by lesion studies. For a particular material, the negative maximum was always found exactly over those cortical areas that are known as functionally specific for processing the related code. This close correspondence supports the assumption that the neuropsychological generators of slow negative shifts are actually located in the underlying cortical tissue. The finding is also consistent with the idea that the negative maximum of a slow wave pattern indicates which cortical areas are in a relatively higher state of activity than others during particular steps of human information processing.

If this interpretation of a slow wave pattern proves to be correct, it will open an interesting perspective for the field of neuropsychological diagnostics. Slow waves could then become a very sensitive tool for monitoring the functional division of the cortex. The method could be used quasi online and it would also allow studying functional distinctions in healthy subjects. Moreover, as slow waves have a much higher temporal resolution than other noninvasive methods, for example, rCBF or PET scans, they could help to disclose very subtle differences of cognitive processes.

The data presented here are already convincing in this respect. If one looks at the behavioral data that were collected in the four experiments—response times and error rates, respectively no material-specific effect would have been detected at all (see Heil, Rösler, and Hennighausen, 1993). Retrieval times were virtually the same irrespective of the kind of associations tested— semantic, spatial, or color associations, respectively. For one thing, this was intended. We wanted to create tasks that were equivalent as closely as possible. There should be no difference in the difficulty of the retrieval processes, just the type of code should be manipulated. However, from a theoretical point of view these behavioral data could also be taken as evidence for the fact that the retrieval processes were exactly the same in all three conditions. Adherents of a theory that assumes a uniform propositional code for all entities stored in long-term memory would be delighted with this finding (e.g., Anderson and Bower, 1973). However, our psychophysiological findings cast some doubt on such a unifying position. Although the retrieval processes may in fact be equivalent with each kind of material, the topographic differences indicate nevertheless that these processes seem to operate on different structures. These structures may constitute different subsets of long-term memory representations. This still leaves open the possibility that everything is transformed into a propositional code, but on accessing this code different transformations seem to be involved.

Of course, monitoring slow event-related brain potentials during memory retrieval does not solve the riddle of the engram: how mnemonic contents are coded, where they are stored and how they are reactivated. However, the findings reported here give at least some clue about the functional role of cortical structures during memory retrieval. Obviously the cortical structures involved during explicit memory retrieval are also those necessary for perception. The engram may be distributed across these structures and, as proposed by Markowitsch (1985, p. 214) "for retrieval to occur, limbic system-related structures, (...) may act as an organ of resonance which induces the (primarily) cortical neuronal network with the particular information to fire in a way which represents the mnemonic event".

Acknowledgments. This work was supported by grants Ro 529/4-1 and 4-2 of the Deutsche Forschungsgemeinschaft. We thank Anette Loell, Peter Puetz, Brigitte Roeder, and Corinna Wende for their support during data acquisition and analysis.

References

Anderson JR (1974): Retrieval of propositional information from long-term memory. *Cognit Psychol* 6:451–474.

Anderson JR, Bower GH (1973): *Human associative memory*. Hillsdale, NJ: Erlbaum.

Birbaumer N, Elbert T, Canavan AGM, Rockstroh B (1990): Slow potentials of the cerebral cortex and behavior. *Physiol Rev* 70:1–41.

Brunia CHM, Damen EJP (1988): Distribution of slow potentials related to motor preparation and stimulus anticipation in a time estimation task. *J Electroencephalogr Clin Neurophysiol* 69:234–243.

Creutzfeldt OD (1983): *Cortex Cerebri: Leistung, strukturelle und funktionelle Organisation der Hirnrinde*. Berlin: Springer.

Deecke L, Uhl F, Spieth F, Lang W, Lang M (1987): Cerebral potentials preceding and accompanying verbal and spatial tasks. In: *EEG Suppl, vol. 40: Current Trends in Event-Related Potential Research*, Johnson RJ, Rohrbaugh JW, Parasuraman R, eds., pp. 17–23. Amsterdam: Elsevier.

Delisle M, Stuss DT, Picton TW (1986): Event-related potentials to feedback in a concept-formation task. In: *Electroencephalography and Clinical Neurophysiology, Suppl. 38: Cerebral Psychophysiology: Studies in Event-Related Potentials*, McCallum WC, Zappoli R, Denoth F, eds., pp. 103–105. Amsterdam: Elsevier.

Elbert T, Rockstroh B (1987): Threshold regulation—a key to the understanding of the combined dynamics of EEG and event-related potentials. *J Psychophysiol* 4:317–333.

Farah MJ, Peronnet F (1989): Event-related potentials in the study of mental imagery. *J Psychophysiol* 3:99–109.

Gratton G, Coles MGH, Donchin E (1983): A new method for off-line removal of ocular artefact. *EEG J* 55:468–484.

Grünewald G, Grünewald-Zuberbier E, Hömberg V, Schuhmacher H (1984): Hemispheric asymmetry of feedback-related slow negative potential shifts in a positioning movement task. In: *Annals of the New York Academy of Sciences, 425: Brain and Information: Event-related potentials*, Karrer R, Cohen J, Tueting P, eds., pp. 470–476. New York: The New York Academy of Sciences.

Haider M, Groll-Knapp E, Ganglberger JA (1981): Event-related slow (DC) potentials in the human brain. *Rev Psychol, Biochem & Pharmacol* 88:126–197.

Hansen JC, Hillyard SA (1983): Selective attention to multidimensional auditory stimuli. *J Exp Psychol: Hum Percept Perform* 9:1–19.

Heil M, Rösler F, Hennighausen E (1990): Slow brain potentials during retrieval of spatial and color representations from long-term memory. *Psychophysiol* 27:S38.

Heil M, Rösler F, Hennighausen E (1993): Dynamics of activation in long-term memory: The retrieval of verbal, pictorial, and color information. *J Exp Psychol Learn Mem Cognit*.

Hennighausen E, Heil M, Rösler F (1993): A simple method for correcting DC-drift artifacts. *Electroencephalogr and Clin Neurophysiol* (in press).

Hillyard SA, Hansen JC (1986): Attention: Electrophysiological approaches. In: *Psychophysiology: Systems, processes, and applications*, Coles MGH, Donchin E, Porges SW, eds., pp. 227–243. Amsterdam: Elsevier.

Horst RL, Ruchkin DS, Munson RC (1987): Event-related potential processing negativities related to workload. In: *Current Trends in Event-Related Potential Research*, EEG Suppl. ed., Johnson R, Rohrbaugh JW, Parasuraman R, eds., pp. 186–190. Amsterdam: Elsevier.

Kornhuber HH, Deecke L (1965): Hirnpotentialänderungen bei Willkürbewegungen und passiven Bewegungen des Menschen: Bereitschaftspotentiale und Reafferente Potentiale. *Pflügers Archiv ges Physiol* 284:1–17.

Kosslyn SM (1987): Seeing and imagining in the cerebral hemispheres: A computational approach. *Psychol Rev* 94:148–175.

Lang M, Lang W, Uhl F, Kornhuber A, Deecke L, Kornhuber HH (1987): Slow negative potential shifts indicating verbal cognitive learning in a concept formation task. *Hum Neurobiol* 6:183–190.

Lang M, Lang W, Podreka I, Steiner M, Uhl F, Suess E, Müller C, Deecke L (1988): DC-potential shifts and regional cerebral blood flow reveal frontal cortex involvement in human visuomotor learning. *Exp Brain Res* 71:353–364.

Lang W, Zilch O, Koska C, Lindinger G, Deecke L (1989): Negative cortical DC shifts preceding and accompanying simple and complex sequential movements. *Experimental Brain Research* 74:99–104.

Lashley KD (1950): In search of the engram. *Symp Soc Exp Biol* 4:454–482.

Looren de Jong H, Kok A, van Rooy JCGM (1987): Electrophysiological indices of visual selection and memory search in young and old subjects. In: *Current Trends in Event-Related Potential Research*, EEG Suppl., Johnson R, Rohrbaugh JW, Parasuraman R, eds., pp. 341–349. Amsterdam: Elsevier.

Markowitsch HJ (1985): Hypotheses on mnemonic information processing by the brain. *Int J Neurosci* 27:191–227.

McCallum WC, Curry SH (eds) (1993): *Proceedings of the NATO ARW on Slow Potential Changes of the Human Brain*, NATO ASI Life Sciences Series. New York: Plenum.

Meadows JC (1974): Disturbed perception of colors associated with localized cerebral lesions. *Brain* 97:615–632.

Mishkin M, Appenzeller T (1987): The anatomy of memory. *Sci Am* 256:80–89.

Peronnet F, Farah MJ (1989): Mental rotation: An event-related potential study with a validated mental rotation task. *Brain Cognit* 9:279–288.

Reder LM, Anderson JR (1980): A partial resolution of the paradox of interference: The role of integrating knowledge. *Cognit Psychol* 12:447–472.

Reder LM, Ross BH (1983): Integrated knowledge in different tasks: The role of retrieval strategy on fan effects. *J Exp Psychol Learn Mem Cognit* 9:55–72.

Risse GL, Rubens AB, Jordan LS (1984): Disturbances of long-term memory in aphasic patients. A comparison of anterior and posterior lesions. *Brain* 107:605–617.

Rösler F, Heil M (1991): Toward a functional categorization of slow waves: taking into account past or future events? *Psychophysiol* 28:344–358.

Rösler F, Heil M, Glowalla U (1993): Memory retrieval from long-term memory by slow event-related brain potentials. *Psychophysiol* 30:170–182.

Rösler F, Heil M, Hennighausen E (1993): Distinct Cortical Activation Patterns During Long-Term Memory Retrieval of Verbal, Spatial, and Color Information (manuscript submitted).

Rösler F, Schumacher G, Sojka B (1990): What the brain tells when it thinks. Event-related potentials during mental rotation and mental arithmetic. *Germ J Psychol* 14:185–203.

Rohrbaugh JW, Gaillard AWK (1983): Sensory and motor aspects of the contingent negative variation. In: *Tutorials in Event-Related Potential Research: Endogenous Components*, Gaillard AWK, Ritter W, eds., pp. 269–310. Amsterdam: North Holland.

Ruchkin DS, Johnson R, Mahaffey D, Sutton S (1988): Toward a functional categorization of slow waves. *Psychophysiol* 25:339–353.

Speckmann EJ, Caspers H, Elger C (1984): Neuronal mechanisms underlying the generation of field potentials. In: *Self-regulation of the Brain and Behavior*, Elbert T, Rockstroh B, Lutzenberger W, Birbaumer N, eds., pp. 9–25. Heidelberg: Springer.

Stuss DT, Leech EE, Sarazin FF, Picton TW (1984): Event-related potentials during naming. In: *Annals of the New York Academy of Sciences, 425: Brain and Information: Event-related potentials*, Karrer R, Cohen J, Tueting P, eds., pp. 278–282. New York: The New York Academy of Sciences.

Stuss DT, Picton TW, Cerri AM (1986): Searching for the names of pictures: an event-related potential study. *Psychophysiol* 23:215–223.

Uhl F, Lang W, Lang M, Kornhuber A, Deecke L (1990): DC potential evidence for bilateral symmetrical frontal activation in non-verbal associative learning. *J Psychophysiol* 4:241–248.

Walter WG, Cooper R, Aldridge V, McCallum WC, Winter AL (1964): Contingent negative variation: An electrical sign of sensorimotor association and expectancy in the human brain. *Nature* 203:380–384.

Wijers AA, Otten LJ, Feenstra S, Mulder G, Mulder LJM (1989): Brain potentials during selective attention, memory search, and mental rotation. *Psychophysiol* 26:452–467.

Chapter 7

Event-Related Potentials Dissociate Immediate and Delayed Memory

L. NIELSEN-BOHLMAN AND R. T. KNIGHT

Evidence from human amnesia suggests that immediate or working memory and long-term memory involve activation of two distinct neural systems, with a transfer of information from immediate to long-term memory occurring from 15 to 60 sec post encoding. Amnesiac patients can correctly repeat six or seven items and carry on apparently normal conversations. This immediate memory process has a limited capacity, so that the addition of new items will impair performance on old items in these patients (Squire, 1986). Further, if they are distracted for a few minutes, the patients will not recall the items or their conversation. Conversely, amnesiac patients are able to recall events that occurred before they sustained hippocampal system damage (Scoville and Milner, 1957). This suggests that while their working memory, long-term memory storage, and recall mechanisms are intact, the hippocampal damage has impaired their ability to transfer information from immediate memory to long-term storage. This transfer problem is referred to as a deficit in short-term memory or anterograde amnesia.

Endogenous event-related potentials (ERPs) have been used to study the chronometry of cognition since it was discovered that these electrophysiological measures were responsive to the psychological state of the subject. First reports of this phenomenon showed that a late positive component of the ERPs was generated at 300 msec after stimulus delivery of unexpected or deviant tones in a tone sequence (Desmet, Debecker, and

Cognitive Electrophysiology
H-J. Heinze, T.F. Münte, and G.R. Mangun, editors
© 1994 Birkhäuser Boston

Manil, 1965; Sutton et al., 1965). It was later shown that this P3 component was independent of the physical parameters of the stimuli, as it was generated by the absence of an expected stimulus in a series (McCarthy and Donchin, 1976). This component has been recorded from several species, including monkeys (Arthur and Starr, 1984; Neville and Foote, 1984) dolphins (Woods et al., 1986), cats (Wilder, Farley, and Starr, 1981), and rats (Yamaguchi and Knight, 1993), indicating that this electrophysiological measure of cognition has a broad ethological range.

Event-related potentials have also been widely employed to assess the biological basis of attention and memory in man. The P3 component has been associated with both stimulus encoding (Fabiani, Karis, and Donchin, 1986; Friedman, Vaughan, and Erlenmeyer-Kimling, 1981; Neville et al., 1986; Paller, Kutas, and Mays, 1987) and recognition processes (Chapman and McCrary, 1981; Gomer, Spicuzza, and O'Donnell, 1976; Neville et al., 1982, 1986). Enhanced P3 amplitude generated by an item presented in an incidental memory paradigm is associated with increased probability of later free recall of that item (Fabiani, Karis, and Donchin, 1986), while enhanced P3 amplitude in a serial recognition paradigm is associated with correct identification of repeated stimuli (Rugg and Nagy, 1989).

Although many studies have examined the increase in P3 latency with memory load (Adams and Collins, 1978; Ford et al., 1979; Gomer, Spicuzza, and O'Donnell, 1976; Marsh, 1975) and the increase in P3 amplitude with sustained mental activity (Johnson et al., 1987; Ruchkin, Johnson, and Sutton, 1988; Stuss et al., 1983) few studies have examined event-related potential correlates of working memory. P3 peak amplitude was shown to decrease and latency increase as the number of items in the memory list increased in a Sternberg (1966) memory paradigm (Ruchkin et al., 1990). When the same memory list stimuli were presented as the target in a cued match-to-sample paradigm, the P3 component peak continued to decrease in amplitude and increase in latency with an increased number of items, although the cue was always a single item. This suggests that the P3 component indexes immediate stimulus processing rather than sustained working memory activation.

Generation of a negativity at −400 msec (N4) has also been associated with memory search processes (Friedman, 1990; Neville et al., 1986; Okita et al., 1985). The N4 generated during recognition memory tasks decreases to recognition stimuli in both linguistic (Rugg and Nagy, 1989) and visuospatial (Puce et al., 1991) tasks. N4 amplitudes decreased to repeated stimuli in a continuous recognition paradigm to words with a repetition lag of more than 18 sec. This reduction was accompanied by a positive shift generated from 300–600 msec by repeated stimuli, which may index an aspect of recognition memory processes (Rugg and Nagy, 1989).

Intracranial recording evidence suggests that this component is generated in entorhinal cortex. The N4 shows large amplitude gradients and polarity reversals in medial temporal structures, and is reduced by anterior temporal lobectomy (Smith and Halgren, 1986; Smith, Stapleton, and Halgren, 1986). Generation of the N4 has been dissociated from generation of the P3 both anatomically in lesion studies (Puce et al., 1991) and by task differences (Kutas and Hillyard, 1980). However, the relationship of the N4 to the temporal dynamics of memory processes has not been defined.

Experimental Design

We assessed whether there was electrophysiological evidence for differential brain mechanisms of immediate versus delayed memory processing, using a task with specific attentional and memory demands (Nielsen-Bohlman and Knight, 1991a, 1991b). Event-related potentials generated during a serial recognition memory paradigm were examined in neurologically normal young (21.5 $\pm$ 2 years of age; $n = 10$) and old (71.2 $\pm$ 7 years; $n = 10$) subjects with normal or corrected vision. Visual images normalized for familiarity (Snodgrass and Vanderwart, 1980) were presented in four blocks (111–114 stimuli per block; 700-msec stimulus duration; 1200 msec inter-stimulus interval). The images were white line drawings presented in the left- or right-upper quadrafield of a black CRT screen; 20% of the images were presented once and 80% were presented twice within a block. The second image occurred at temporal delays of 1.2, 4–12, 22–58, or 100–158 seconds. Thus, the second image was presented immediately following the first image or with 2–6, 11–29, or 50–79 intervening images. The subjects had to indicate whether they had seen the stimulus before by pressing a "yes" button with the first finger or a "no" button with the second finger of their right hand. The EEG was recorded from electrodes placed at Fpz, F3, Fz, F4, T3, C3, Cz, C4, T4, T5, P3, Pz, P4, T6, O1, Oz, and O2 (Jasper, 1958), references to a balanced noncephalic site (Stephenson and Gibbs, 1951). A below-eye electrode placed 2 cm below the eye and a lateral-eye electrode placed on the outer canthus were used to monitor eye movements. The N4 and P3 were identified from grand averaged ERPs and waveforms from each subject. Peak and mean amplitude and latency measures were referenced to a 200-msec prestimulus baseline.

In young subjects, the first stimulus presentation generated both a parietal maximum P3 (9.2 μV; latency, 601 msec at Pz) and a central N4 (-2.4 μV; latency, 404 msec at Cz). Reaction time to the initial stimuli was 711 msec, with 82% target detection accuracy. Immediate recognition stimuli (1.2-sec delay) generated a parietal maximal P3 (Pz $= 14.9$ μV) and a reduced N4 component. Mean reaction time to immediate recognition stimuli was 625 msec, with a target detection accuracy of 95%.

At delayed recognition stimulus lags of 4–12 and 22–58 sec, P3 amplitude was not reduced (at Pz, 4–12 $= 12.8$ μV; 22–58 $= 13.7$ μV). However, P3 amplitude decreased at the longest stimulus lag of 100–158 sec (10.2 μV at Pz, $p < .005$; Figures 1 and 2). An N4 component was generated by all delayed recognition lags (4–12 $= -1.1$ μV; 22–58 $= -0.6$ μV; 95–105 $=

$-2.6\ \mu$V; $p < .005$ for each long lag versus 1.2-sec lag; see Figure 2C). Delayed recognition processes generated increased reaction times at all longer lags versus immediate recognition (4–12 = 733 msec; 22–58 = 679 msec; 100– 158 = 682 msec; $p < .005$ for each long lag versus 1.2-sec lag). Accuracy decreased from the immediate to the delayed lags (4–12 = 85%, $p = .005$; 22–58 = 77%, $p < .005$; 100–158 = 79%, $p < .05$).

Recognition memory processes showed effects of both recognition interval and age. In the older subjects, the first stimulus presentation generated a parietal maximal P3 (8.1 μV; latency, 606 msec at Pz) comparably to the younger subjects. The central N4 (3.2 μV; latency, 388 msec at Cz) was reduced in comparison to the young group ($p < .05$). Reaction time to initial stimuli was 766 msec, with 84% target detection accuracy. In the older subjects, a large (14.5 μV) frontal positivity was generated by all stimuli.

Immediate recognition stimuli (1.2-sec interstimulus interval) generated a parietal maximal P3 (Pz = 13.3 μV). Similarly to the young subjects, the N4 component was absent. Mean reaction time to immediate recognition stimuli was 664 msec, with a target detection accuracy of 97%. The immediate recognition stimuli generated comparable electrophysiological and behavioral responses in the old and young subjects.

P3 amplitudes generated by old subjects were reduced at all delayed recognition lags versus the immediate recognition stimuli (at Pz, 4–12 = 10.5 μV; $p = .058$; 22–58 = 8.9 μV, $p < .05$; 100–158 = 9.2 μV, $p < .05$). The P3 generated by the old group was decreased from that of the young group at the delayed recognition lag of 22–58 sec (old = 8.9, young = 13.7, $p < .05$). The old group also tended to generate decreased N4 amplitudes to the delayed recognition stimuli in comparison to young; this decrease was significant at the delays greater than 100 seconds (old = 3.9 μV, young = $-2.6\ \mu$V, $p < .05$).

Delayed recognition processes generated increased reaction times at all longer lags versus recognition (4–12 = 856 msec; 22–58 = 833 msec; 100– 158 = 844 msec; $p < .001$ for each long lag versus the 1.2-sec lag). Accuracy decreased from the immediate to the delayed lags (4–12 = 66%; 22–58 = 60%; 100–158 = 54%; $p < .001$ for each long lag versus 1.2-sec lag). Reaction times in the old group showed a greater increase at long recognition stimulus lags (4–12, young = 733 msec, old = 856 msec; 22–58, young = 679 msec, old = 833 msec; 100–158, young = 682 msec, old = 843 msec; $p < .05$ for each old versus young comparison).

Discussion

In both young and old subjects a different event-related potential pattern was generated by stimuli recognized at a lag of 1.2 sec in comparison to stimuli recognized at lags greater than 4 sec. The immediate recognition

Initial Presentation

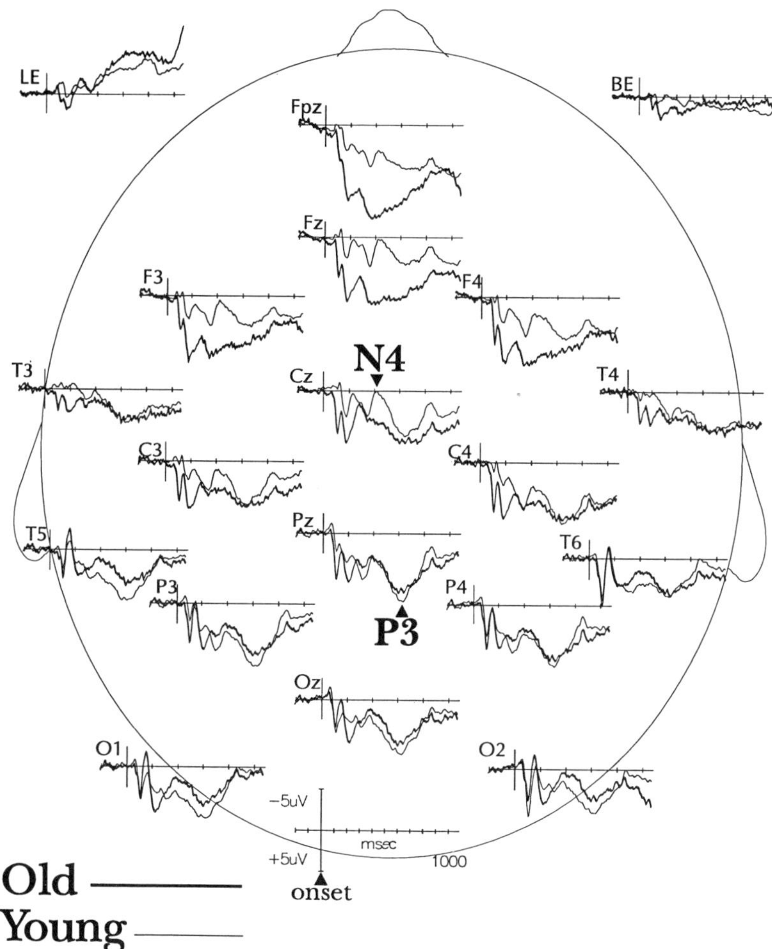

Figure 1.   Grand average event-related potentials generated by old (*thick line*) and young (*thin line*) subjects to correctly recognized stimuli across four lag periods. A. ERPs generated to initial presentation of stimuli. Note presence of both N4 and P3 components in young group and enhanced frontal positivity in old group.

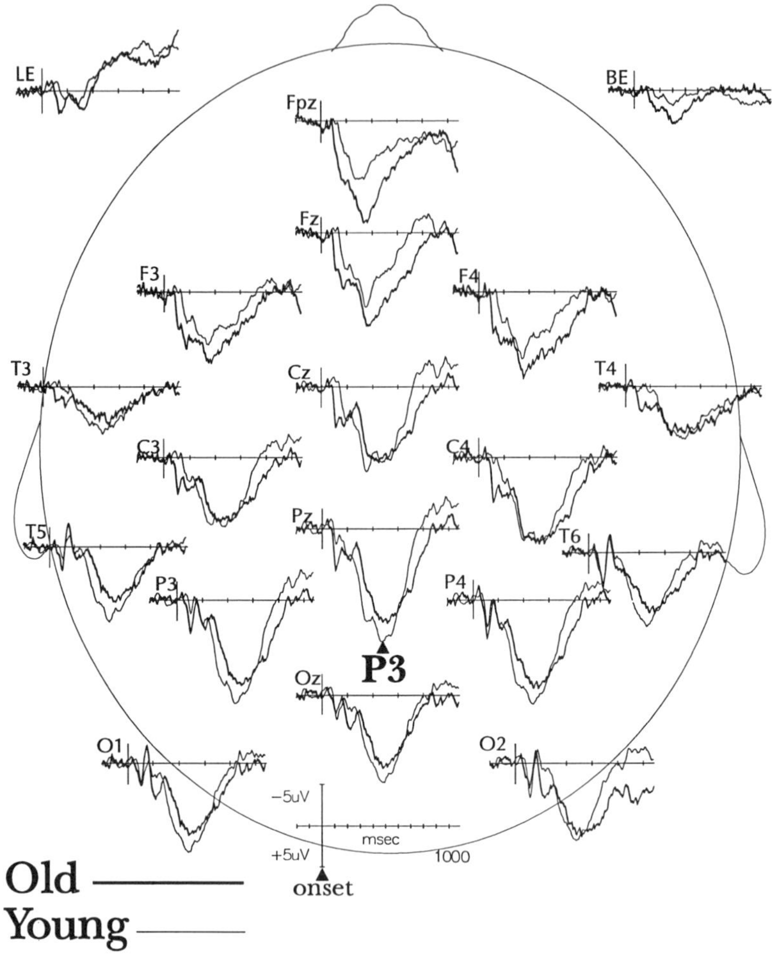

Figure 1. *(continued)* **B.**ERPs generated to recognition stimuli presented at lag of 1.2 sec. Note large P3 amplitude in both groups and absence of negativity at 400 msec.

Delayed Recognition: 4-12 sec.

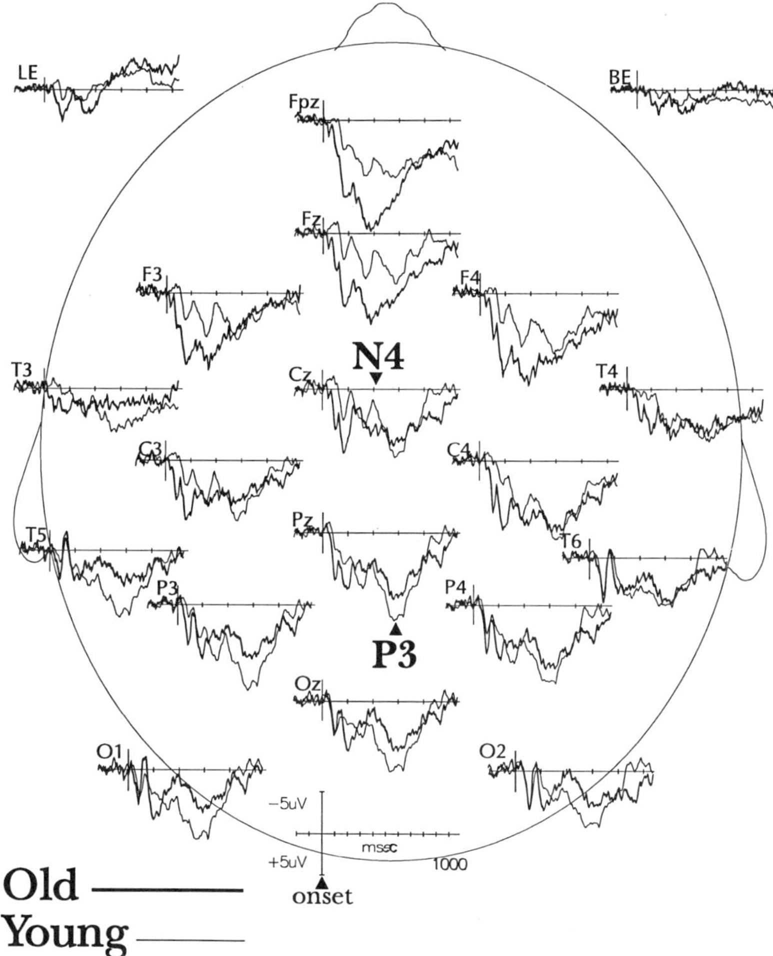

Figure 1.   (*continued*) **C.** ERPs generated to recognition stimuli presented at lag of 4–12 sec. Note decrease in P3 amplitude in old group and generation of additional negative component (N4) at 400 msec after stimulus delivery.

Delayed Recognition: 20-58 sec.

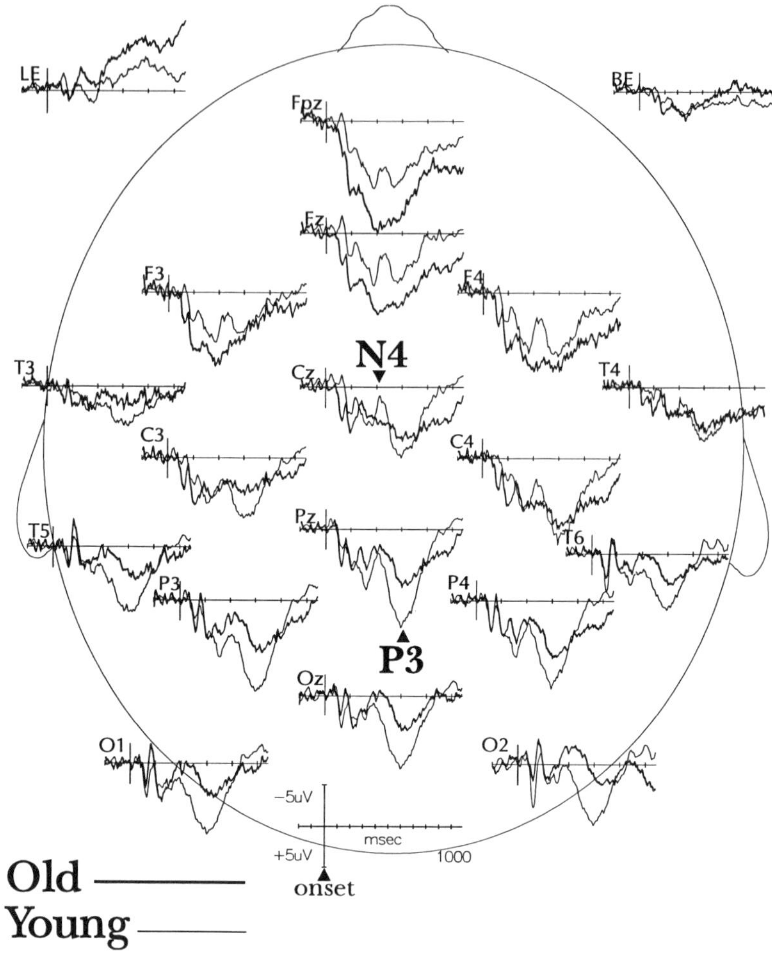

Figure 1.   (*continued*) **D.** ERPs generated to recognition stimuli presented at lag of 25–58 sec. P3 amplitude generated by old group is decreased both from 1.2-sec lag and from young group.

Delayed Recognition:100-158 sec.

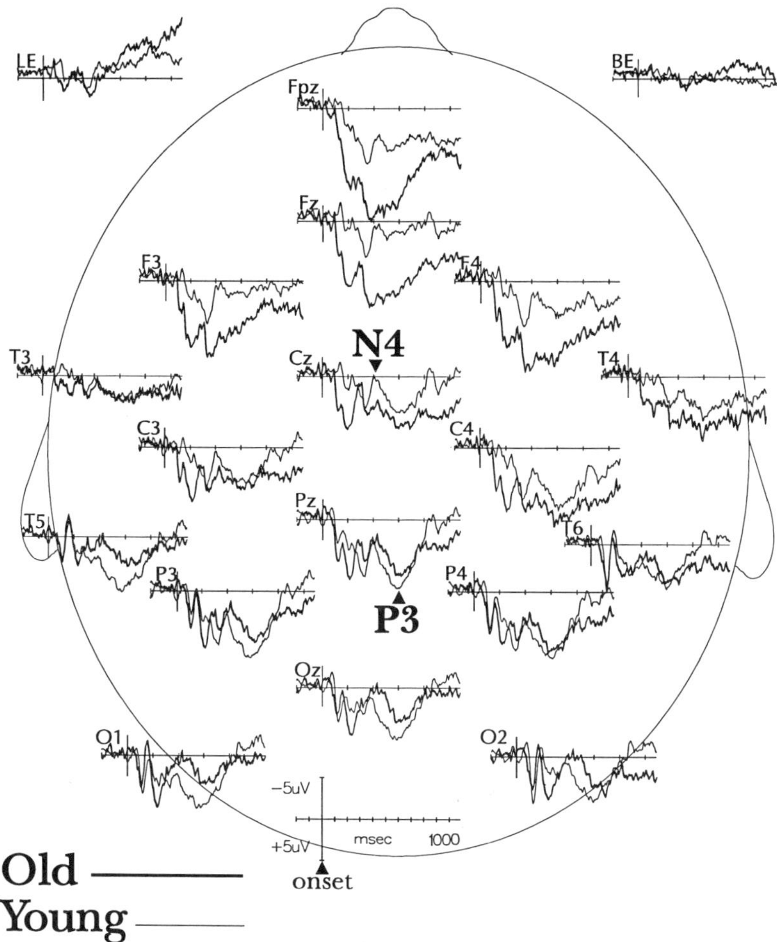

Figure 1.   (*continued*) **E.** ERPs generated to stimuli recognized at more than 100 sec. Note that both groups now have decreased P3 amplitudes from 1.2-sec lag.

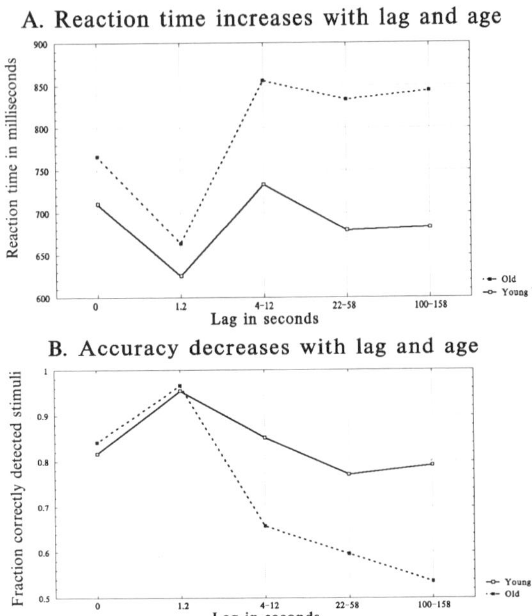

Figure 2.   Performance is affected by both lag and age. **A.** Reaction times increase in both young (*solid line*) and old (*dashed line*) groups with increases in stimulus presentation lag. Reaction times of both groups are significantly increased at lags greater than 4 sec from 1.2-sec lag. Although both groups showed comparable reaction times to first presentation and 1.2-sec recognition lag, reaction times in old group showed greater increase at long recognition stimulus lags. **B.** Response accuracy increases in both young and old subjects with increases in stimulus presentation lag. Again, both groups showed comparable performance at the first presentation and immediate recognition (lags, 0 and 1.2 sec). Older subjects showed reduced accuracy at all lags greater than 4 sec; young subjects showed reduced accuracy at lags greater than 22 sec.

stimuli generated an enhanced P3 component to correctly identified images in both young and old adults, which decreased in amplitude at delayed recognition lags, most rapidly in the older subjects. Both young and old subjects generated an additional frontocentral negative (N4) component at delayed recognition lags greater than 4 seconds, with an associated increased reaction time. These data indicate that different memory processes are engaged during immediate versus delayed recognition memory. While the P3 may index immediate memory processes, the N4 may index a neural process involved in transfer of information from a rapid sensory trace to long-term storage.

Topography at 100 sec. Lag

YOUNG

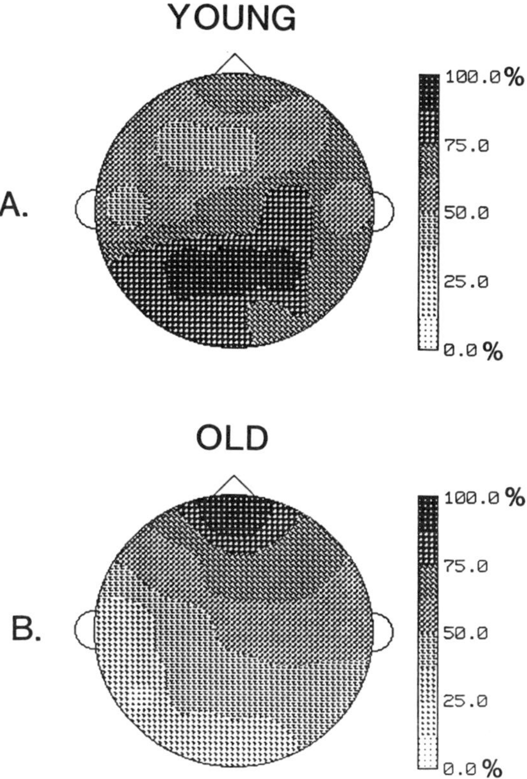

OLD

Figure 3.   Topographic maps show changes in ERP topography with age (window, 500–600 msec; lag, 100–158 sec). **A.** Young group shows maximal positivity at parietal sites (100% at Pz = 10.2 μV). **B.** Old group shows maximal positivity at frontal sites (100% at Fpz = 20.6 μV).

The old and young subjects generated comparable P3 amplitudes and shortened reaction times to the immediate recognition stimuli, supporting the notion that immediate recognition memory is preserved in aging. In contrast, at the delayed recognition lags the older group generated disproportionally longer reaction times than the young group, with an associated decrease in parietal P3 amplitude. This was associated with an enhanced frontal positivity to all stimuli, indicating that prefrontal changes are associated with memory search deficits in aging.

This and prior studies (Friedman, 1990; Neville et al., 1986) support the theory that N4 generation indexes a memory search process. A memory search process must be engaged to evaluate both the first and delayed second stimuli. However, the immediate second stimuli can be processed without a memory search, and the N4 component is absent following these stimuli, because attentional processes engaged by the stimulus are uninterrupted between the first and second presentation. Therefore, in this priming-like process the stimuli may be evaluated without a memory search. The P3 component is maximal to these immediately repeated stimuli. This and prior studies (Knight, 1984; Knight et al., 1989; Yamaguchi and Knight, 1991) suggest that the P3 component indexes an attentional or working memory process. Although this process does not involve long-term memory search, it may be associated with the degree of input to long-term memory processes (Fabiani, Karis, and Donchin, 1986).

References

Adam N, Collins GI (1978): Late components of the visual evoked response to search in short-term memory. *Electroencephalogr Clin Neurophysiol* 44:147–156.

Arthur DL, Starr A (1984): Task-relevant late positive component of the auditory event-related potential in monkeys resembles P300 in humans. *Science* 223:186–188.

Chapman RM, McCrary JW (1981): Memory processes and evoked potentials. *Can J Psychol* 35:201–212.

Desmet JE, Debecker J, Manil J (1965): Mise en evidence d'un signe electrique cerebral associe a la detection par le suject d'un stimulus sensoriel tactile. *Bull Acad R Med Belg* 5:887–936.

Fabiani M, Karis D, Donchin E (1986): P300 and recall in an incidental memory paradigm. *Psychophysiology* 23(3):298–308.

Ford JM, Roth WT, Mohs RC, Hopkins WF, Kopell BS (1979): Event-related potentials recorded from young and old adults during a memory retrieval task. *Electroencephalogr Clin Neurophysiol* 47:450–459.

Friedman D (1990): Cognitive event-related potential components during continuous recognition memory for pictures. *Psychophysiology* 27(2):136–148.

Friedman D, Vaughan HG, Jr., Erlenmeyer-Kimling L (1981): Multiple late positive potentials in two visual discrimination tasks. *Psychophysiology* 18(6):635–649.

Gomer FE, Spicuzza RJ, O'Donnell RD (1976): Evoked potential correlates of visual item recognition during memory scanning tasks. *Physiol Psychol* 4:61–65.

Jasper HH (1958): Report of the committee on methods of clinical examination in electroencephalography. *Electroencephalogr Clin Neurophysiol* 10:371–375.

Johnson R, Jr., Cox C, Fedio P (1987): Event-related potential evidence for individual differences in a mental rotation task. In: *Current Trends in Event-Related Potential Research*, Johnson R, Jr., Rohrbaugh W, Parasuraman R, eds., *Electro-encephalogr Clin Neurophysiol* (Suppl.) 40:191–197.

Knight RT (1984): Decreased response to novel stimuli after prefrontal lesions in man. *Electroencephalogr Clin Neurophysiol* 59:9–20.

Knight RT, Scabini D, Woods DL, Clayworth CC (1989): Contributions of temporal-parietal junction to the human auditory P3. *Brain Res* 502:109–116.

Kutas M, Hillyard SA (1980): Reading senseless sentences: Brain potentials reflect semantic incongruity. *Science* 211:77–80.

Marsh GR (1975): Age differences in evoked potential correlates of a memory scanning process. *Exp Aging Res* 1:3–16.

McCarthy G, Donchin E (1976): The effects of temporal and event uncertainty in a vigilance situation determining the waveform of the auditory event related potential. *Psychophysiol* 13:581–590.

Neville HJ, Foote SL (1984): Auditory event-related potentials in the squirrel monkey: Parallels to human late wave responses. *Brain Res* 298:107–116.

Neville H, Snyder E, Woods DL, Galambos R (1982): Recognition and surprise alter the human visual evoked response. *Proc Natl Acad Sci USA* 79:2121–2123.

Neville H, Kutas M, Chesney G, Schmidt A (1986): Event-related brain potentials during initial encoding and recognition memory of congruous and incongruous words. *J Mem Lang* 25:75–92.

Nielsen-Bohlman L, Knight RT (1991a): Electrophysiological measures of recognition memory. *IBRO Abstr* 164.

Nielsen-Bohlman L, Knight RT (1991b): Immediate and delayed memory are differentially affected by aging. *Soc Neurosci Abstr* 17(1):661.

Okita T, Wijers A, Mulder G, Mulder LJM (1985): Memory search and visual spatial attention: an event-related brain potential analysis. *Acta Psychol* 60:263–292.

Paller KA, Kutas M, Mayes A (1987): Neural correlates of encoding in an incidental learning paradigm. *Electroencephalogr Clin Neurophysiol* 67:360–371.

Puce A, Andrewers DG, Berkovic SF, Bladin PF (1991): Visual recognition memory. *Brain* 114:1647–1666.

Ruchkin DS, Johnson R, Jr., Sutton S (1988): Toward a functional categorization of slow waves. *Psychophysiol* 25:339–353.

Ruchkin DS, Johnson R, Jr., Canoune H, Ritter W (1990): Short-term memory storage and retention: an event-related brain potential study. *Electroencephalogr Clin Neurophysiol* 76:419–439.

Rugg MD, Nagy ME (1989): Event-related potentials and recognition memory for words. *Electroencephalogr Clin Neurophysiol* 72:395–406.

Scoville WB, Milner B (1957): Loss of recent memory after bilateral hippocampal lesions. *J Neurology, Neurosurg and Psychiatry* 20:11–21.

Smith ME, Halgren E (1986): Attenuation of a sustained visual processing negativity after lesions that include the inferotemporal cortex. *Electroencephalogr Clin Neurophysiol* 70:366–369.

Smith ME, Stapleton JM, Halgren E (1986): Human medial temporal lobe potentials evoked in memory and language tasks. *Electroencephalogr Clin Neurophysiol* 63:145–159.

Snodgrass JG, Vanderwart M (1980): A standardized set of 260 pictures: Norms for name agreement, image agreement, familiarity, and visual complexity. *J Exp Psychol* 6(2):174–215.

Squire LR (1986): Mechanisms of memory. *Science* 232:1612–1619.

Stephenson WA, Gibbs FA (1951): A balanced noncephalic reference electrode. *Electroencephalogr Clin Neurophysiol* 3:237–240.

Sternberg S (1966): High-speed scanning in human memory. *Science* 153:652–654.

Stuss DT, Sarazin FF, Leech EE, Picton TW (1983): Event-related potentials during naming and mental rotation. *Electroencephalogr Clin Neurophysiol* 56:133–136.

Sutton S, Braren M, Zubin J, John ER (1965): Evoked-potential correlates of stimulus uncertainty. *Science* 155:1436–1439.

Woods DL, Ridgway SH, Carder DG, Bullock TH (1986): Middle- and long-latency auditory event-related potentials in dolphins. In: *Dolphin Cognition and Behavior: A Comparative Approach*, Schusterman RJ, Thomas JA, Wood FG, eds., pp. 61–78. Hillsdale, NJ: Erlbaum.

Wilder MB, Farley GR, Starr A (1981): Endogenous late positive component of the evoked potential in cats corresponding to P300 in humans. *Science* 211:605–607.

Yamaguchi S, Globus H, Knight RT (1993): P3a-like potentials in rats. *Electroencephalogr Clin Neurophysiol* 88:151–154.

Yamaguchi S, Knight RT (1991): Age effects on the P300 to novel somatosensory stimuli. *Electroencephalogr Clin Neurophysiol* 78:297–301.

Chapter 8

What Is Who Violating? A Reconsideration of Linguistic Violations in Light of Event-Related Brain Potentials

Marta Kutas and Robert Kluender

It is often a worthwhile exercise to step back from one's field of endeavor and to think in broad terms about what its proper goals are, or perhaps what they should be. It seems to us that the proper goals of event-related brain potential (ERP) research into language comprehension should at minimum include enquiries into the following questions:

1. How many independent factors contribute to our understanding of natural language?
2. What function relates all these factors to comprehension?
3. Are these factors specific to language, or do they cut across cognitive domains?
4. Which consequences of these factors can be consciously modified (e.g., stopped or gated), and which are not under conscious control?

Although these remain at present long- range objectives of ERP language research, there is nonetheless a slowly accumulating body of evidence that has begun to address issues in each of these areas. At the same time, more than 10 years of research have established beyond a doubt that ERPs are effective tools for the investigation of factors and processes involved in language comprehension. A host of ERP measures, including amplitude, latency, and distribution, have been shown to vary, often systematically, with factors relevant to understanding. Most of these studies have taken advan-

Cognitive Electrophysiology
H-J. Heinze, T.F. Münte, and G.R. Mangun, editors
© 1994 Birkhäuser Boston

tage of the fact that ERPs are particularly sensitive indices of violations of expectancy.

Kutas and Hillyard (1980) first observed that, following a violation of a semantic expectation at the end of a sentence, there was a negative component in the averaged waveform that was larger over posterior regions than over anterior regions and larger over the right hemisphere than over the left hemisphere. This monophasic negative wave onset about 200 msec and was between 300 and 400 msec in duration; it was called the N400 to reflect the fact that it was negative going and peaked around 400 msec post stimulus (Fig. 1). Since then, a number of studies have shown the reliability and robustness of this effect in a variety of paradigms; some of these are reviewed in this chapter (for reviews, see Fischler, 1990; Halgren, 1990; Kutas and Van Petten, 1988).

Violations are thus known to be good vehicles for eliciting ERP effects. These can in turn be modulated by other factors in a way that makes it possible to ask questions about various processing models of language. However, the fact that this is the case does not imply that language ERP components are specific or unique reflections of violation. For example, the

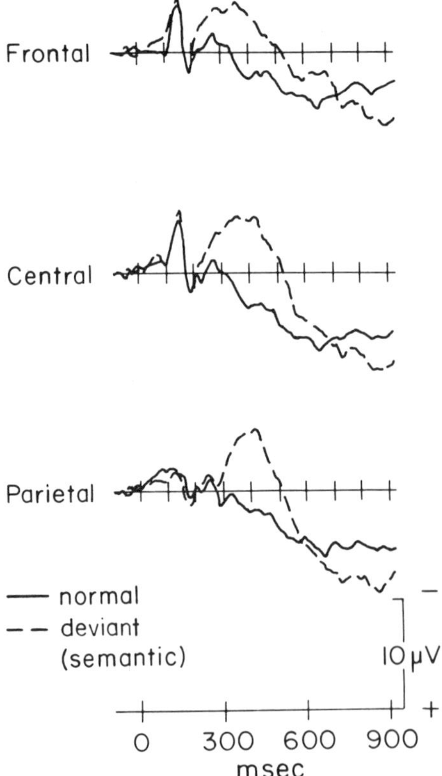

Figure 1.   Grand average ERPs elicited by semantically congruous and incongruous sentence-terminal words from frontal, central, and parietal midline locations. Negative is up in this and all subsequent figures. (Data from Kutas and Hillyard, 1980.)

N400 was originally viewed as a semantic violation detector. That perception has persisted to the present day, even though several studies have since demonstrated that while the N400 is largest in amplitude following semantic violations, violations are neither necessary nor sufficient to yield an N400 component. In this chapter, we highlight those studies that have demonstrated the sensitivity of the N400 to factors other than semantic violation.

Additionally, we address the question of just what constitutes a violation in the first place. Although the utility of violations in helping to decide between models of language processing has been amply demonstrated, this in no way implies or guarantees that researchers know exactly what is "expected" and therefore what is "violated." Nor can we overlook the fact that even the description of the domain of a violation is theory dependent. We would therefore like to raise the possibility that violations are epiphenomenal, that is, reducible to more fundamental properties of language or even of cognition in general.

We will do this in the following way. First, we review the more general biological and cognitive factors known to influence N400 amplitude, latency, and distribution. Then we consider some recent studies that have investigated the effects of violations of various sorts on the ERP record of sentence processing. Finally, we review studies indicating that the violations in question are perhaps best viewed as emerging from the interaction of a number of different variables, some of which may not be language specific.

The Influence of Biological and Cognitive Factors on the N400

There is ample evidence indicating that the N400 and the cognitive function which underlies its elicitation are modality independent. N400 effects have been observed in the visual modality in a variety of languages, including written English (Kutas, Van Petten, and Besson, 1988). Dutch (Gunter, Jackson, and Mulder, 1992). French (Besson and Macar, 1987; Ardal et al., 1990), Spanish (Kutas et al., in manuscript), and Japanese (Koyama et al., 1991), for visually presented signs of American Sign Language in congenitally deaf adults (Kutas, Neville, and Holcomb, 1987; Neville, 1985), for line drawings terminating sentences presented visually (Kutas and Van Petten, 1990; Nigam, Hoffman, and Simons, 1990), and in the auditory modality for both natural and metered speech (Bentin, Kutas, and Hillyard, in press; Connolly, Stewart, and Phillips, 1990; Herning, Jones, and Hunt, 1987; Holcomb and Neville, 1990; McCallum, Farmer, and Pocock, 1984; O'Halloran et al., 1988). In all these situations, the relative difference between the ERP to a word or picture that fits with its context and the ERP to one which does not is a monophasic negativity.

Modality independence, however, does not imply an insensitivity to the eliciting modality. Indeed, the onset latency, peak latency, duration, and laterality of both the raw N400 elicited by one condition as well as the N400 effect itself (i.e., the difference between two conditions) seem to vary systematically as a function of modality. For example, the N400 effect in comprehension of speech generally onsets earlier and lasts longer than that observed during reading. Insofar as asymmetries have been obtained, N400s in the visual and auditory modalities seem to exhibit reversed lateralities; that is, the effect is larger over the right hemisphere than over the left hemisphere during reading, but reversed during speech processing (Holcomb and Neville, 1990).

Age and family history of handedness are other biologically determined factors that influence the N400. For the same stimulus sets, young adults generate larger and more peaked N400 waves than do older adults (Gunter, Jackson, and Mulder, 1992; Hamberger and Friedman, 1989; Harbin, Marsh, and Harvey, 1984); there appears to be a linear reduction in N400 amplitude with age (Kutas, Mitchiner, and Iragui, 1992). Although there have been no studies of the N400 specifically and systematically devoted to determining the relationship between its *laterality* and handedness, some patterns have emerged. During reading, the N400 seems to be larger over the right hemisphere so long as the subjects do not have a left-handed relative in their immediate family. The N400s of subjects with left-handed family members are markedly less asymmetric (Kutas, Van Petten, and Besson, 1988; see Fig. 2). As mentioned, this right predominance of the visual N400 appears to be reversed in the auditory modality. Whether the laterality of the auditory N400 is also sensitive to family history of left-handedness is unknown.

N400 amplitude is also influenced by a number of cognitive manipulations. Within an experimental session, N400 amplitudes are larger for the first occurrence of a word or picture relative to the second occurrence (for a review see Van Petten et al., 1991). This reduction in N400 amplitude with repetition interacts with frequency and congruity. N400 repetition effects are more pronounced for low-frequency than high-frequency words (Rugg, 1990; Van Petten et al., 1991) and for incongruous than congruous sentence-terminal words (Besson, Kutas, and Van Petten, in press). Both the semantic priming and repetition priming effects on the N400 are further subject to attentional manipulations. Priming effects are either reduced or absent for words occurring outside the focus of attention (Holcomb, 1986; Nobre and McCarthy, 1987; Otten, Rugg, and Doyle, 1992). On average, N400 amplitudes are also smaller for the less fluent language of a bilingual (Ardal et al., 1990; Kutas et al., in manuscript).

Although most of the studies utilizing the N400 effect have focused on modulations of its amplitude, some data concerning its timing have also accumulated. Specifically, the onset and peak latencies—and on occasion the duration—of the N400 have similarly proven to be susceptible to biological and cognitive influences. Both onset and peak latencies are delayed with increasing age (Gunter, Jackson, and Mulder, 1992; Kutas, Mitchiner, and Iragui, 1992). The rate of presentation of words in sentences also affects N400 latency. Semantically anomalous words completing sentences pre-

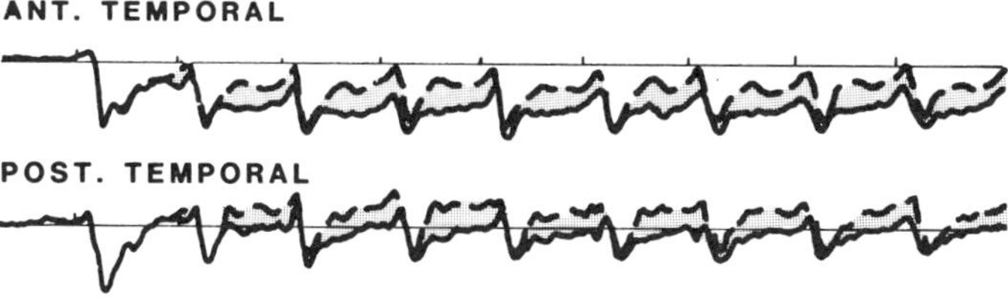

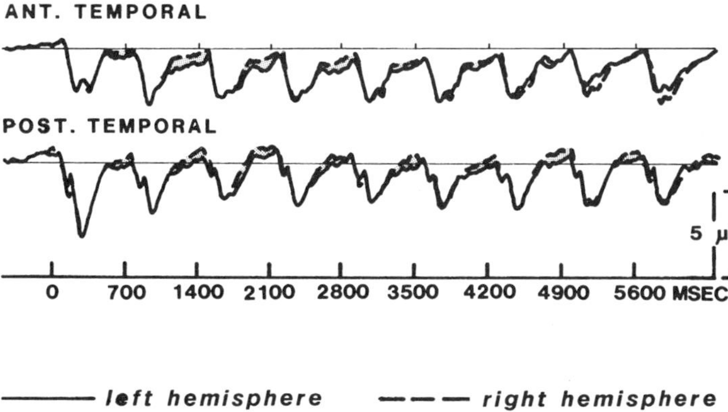

Figure 2.   Grand average ERPs for first seven words of sentences that subjects read for meaning; averaged separately for nine right-handed subjects without and five right-handed subjects with left-handed relatives. Right and left hemisphere sites are overlapped. Note absence of asymmetry in 300- to 600-msec range for right-handers with left-handed family members. Reprinted from Kutas, Van Petten, and Besson, 1988, by permission of Elsevier Publishers).

sented one word at a time (also known as rapid serial visual presentation or "RSVP") elicit N400s with a peak latency about 400 msec when the presentation rate is one per 250 msec or slower. At a presentation rate of one word every 100 msec, both the onset and peak of the N400 effect are delayed by between 80 and 100 msec (Kutas, 1987). We have also found a relationship between the timing of the N400 effect and a bilingual's fluency in his/her two languages; the N400 in the less fluent language peaks later and lasts longer than that in the more fluent language (Kutas et al., in manuscript; see also Ardal et al., 1990). The size of the set of semantically related words from which a word is chosen appears to influence N400 latency as well. In semantic verification tasks, the N400s elicited by words that do not match a specified definition both start and peak earlier if the definition mentions a polar opposite rather than a category with a number of other members. A possibly related effect is that involving the ordinal position of the eliciting word in a sentence. The N400 elicited by a semantically anomalous word in the middle of sentences, where context does not fully constrain the occurrence of an ensuing word, peaks approximately 30 msec later than that elicited by an anomalous word at sentence end, where contextual constraint is greater (Kutas and Hillyard, 1983).

ERP Studies of the Effects of Violation on Sentence Processing

Let us now examine some ERP consequences of various violations within the domain of language, Kutas and Hillyard (1983) investigated the ERP consequences of morphological violations embedded within written text. Specifically, subjects were asked to read a number of different passages one word at a time in order to answer a series of multiple choice questions. In addition to various semantic violations, the text included sentences with morphological violations of the following kinds:

(1)   (a) ... as a turtle grows its shell *grow* too ...
       (b) ... all turtles have four *leg* and a tail ...
       (c) Turtles will spit out things they *does* not like to eat.
       (d) ... they tend to *been* as cool as ... .

These morphological violations did not yield an N400 with the same amplitude and distribution as that elicited by the semantic violations, but there was nonetheless an enhanced negativity in the 300- to 500-msec range relative to control words, as well as a hint of an enhanced late positivity that carried over into the next word. A replication (Kutas et al., in manuscript) with another group of monolingual subjects showed that both the enhanced negativity and positivity were real effects and that the different morphological violations were associated with different ERP changes. Bilingual Spanish speakers reading similar materials in Spanish showed no negativity but an even more pronounced positivity than monolingual subjects in response to verb tense violations (Fig. 3). Osterhout (1990) examined the ERPs associated with a number of different syntactic violations. In one of his paradigms, ERPs were recorded while subjects read sentences containing "subcategorization violations"; after each sentence, subjects were asked to make a decision as to its well-formedness (a so-called grammaticality judgment).

On the assumption that words in the lexicon are assigned to various categories (e.g., nouns, verbs, adjectives, adverbs, etc.), verbs as a class can be further subdivided into subcategories on the basis of the complements that they take. For example, some verbs take direct object complements while others do not. Thus the verb *persuade* requires a direct object but cannot take a prepositional complement (e.g., *John persuaded me.* vs. **John*

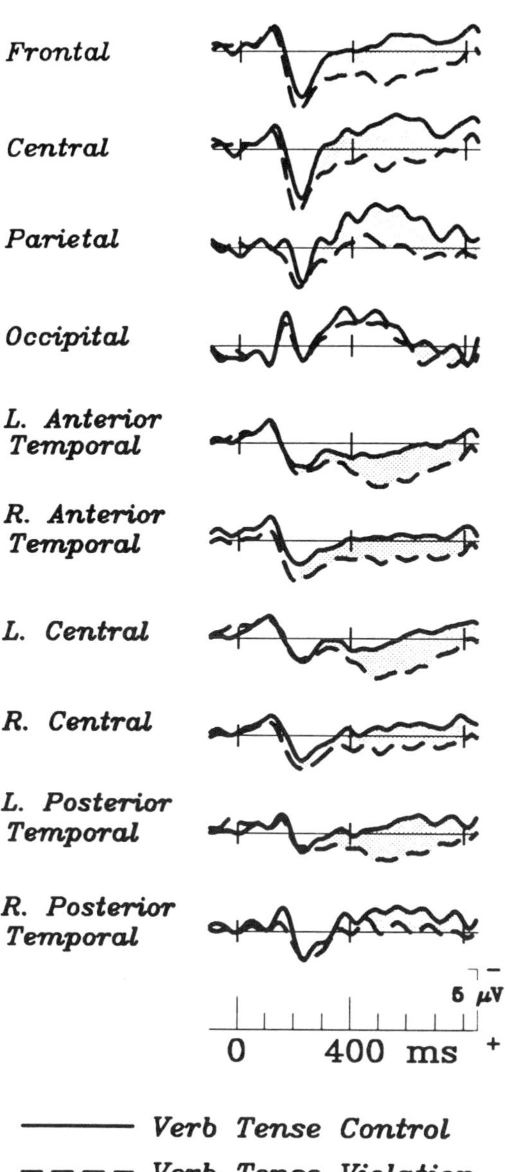

Figure 3. Grand average ERPs ($n = 15$) from Spanish/ English bilinguals reading simple Spanish text for comprehension. Compared are ERPs elicited by verb tense violations and control words matched on word position, length, and lexical class. Spanish was primary language of these bilinguals who were first exposed to English after age 8. Note enhanced late positivity elicited by verb tense violations.

persuaded to me.), *while the verb hope* never takes a direct object but can take a prepositional complement (e.g., **They hoped a solution.* vs. *They hoped for a solution.*). More generally, it can be said that *persuade* belongs to that subcategory of verbs which take a direct object but not a prepositional complement, while *hope* belongs to the subcategory of verbs that do not take a direct object but do take a prepositional complement. Another way of saying this is that *persuade* "subcategorizes" for a direct object while *hope* "subcategorizes" for a prepositional complement. Hence, a verb subcategorization violation refers to the state of affairs that pertains when an expectancy set up by the verb for a certain constituent (such as a noun-phrase direct object or a prepositional phrase) is not fulfilled, or in other words, is violated.

None of the sentences that Osterhout used in his experiment contained direct objects, but half the sentences contained verbs that require a direct object and thus constituted subcategorization violations, while the other half contained verbs that have no such requirement and were thus well formed. These sentences were of the following types.

(2)   (a) The broker hoped *to* sell the *stock.*
       (b) The broker persuaded *to* the sell the *stock.*

Although both *hope* and *persuade* also take infinitival complements (i.e., to *sell the stock*), *persuade* still requires an immediately following direct object (e.g., *The broker persuaded* **me** *to sell the stock.*) while *hope* does not (e.g., **The broker hoped* **me** *to sell the stock.*). Osterhout and Holcomb recorded ERPs to the preposition *to*, which can under no circumstances introduce a direct object noun phrase and is thus unambiguously wrong when a direct object is required, as in the case of *persuade*. Thus, the ERP to the *to* following *persuaded* could be expected to show some sign of the subcategorization violation as soon as subjects became aware that the required direct object was missing, relative to the ERP elicited by the *to* following *hoped*, which requires no direct object.

And, indeed, the data in Figure 4 reveal that the ERP to the subcategorization violation in the *persuade* condition is more positive from about 400 msec on than the corresponding ERP in the control *hope* condition; Osterhout and Holcomb labeled this response the P600. Notice that the ERP elicited by the final word of the *persuade* condition containing the subcategorization violations sports an N400 relative to the control *hope* condition that contains no syntactic violations. It remains an open question whether this N400 reflects the inability of subjects to find a suitable syntactic structure to assign to such sentences, an inability to make sense of them semantically, or both. Also worth noting is the similarity between the P600 observed by Osterhout and the enhanced late positivity observed by Kutas et al. (in manuscript) following morphological violations. More recording sites and greater scrutiny will be required to ascertain whether or not these two positivities are in fact equivalent.

Neville et al. (1991) reported on the ERPs associated with violations of certain syntactic constraints posited by extended standard theory (Chomsky, 1973, 1977, 1981, 1986a, 1986b). In one condition, Neville et al. included phrase structure violations, which for present purposes can best be viewed as violations of normal word order. Subjects were asked to make grammaticality

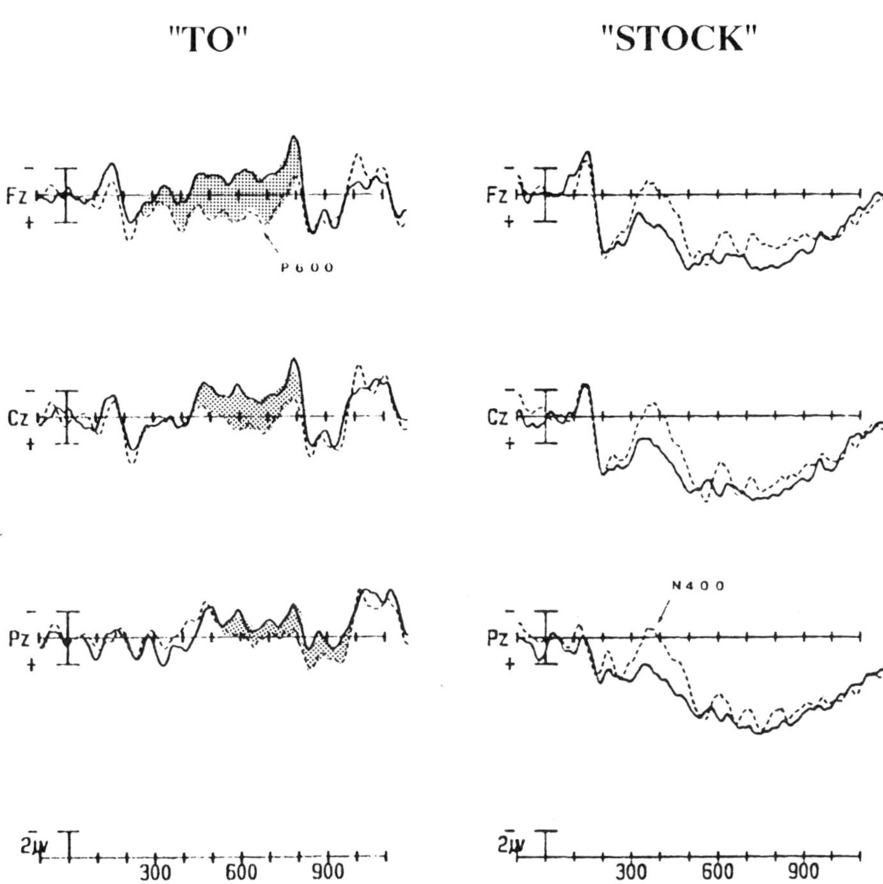

Figure 4. Grand average ERPs ($n = 15$) to subcategorization violations and control words (*to*) and to final words of sentences containing such violations and their control words (*stock*). Note late positivity (P600) elicited by subcategorization violations and negativity (N400) elicited by sentence terminal words in sentences containing subcategorization violations. (Data from Osterhout, 1990.)

judgments on sentences such as

(3)   (a) The scientist criticized a proof *of* the theorem.
      *(b) The scientist criticized Max's *of* proof the theorem.

The ERPs in response to the preposition *of* were compared across the two

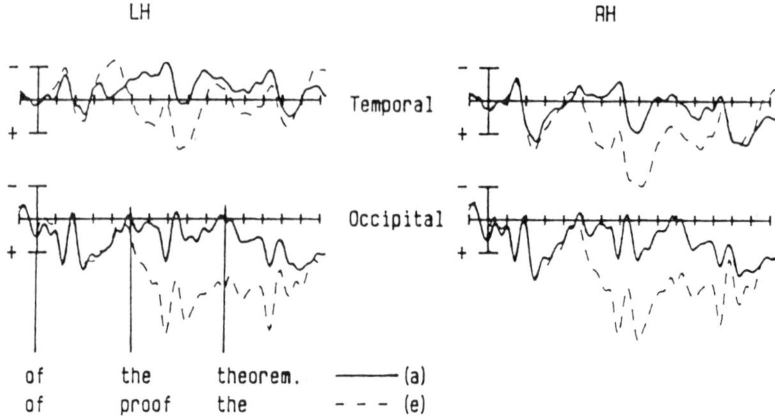

Figure 5.   Grand average ERPs ($n = 40$) elicited by phrase structure violations and two subsequent words (*dashed line*) versus those elicited by control words (*solid line*). Recordings are from temporal and occipital regions of left and right hemispheres. (Adapted from Neville et al., 1991.)

conditions. The preposition violates the expectancy for a noun or adjective to follow *Max's* in 3b, but is acceptable following the noun *proof* in 3a. The violation in 3b was associated with a number of ERP effects relative to 3a, including a pronounced negativity between 300 and 500 msec over temporal regions of the left hemisphere and a more broadly and bilaterally distributed late positivity (see Fig. 5). Neville et al. also investigated violations of a constraint referred to as the "specified subject condition" (Chomsky, 1973, 1977, 1981, 1986a, 1986b; Huang, 1982). Without going into the details and history of this constraint within syntactic theory, we can simply say that it is considered responsible for the ungrammaticality of 4b relative to a grammatical sentence such as 4a.

(4)   (a)   What did the scientist criticize a *proof* of ____?
      *(b)   What did the scientist criticize Max's *proof* of ____?

The ERPs in Figure 6 show that the response to the word *proof* following *Max's* in 4b is associated with an enhanced negativity in the region of the

Figure 6.   Grand average ERPs ($n = 40$) elicited by specified subject condition violations (*dashed line*) and control words (*solid line*) as subjects made grammaticality judgments. Words were presented once every 500 msec. (Data from Neville et al., 1991.)

Wh-movement (Specified subject constraint)

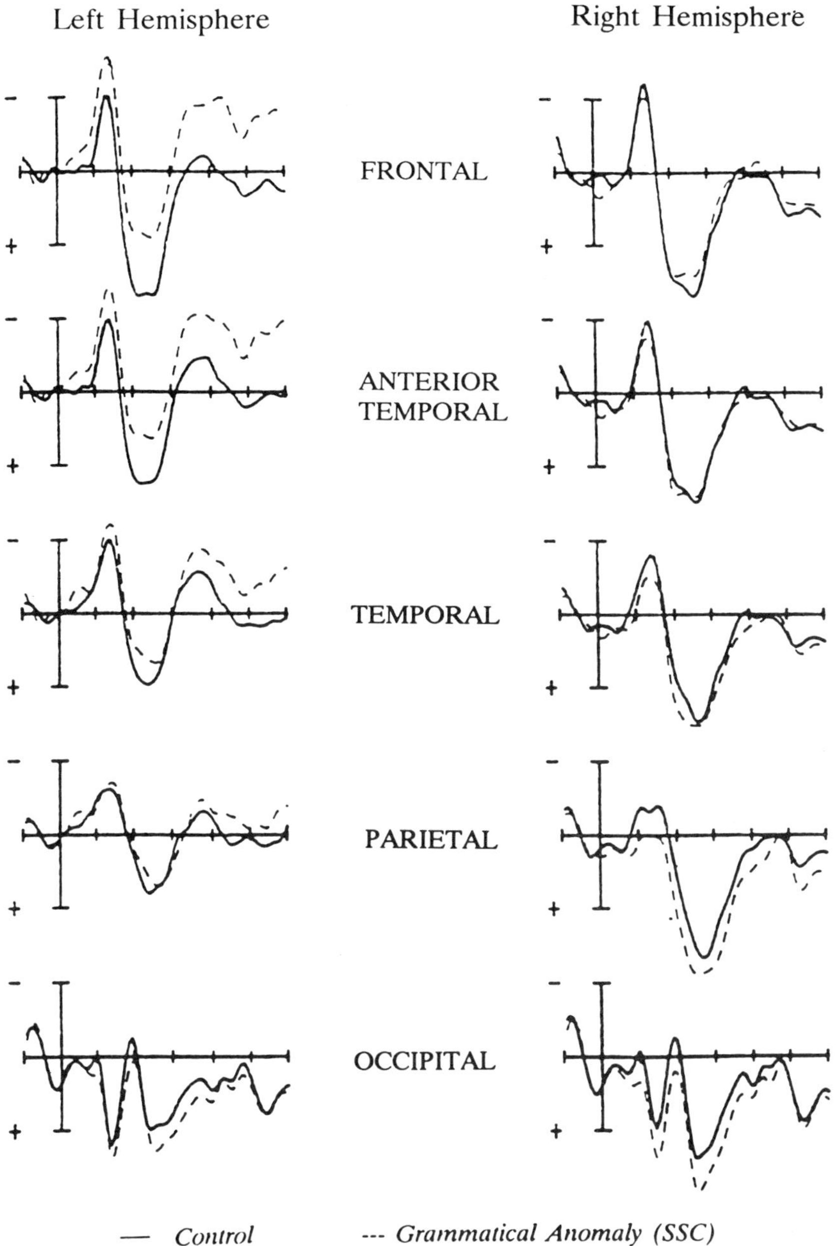

The scientist criticized Max's <u>proof</u> of the theorem.
What did the scientist criticize Max's <u>proof</u> of?

N1 component (about 100 msec) as well as between 300 and 500 msec post stimulus, primarily over anterior regions of the left hemisphere. Although the earlier effect occurs perhaps too early to be an enhancement of the N1 per se, and may instead be a remnant of differential processing of the previous word, Neville et al. take both ERP effects to be specific indices of the violation of the specified subject condition. They looked at other violations as well, but these two are sufficient to show the diversity of ERP componentry elicited by various "syntactic" violations.

So far we have simply described the various ERP components apparently associated with violations of expectancies or constraints within different language domains. Within the domain of syntax, the variety of ERP responses that have been observed suggests that there is no unique index of syntactic violation. Although in the next section we show that the N400 is not a specific response to semantic anomaly, we first demonstrate how the fact that an N400 is reliably elicited by semantic anomaly has been used to yield information about syntactic processing.

An example of this can be seen in the work of Garnsey and colleagues (Garnsey, Tanenhaus, and Chapman, 1989). Garnsey exploited the fact that N400s reliably occur in response to semantically incongruous words to test alternative views of how people parse sentences, that is, how they figure out the syntactic relations between words in a sentence. In English, determining the grammatical functions (e.g., subject, direct object, indirect object, etc.) of individual words in a sentence is usually fairly simple, as this information is explicitly encoded in word order. However, for certain constructions word order cannot be used to assign grammatical functions. For instance, in *wh*-questions, where the questioned element occurs at the beginning of a sentence (e.g., *Which customer did the secretary call?*), it is impossible to know what the grammatical function of the questioned element (*Which customer*) is until later in the sentence. One cannot know, immediately on reading or hearing *Which customer*, whether it serves as the subject, object of the verb, or object of a preposition, etc. To understand this sentence the reader must keep this questioned element in working memory until its grammatical function becomes clearly specified.

Wh-questions represent one type of "filler-gap" construction. The questioned element or phrase (*Which customer*) is known as the "filler." The empty position in the sentence where the filler would occur in an "echo" question (*The secretary called **which customer**?*) is called the "gap," often indicated by a blank line (*Which customer did the secretary call _____?*). To understand a sentence containing a filler-gap construction, a hypothetical parser must locate the gap and assign the filler to it. This procedure is occasionally difficult or costly in terms of the allocation of mental resources because there are no cues about gap location until some time after the filler is presented, and because sometimes there is more than one possible gap location. For these reasons, there exists a temporary ambiguity as to the grammatical function of the filler while the sentence is being processed.

Depending on one's theoretical bent, one can choose among several different procedures that a parser might follow to deal with the ambiguities of filler-gap constructions (Fodor, 1978). At one extreme is the view that an attempt is made to assign the filler to the first possible gap location; this approach is called the "first resort" strategy. At the other extreme is the "last resort" strategy, which requires the parser to leave the filler unassigned until there is unambiguous information about where the true gap is, and only then assign the filler to the gap. For the sentence *Which customer did the secretary call [____] about ____?*, the first "potential" gap location occurs after *call* in square brackets, whereas the actual gap occurs after the word *about*. These first and last resort strategy views of how a *wh*-question is parsed make very different predictions as to what mental processes occur immediately following the word *call*.

Garnsey, Tanenhaus, and Chapman (1989) devised an electrophysiological test of these two alternative positions by constructing sentences with embedded *wh* questions wherein the filler was either plausible or implausible:

(5)　(a)　The businessman knew which customer the secretary
　　　　　　called ____at home.
　　　(b)　The businessman knew which article the secretary
　　　　　　called ____at home.

As a control, Garnsey also included simple declarative sentences which either did or did not contain a semantically anomalous word, thereby providing a baseline against which N400 effects in the embedded *wh*-question sentences could be compared.

(6)　(a)　The businessman knew whether the secretary
　　　　　　called the *customer* at home.
　　　(b)　The businessman knew whether the secretary
　　　　　　called the *article* at home.

The beauty of this design is that while the filler *which article* in 5b is anomalous and will therefore elicit an enhanced N400 component according to both resort parsing strategies, the two make very different predictions as to exactly where in the sentence the N400 should appear. If the first resort strategy is in effect, then *which article* violates a semantic expectancy after *called* and should elicit an N400 at that point. On the other hand, if the parser is using a last resort strategy, then the filler *which article* should not elicit an N400 until the sentence ends and it is clear tht there is no gap to which that filler can meaningfully be assigned.

Because the ERP elicited by *called* does indeed contain an N400 when preceded by an implausible filler (*which article*) but not by a plausible one (*which customer*) (Fig. 7), the results support a first resort strategy and argue against the last resort strategy outlined previously. They also argue against the existence of an autonomous syntactic processing module: in such a model the parser would make the filler-gap assignment at the first possible gap,

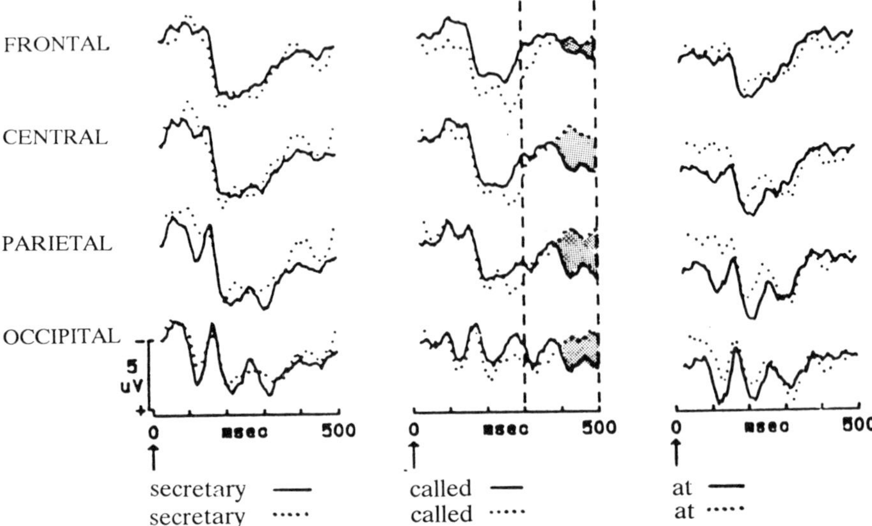

Figure 7.   Grand average ERPs ($n = 32$) for plausible (*solid line*) and implausible (*dashed line*) filler-gap sentences. At top of figure are sentence contexts preceding words below to which ERPs were recorded. *Vertical dashed lines* indicate 200-msec range across which N400 amplitude was quantified. Shading indicates plausible–implausible difference. (Modified from Garnsey, Tanenhaus, and Chapman, 1989.)

but would not evaluate its semantic plausibility until some later stage of processing when all potential gap sites had been eliminated from consideration.

It should be emphasized that the N400 is not a direct reflection of gap-filling, but rather a reflection of the incongruity that is either a consequence of making an implausible filler-gap assignment or at least of evaluating the possibility of that assignment. There are many potential sources of information that could help the parser in making filler-gap assignment, and Garnsey's present research is aimed at determining whether the parser makes use of these different sources of information, and if so, when.

Are Linguistic Violations Reducible to Other Factors?

Numerous experiments attest to the fact that while semantic anomalies may elicit the largest amplitude N400s, the response is not unique or specific to the presence of a semantic anomaly. As early as 1984, Kutas and colleagues (Kutas and Hillyard, 1984; Kutas, Lindamood, and Hillyard, 1984) showed in a series of studies that semantically congruous words of low cloze

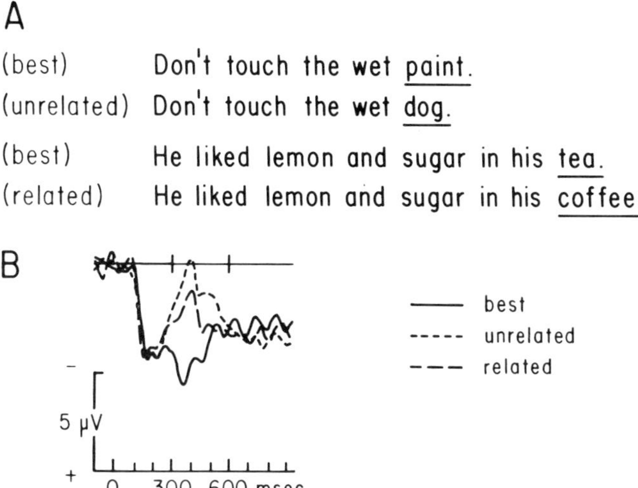

Figure 8.   A. Examples of sentences used in this study. Each pair of sentences exhibits high contextual constraints, but whereas the first sentence in each pair is terminated by its "best completion" (as determined by a sentence completion norming study), the second sentence in each pair is completed by a low close probability word. In first pair, low cloze probability ending is semantically unrelated to best completion; in second pair, it is semantically related. **B.** Grand average ERPs from Pz for best completions (*solid line*), semantically related (*large dashed line*), and semantically unrelated (*small dashed line*), low-cloze probability words. (Reprinted from Kutas and Hillyard, 1984.)

probability elicit a larger N400 than semantically congruous words of high cloze probability. This finding indicated that semantic anomaly is not a necessary condition for the elicitation of an N400; instead, N400 amplitude varies as an inverse function of word expectancy in a given context. In addition, Kutas and her colleagues showed that N400 amplitude is attenuated in response both to semantically anomalous words and to low cloze probability words when these are semantically related to high cloze probability words "best completions" (Fig. 8). This finding suggested that N400 amplitude is modulated by the degree to which a word is semantically primed by prior context.

More recent studies have shown that semantic anomaly is not even a sufficient condition for the elicitation of an N400 response. For example, Besson, Kutas, and Van Petten (1992) have shown that by the third presentation of a sentence containing a semantic anomaly, the N400 effect is eliminated (Fig. 9). Although a particular word may render a sentence anomalous in isolation, the mere act of repetition increases the expectancy for that particular word to occur in the same context, thereby reducing N400

REPETITION EFFECTS

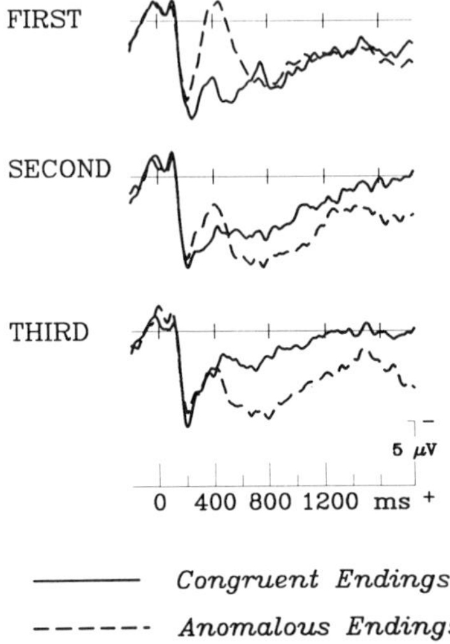

Figure 9.   Grand average ERPs ($n = 17$) elicited by semantically congruous and incongruous sentence endings after three different presentations separated by more than 15 min. Note reduction of N400 amplitude with repetition. (Data from Besson, Kutas, and Van Petten, 1992.)

amplitude. Clearly it is the variation in expectancy and not the violation per se that determines whether an N400 will be elicited by an anomalous word.

N400s are in fact elicited by all words to varying degrees as determined by a number of factors; more general biological and cognitive factors were outlined in the second section of this chapter. We now turn to factors more closely tied to sentence processing. Among the most robust of these observations concerning N400 amplitude are these:

1.   N400s are larger in response to open-class words than to closed-class words (Besson, Kutas, and Van Petten, 1992; Kutas and Hillyard, 1983; Kutas, Van Petten, and Besson, 1988; Van Petten and Kutas, 1991a, 1991b).
2.   N400s are larger in response to open-class words of low frequency than to open-class words of high frequency when such words occur in word lists or at the beginning of sentences. This effect of frequency on the N400 interacts with semantic constraints (Van Petten and Kutas, 1991a, 1991b), semantic congruity (Besson, Kutas, and Van Petten, in press), and word repetition (Besson, Kutas, and Van Petten, in press; Rugg, 1990; Van Petten et al., 1991).
3.   N400 amplitudes are not only reduced by the prior occurrence of a semantic associate or semantically related word, but increasingly dampened by the accumulation of semantic constraints such as those that build up during the course of a sentence (Van Petten, 1989; Van Petten and Kutas, 1991a, 1991b). The latter effect is manifest in a relative reduction in the amplitude of N400s elicited by open-class words as a function of word

position in isolated but meaningful sentences. In two studies (Van Petten, 1989; Van Petten and Kutas, 1991a, 1991b), it has also been shown that this effect of contextual constraint on N400 amplitude interacts with the eliciting word's frequency of occurrence in the language such that the accumulation of prior context overrides the initial difference in N400 amplitude between high and low frequency words.

It was with some of these observations in hand that Van Petten used the N400 to focus on interactions between lexical and sentence-level factors in an attempt to resolve the controversy that revolves around the time course of sentential context effects. In a series of experiments, Van Petten and Kutas (1987, 1991a, 1991b) tested the timing predictions of two different views of language processing. One view assumes that each word has a predefined lexical representation that is first accessed and only subsequently subject to higher level (e.g., sentential or discourse) constraints (Fodor, 1983; Forster, 1981; Garrett, 1978); the other view is based on the assumption that a reader's primary goal is comprehension, and that the brain substrates for language are organized so that this goal of comprehension pervades and impacts other levels of analysis at a number of different time points, both early and late (Marslen-Wilson, 1987; Marslen-Wilson and Tyler, 1980, 1987; Tyler and Marslen-Wilson, 1977; Tyler and Wessels, 1983). On the whole, the results of the Van Petten and Kutas studies cannot be accommodated easily by word processing models that rely strictly on static representations of words in a mental lexicon slowly built over years of use, and have led to our belief that comprehension is a product of converging constraints that operate in parallel at multiple levels of linguistic analysis.

The main conclusions that can be drawn from these experiments include:

1. the process that yields frequency effects for words presented in isolation is not mandatory or immune to sentence-level context;
2. the influence of sentence-level context can be as powerful and act as early as that of a single lexically associated word; and
3. sentence context can be used to pick out the appropriate core meaning of an ambiguous word without first passing through an early stage of indiscriminate semantic activation.

Let us examine the logic of one of these experiments. The materials for this experiment consisted of four types of sentences, each containing a critical pair of words. The same associated pairs of words (such as *salt–pepper*) were embedded in both congruent and syntactically legal but semantically anomalous sentences ("Congruent Associated" and "Anomalous Associated," respectively). Likewise, pairs of words that were not particularly related to one another outside of a sentence context were embedded in both congruent and anomalous sentences ("Congruent Unassociated" and "Anomalous Unassociated"). With this design, the second word of a pair could thus benefit from both sentential and lexical context as in the Congruent Associated condition, lexical context alone as in the Anomalous Associated condition, sentential context alone as in the Congruent Associated condition, and neither as in the Anomalous Unassociated condition.

The ERPs elicited by the critical word pairs are shown in Figure 10. Based on what we know about the sensitivity of the N400 to lexical associative effects and to contextual effects, all except the Anomalous Unassociated condition should show an N400 effect. The critical question, however, concerns the timing of the N400 effect. Does the lexical N400 effect start earlier than the sentential N400 effect as the autonomous lexical view would predict, or is there some other pattern of timing relations? The answer is that there is no indication that the lexical effect precedes the sentential

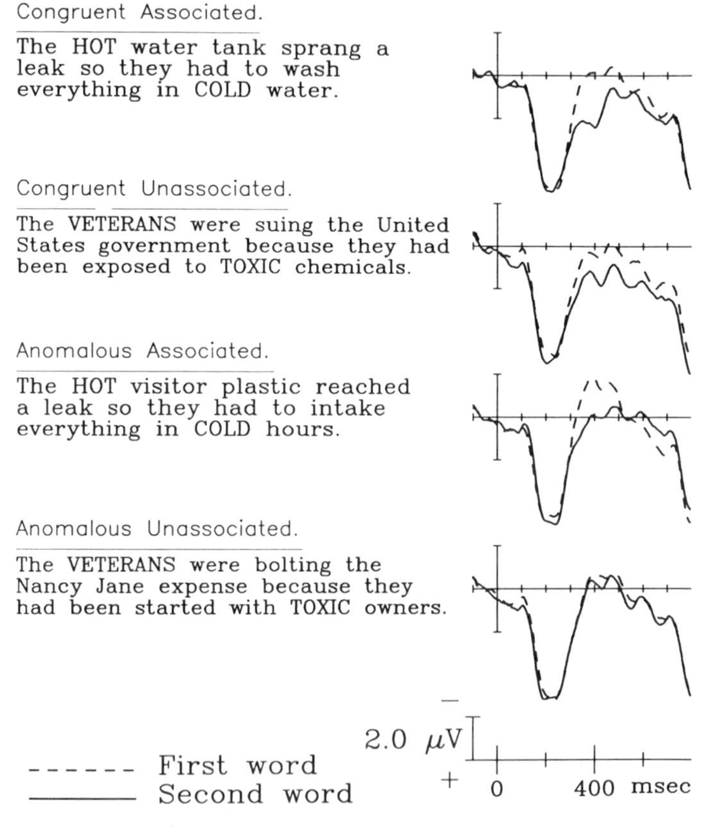

Figure 10.   Grand average ERPs to critical pairs of words (shown capitalized) in each of four sentence types used to contrast sentential and lexical contexts. (Reprinted from Van Petten and Kutas, 1991.)

effect, for the N400 effects appear to start at the same time. Thus we find no evidence for the view that there is a distinct stage in the processing of a word that is influenced by associative links but blind to sentence-level context.

These ERP data indicate that lexical and sentential effects are remarkably similar. Indeed, there seems to be no reason to assume that the mechanisms that underlie the two are completely orthogonal. The sensitivity of the N400 to lexical factors such as word class membership and frequency as well as to sentential influences underscores its independence from semantic violations.

This raises the question of whether the ERP effects seen in response to so-called syntactic violations in the preceding section are actually specific responses to violations. If they are, then of course it is necessary to figure out exactly what constraints are being violated, and to show that a particular violation will always elicit a specific ERP pattern. But consider the possibility, based on what we have learned about the N400 and its relation to semantic violations, that these ERPs are not specific to syntactic violations per se but rather reflect the interactions of various lexical, semantic, and pragmatic factors with the surface structural properties of language. On this view, syntactic violations merely represent one end of a continuum of such interactions. At the opposite end of the continuum one would find interactions of these properties that produce completely grammatical sentences; in between these two extremes would exist an entire range of sentences whose grammaticality would be determined by the confluence of these factors in various combinations.

An example of this perspective on the relationship between ERPs and grammaticality is provided by the work of Kluender (Kluender, 1991; Kluender and Kutas, 1993). Specifically, Kluender focused on ERPs elicited by words in yes/no and *wh* questions, making direct comparisons between sentences of both types that differed in the closed-class item that introduced an embedded clause (e.g., *that*, *if*, or *who/what*).

(7)   (a) Can't you remember *that* he advised them against it on previous occasions?
(b) Can't you remember *if* he advised them against it on previous occasions?
(c) Can't you remember *who* he advised ____ against it on previous occasions?
(d) What did you remember *that* he advised them against ____on previous occasions?
(e) What can't you remember *if* he advised them against ____on previous occasions?
(f) What can't you remember *who* he advised ____ against ____ on previous occasions?

This work was based on the notion that the membership of a word in the

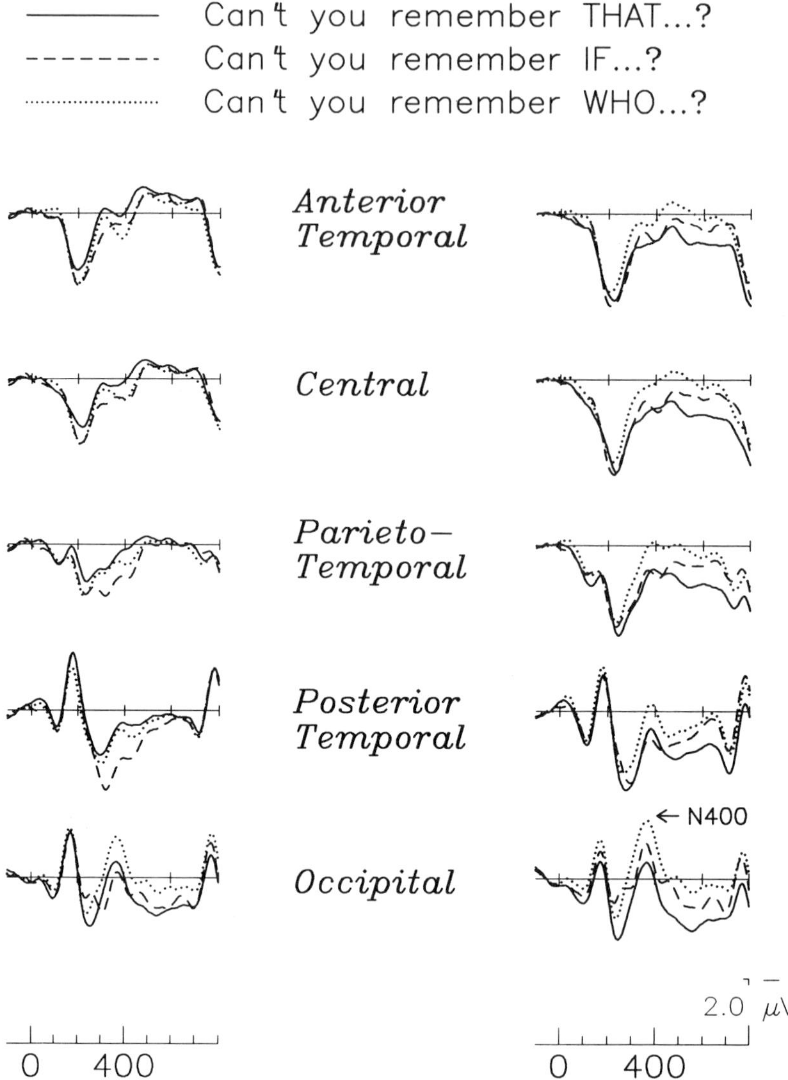

Figure 11.   Grand average ERPs ($n = 30$) elicited by *that-* (*solid line*) and *if*-complementizers (*dashed line*) and by the *wh*-fillers *who* and *what* (*dotted line*) in initial position of embedded clauses in *yes/no* questions. Note typical distribution of monotonic N400 (also in 500–700 msec range): greater over right hemisphere than over left, and greater over posterior regions than over anterior regions. Words were presented every 600 msec; subjects were given probe word task at end of each sentence. (Data from Kluender, 1991.)

open or closed class is highly correlated with its frequency of occurrence in the language (Gentner, 1981), and with the extent to which it refers to entities or relationships in the real word, that is, its referentiality (Kluender, 1992). Kluender predicted that closed-class items would elicit N400s of different amplitude to the extent that they varied along these dimensions. And indeed, as seen in Figure 11, which compares the yes/no-question conditions 7a through 7c, *who*, the least frequent and most referential of these closed-class words, elicits the largest N400; *if*, which is intermediate on both dimensions, elicits an N400 intermediate in amplitude; and *that*, the most frequent and most semantically neutral item, elicits the smallest N400.

In addition to these strictly lexical ERP effects, Kluender and Kutas (1993) observed an ERP effect that reflected the structural properties of sentences containing filler-gap constructions. Specifically, any condition containing a filler-gap relationship elicits a larger negativity over left-anterior regions when compared to conditions in which this filler-gap relationship is absent. This can be seen in comparisons made at the subject *he* in the embedded clause of all three yes/no questions (conditions 7a, 7b, and 7c): when *he* is preceded by the embedded filler *who* as in condition 7c, a left-anterior negativity is seen (Fig. 12). Note that all three sentences are perfectly grammatical, that is, no syntactic violation is involved in this comparison.

Figure 13 shows similar effects seen at three different positions within the embedded clauses of *wh* questions, conditions 7d, 7e, and 7f. Although all three conditions contain main clause *wh* fillers associated with prepositional object gaps in the embedded clause, condition 7f contains an additional filler-gap relationship in the embedded clause that does not exist in the other two conditions. Left-anterior negativity is seen at the embedded subject *he* in condition 7f, the *wh* question with the embedded filler *who*, relative to conditions 7d and 7e, which contain embedded *that* and *if* clauses. Note that although the eliciting condition 7f is in this case ungrammatical, the response does not differ in latency or distribution from that seen at the same position in the grammatical yes/no question 7c in Figure 12. Further, for some speakers condition 7e is also considered ungrammatical, but here no left-anterior negativity emerges in the ERP.

The second position where this effect is seen is the embedded preposition *against*; in condition 7f, which contains the embedded filler-gap relationship, this preposition is preceded by a direct object gap, while in conditions 7d and 7e the preposition is preceded by a lexical noun phrase, in this example the pronoun *them*. Once again, condition 7f shows increased left-anterior negativity relative to the other two conditions, this time in response to gap location. Although the eliciting condition is again the ungrammatical one, a similar effect is seen in response to the embedded preposition in condition 7c, that is, in grammatical yes/no questions with embedded direct object gaps.

The third position is the preposition *on* in the embedded clause. In all three conditions, 7d, 7e, and 7f, this preposition is preceded by a prepositional

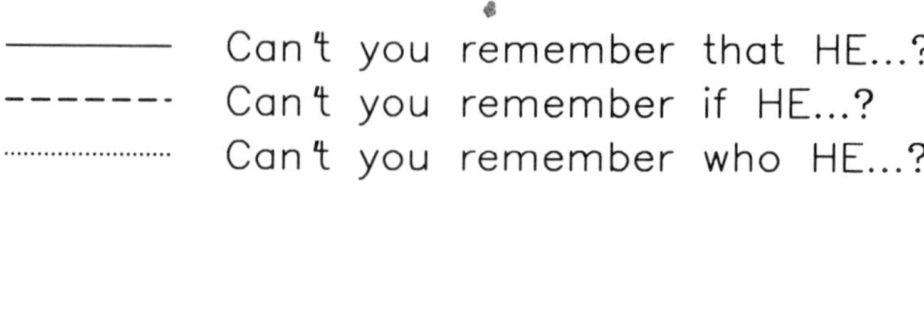

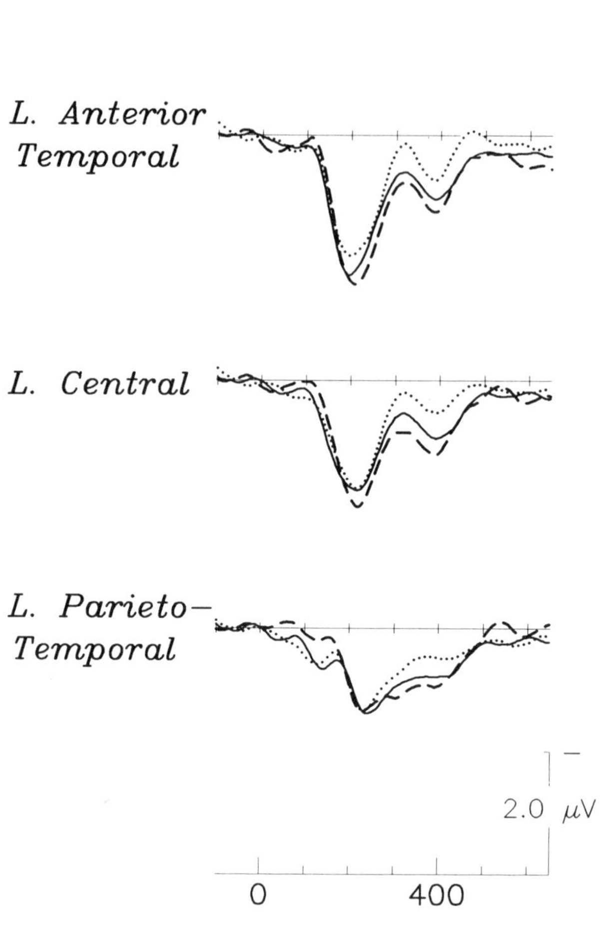

Figure 12.   Grand average ERPs ($n = 30$) recorded at three anterior left hemisphere sites to function words (*capitalized*) immediately following embedded complementizers and *wh* fillers in *yes/no* questions. Note enhanced negativity between 300 and 500 msec associated with existence of filler-gap relationship in condition containing embedded *wh* question. (Data from Kluender, 1991.)

object gap associated with the main clause filler *What*. When all three conditions are thus equivalent in the presence of an immediately preceding gap, lexical effects reflecting the occurrence of *who*, *if*, or *that* earlier in the sentence modulate the amplitude of the negativity indexing the filler-gap

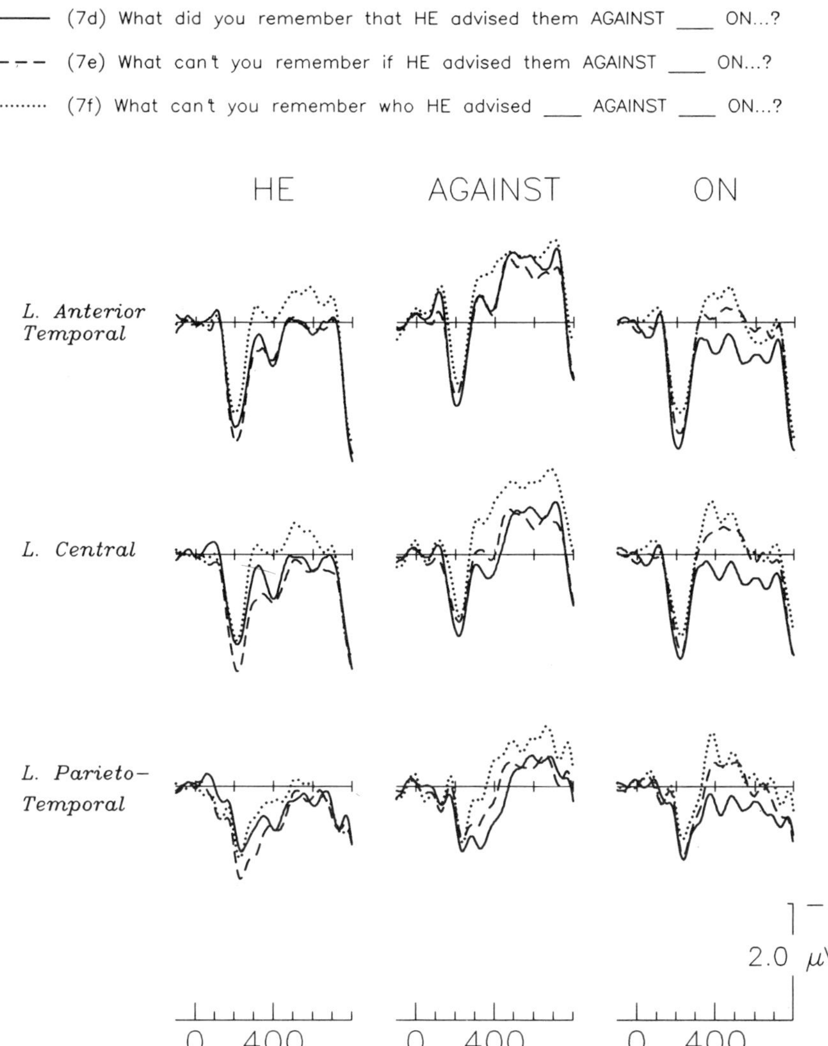

Figure 13.   Grand average ERPs ($n = 30$) from three anterior left-hemisphere sites elicited by three different positions in *wh* questions: function word immediately following embedded complementizers and *wh* fillers (*HE*), preposition immediately following embedded direct object position (*AGAINST*), and initial function word of sentence-ending adverbial immediately following prepositional object gap in all three conditions (*ON*). Note that at latter position, left-anterior negativity is modulated in much the same way as monotonic N400 seen in Figure 11. (Data from Kluender, 1991.)

relationship (the possibility that the effect is caused by the difference in the main clause auxiliaries *did* vs. *can't* is relatively remote; see Kluender, 1991, for discussion). The amplitude of the left-anterior response is larger in conditions 7e and 7f, which contain *who* and *if* embeddings, respectively, than in condition 7d, which contains a *that* embedding. The important points to note are (1) that this negativity has persisted throughout the extent of the filler-gap domain, that is, until the filler is assigned and the dependency released, and (2) that this structurally influenced sentential effect also responds to lexical influences. Lexically induced modulations of the response to sentence-final words (*occasions*) were also observed.

Traditionally, the differences in the grammaticality of these sentences have been explained in terms of violations of syntactic constraints. However, it may be possible to explain them in terms of the processing requirements of a filler-gap construction and the demands of processing intervening lexical items. For example, these processing difficulties may reflect strain on the limited capacity of a working memory that must store a filler along with the intermediate products of comprehension.

Conclusion

We hope to have convincingly laid to rest the misconception that the N400 is exclusively associated with semantic anomaly; we reiterate for emphasis that semantic anomaly is neither a necessary nor a sufficient condition for the elicitation of an N400. Our reading of the currently available research is that sentential processes can and do influence lexical processing, and conversely, that lexical processes influence sentential processing. In other words, we do not see ERP evidence for a strict modularist position in these domains. In addition, we have suggested that the canonical construal of linguistic violation may be fundamentally flawed. Just as semantic anomalies have proved to be derivable from an interaction of lexical properties (such as frequency of occurrence and lexical class membership) with contextual constraints at the sentence and discourse levels, so too it may prove possible to derive syntactic violations in the traditional sense from the interaction of the sequential structure of language with lexical and semantic properties such as frequency of occurrence and referentiality. If this is the case, then the next logical question is whether and to what extent such interactions can be found in other cognitive domains as well. Working memory has frequently been invoked as a factor influencing language processing (Carpenter and Just, 1992; King and Just, 1991, in manuscript). Thus, if specified subject condition violations can be shown to fall out from the interaction of short-term memory constraints with the need to access mental representations in simultaneous lexical processing, one would want to know if similar effects can be found in other memory tasks that do not involve language, as for example in perceptual monitoring tasks. So, in the long run these data can be used to assess the common notion that language processing is somehow special, unique among cognitive processes, different in the nature of the mental operations involved, and subject to very different processing constraints than for example face recognition or comprehension of visual scenes. We hope that this is an empirical question to which answers can be found.

Acknowledgments. Marta Kutas and some of the research herein were supported by an RSDA from NIMH (MH00322) and grants from NICHD (HD22614) and NIA (AG08313). Robert Kluender was supported by a McDonnell-Pew postdoctoral fellowship in cognitive neuroscience. We would like to thank Claudia Brugman, Heather McIsaac, Ken Paller, and Cyma Van Petten for comments on earlier versions of this chapter.

References

Ardal S, Donald MW, Meuter R, Muldrew S, Luce M (1990): Brain responses to semantic incongruity in bilinguals. *Brain Lang* 39:187–205.

Bentin S, Kutas M, Hillyard SA (1993): Electrophysiological evidence for task effects on semantic priming in auditory word processing. *Psychophysiology* 30(2):161–169.

Besson M, Macar F (1987): An event-related potential analysis of incongruity in music and other non-linguistic contexts. *Psychophysiology* 24:14–25.

Besson M, Kutas M, Van Petten C (1992): An event-related potential (ERP) analysis of semantic congruity and repetition effects in sentences. *J Cognit Neurosci* 4(2):132–149.

Carpenter PA, Just MA (1992): A capacity theory of comprehension: Individual differences in working memory. *Psychol Rev* 99(1):122–149.

Chomsky N (1973): Conditions on transformations. In: *A Festschrift for Morris Halle*, Anderson S, Kiparsky P, eds. New York: Holt, Rinehart and Winston.

Chomsky N (1977): On wh-movement. In: *Formal Syntax*, Culicover P, Akmajian A, Wasow T, eds. New York: Academic Press.

Chomsky N (1981): *Lectures on Government and Binding*. Dordrecht: Foris.

Chomsky N (1986a): *Barriers*. Cambridge: MIT Press.

Chomsky N (1986b): *Knowledge of Language: Its Nature, Origin, and Use*. New York: Praeger.

Connolly JF, Stewart SH, Phillips NA (1990): The effects of processing requirements on neurophysiological responses to spoken sentences. *Brain Lang* 39:302–318.

Fischler I (1990): Comprehending language with event-related potentials. In: *Event-Related Brain Potentials: Issues and Applications*, Rohrbaugh JW, Parasuraman R, Johnson RE, Jr., eds. New York: Oxford University Press.

Fodor JA (1983): *The Modularity of Mind*. Cambridge: MIT Press.

Fodor JD (1978): Parsing strategies and constraints on transformations. *Linguist Inquiry* 9:427–473.

Forster KI (1981): Priming and the effects of sentence and lexical contexts on naming time: Evidence for autonomous lexical processing. *Q J Exp Psychol* 33A:465–495.

Garnsey SM, Tanenhaus MK, Chapman RM (1989): Evoked potentials and the study of sentence comprehension. *J Psycholinguist Res* 18:51–60.

Garrett M (1978): Word and sentence perception. In: *Handbook of Sensory Physiology 8, Perception*, Held R, Leibowitz HW, Teuber HL, eds. Berlin: Springer-Verlag.

Gentner D (1981): Some interesting differences between verbs and nouns. *Cognit Brain Theory* 4:161–178.

Gunter TC, Jackson JL, Mulder G (1992): An electrophysiological study of semantic processing in young and middle aged academics. *Psychophysiology* 29(1):38–54.

Halgren E (1990): Insights from evoked potentials into the neuropsychological mechanisms of reading. In: *Neurobiology of Higher Cognitive Function*, Scheibel AB, Wechsler AF, eds. New York: Guilford Press.

Hamberger MJ, Friedman D (1989): Age-related changes in semantic activation: Evidence from event-related potentials. *Proc EPIC IX, Noordwijk, The Netherlands* III:38–39.

Harbin TJ, Marsh GR, Harvey MT (1984): Differences in the late components of the event-related potential due to age and to semantic and nonsemantic tasks. *Electroencephalogr Clin Neurophysiol* 59:489–496.

Herning RI, Jones RT, Hunt JS (1987): Speech event-related potentials reflect linguistic content and processing level. *Brain Lang* 30:116–129.

Holcomb PJ (1986): ERP correlates of semantic facilitation. In: *Electroencephalography and Clinical Neurophysiology Supplement 38, Cerebral Psychophysiology: Studies in Event-Related Potentials*, McCallum WC, Zappoli R, Denoth F, eds. Amsterdam: Elsevier.

Holcomb PJ, Neville HJ (1990): Auditory and visual semantic priming in lexical decision—a comparison using event-related brain potentials. *Lang Cognit Processes* 5:281–312.

Huang CTJ (1982): *Logical Relations in Chinese and the Theory of Grammar*. Doctoral Dissertation, Massachusetts Institute of Technology, Cambridge.

King J, Just MA (1991): Individual differences in syntactic processing: The role of working memory. *J Mem Lang* 30:580–602.

King J, Just MA: Individual differences in working memory and the use of contextual information during parsing (in manuscript).

Kluender R (1991): *Cognitive Constraints on Variables in Syntax*. Doctoral Dissertation, University of California, San Diego.

Kluender R (1992): Deriving island constraints from principles of predication. In: *Island Constraints: Theory, Acquisition, and Processing*. Goodluck H, Rochemont M, eds. Dordrecht: Kluwer Academic Press (in press).

Kluender R, Kutas M (1993): Bridging the gap—Evidence from ERPs on the processing of unbounded dependencies. *J Cognit Neurosci* 5:196–214.

Koyama S, Nageishi Y, Shimokochi M, Hokama M, Miyazato Y, Miyatani M, Ogura C (1991): The N400 component of event-related potentials in schizophrenic patients: A preliminary study. *Electroencephalogr Clin Neurophysiol* 78:124–132.

Kutas M (1987): Event-related brain potentials (ERPs) elicited during rapid serial visual presentation of congruous and incongruous sentences. In: *Electroencephalography and Clinical Neurophysiology Supplement 40, Current Trends in Event-Related Brain Potential Research*, Johnson R, Jr., Parasuraman R, Rohrbaugh JW, eds. Amsterdam: Elsevier.

Kutas M, Hillyard SA (1980): Reading senseless sentences: Brain potentials reflect semantic incongruity. *Science* 207:203–205.

Kutas M, Hillyard SA (1983): Event-related brain potentials to grammatical errors and semantic anomalies. *Mem Cognit* 11:539–550.

Kutas M, Hillyard SA (1984): Brain potentials during reading reflect word expectancy and semantic association. *Nature* (London): 307:161–163.

Kutas M, Van Petten C (1988): Event-related brain potential studies of language. In: *Advances in Psychophysiology 3*, Ackles PK, Jennings JR, Coles MGH, eds. Greenwich, CT: JAI Press.

Kutas M, Van Petten C (1990): Electrophysiological perspectives on comprehending written language. In: *Electroencephalography and Clinical Neurophysiology Supplement 41, New Trends and Advanced Techniques in Clinical Neurophysiology*, Rossini PM, Mauguiere F, eds. Amsterdam: Elsevier.

Kutas M, Lindamood T, Hillyard SA (1984): Word expectancy and event-related brain potentials during sentence processing. In: *Preparatory States and Processes*, Kornblum S, Requin J, eds. Hillsdale, NJ: Erlbaum.

Kutas M, Mitchiner M, Iragui V (1992) (in manuscript): Effects of aging on the N400 component of the ERP in a semantic categorization task. *Psychophysiology* 29:547.

Kutas M, Neville HJ, Holcomb PJ (1987): A preliminary comparison of the N400 response to semantic anomalies during reading, listening, and signing. In: *Electroencephalography and Clinical Neurophysiology Supplement 39, The London Symposia*, Ellingson RJ, Murray NMF, Halliday AM, eds. Amsterdam: Elsevier.

Kutas M, Van Petten C, Besson M (1988): Event-related potential asymmetries during the reading of sentences. *Electroencephalogr Clin Neurophysiol* 69:218–233.

Kutas M, Bates E, Kluender R, Van Petten C, Clark V, Blesch F: What's critical about the critical period? Effects of early experience in the language processing of bilinguals (in manuscript).

Marslen-Wilson WD (1987): Functional parallelism in spoken word-recognition. *Cognition* 25:71–102.

Marslen-Wilson WD, Tyler LK (1980): The temporal structure of spoken language understanding. *Cognition* 8:1–71.

Marslen-Wilson WD, Tyler LK (1987): Against modularity. In: *Modularity in Knowledge Representation and Natural-Language Understanding*, Garfield JL, ed. Cambridge: MIT Press.

McCallum WC, Farmer SF, Pocock PK (1984): The effects of physical and semantic incongruities on auditory event-related potentials. *Electroencephalogr Clin Neurophysiol* 59:477–488.

Neville HJ (1985): Biological constraints on semantic processing: A comparison of spoken and signed language (abstract). *Psychophysiology* 22:576.

Neville HJ, Nicol JL, Barss A, Forster KI, Garrett MF (1991): Syntactically based sentence processing classes: Evidence from event-related brain potentials. *J Cognit Neurosci* 3:151–165.

Nigam A, Hoffman JE, Simons RF (1992): N400 to semantically anomalous pictures and words. *J Cognit Neurosci* 4:15–22.

Nobre AC, McCarthy G (1987): Visual selective attention to meaningful text. An analysis of event-related potentials. *Soc Neurosci Abstr* 13:852.

O'Halloran JP, Isenhart R, Sandman CA, Larkey LS (1988): Brain responses to semantic anomaly in natural, continuous speech. *Int J Psychophysiol* 6:243–254.

Osterhout L (1990): *Event-Related Brain Potentials Elicited During Sentence Comprehension*. Doctoral Dissertation, Tufts University, Massachusetts.

Otten LJ, Rugg MD, Doyle MC: Modulation of event-related potentials by word repetition: The role of selective attention (in manuscript).

Rugg MD (1990): Event-related brain potentials dissociate repetition effects of high- and low-frequency words. *Mem Cognit* 18:367–379.

Tyler LK, Marslen-Wilson WD (1977): The on-line effects of semantic context on syntactic processing. *J Verb Learn Verb Behav* 16:645–659.

Tyler LK, Wessels J (1983): Quantifying contextual contributions to word recognition processes. *Percept Psychophys* 34:409–420.

Van Petten C (1989): *Context Effects in Word Recognition: Studies with Event-Related Brain Potentials.* Doctoral Dissertation, University of California, San Diego.

Van Petten C, Kutas M (1987): Ambiguous words in context: and event-related potential analysis of the time course of meaning course of meaning activation. *J Mem Lang* 26:188–208.

Van Petten C, Kutas M (1991a): Interactions between sentence context and word frequency in event-related brain potentials. *Mem Cognit* 18:380–393.

Van Petten C, Kutas M (1991b): Influences of semantic and syntactic context on open and closed class words. *Mem Cognit* 19:95–112.

Van Petten C, Kutas M (1991c): Electrophysiological evidence for the flexibility of lexical processing. In: *Word and Sentence*, Simpson G, ed. Amsterdam: North Holland.

Van Petten C, Kutas M, Kluender R, Mitchiner M, McIssac H (1991): Fractionating the word repetition effect with event-related potentials. *J Cognit Neurosci* 3:131–150.

Chapter 9

ERP Negativities During Syntactic Processing of Written Words

T. F. MÜNTE AND H.-J. HEINZE

The observation of a negative component in the event-related potential (ERP) in response to semantic anomalies by Kutas and Hillyard in 1980 represents a hallmark in the use of electrophysiological measures for the investigation of language processing. Subsequent work showed that this negative component can be reliably recorded with an onset latency of about 250 msec and a peak latency of 400 msec, thus leading to the label N400. Most of the research of the past decade has been devoted to the delineation of the factors that influence the N400. This line of experiments was reviewed recently by Kutas and Van Petten (1988), Halgren (1990), and Fischler (1990). It is beyond the scope of this chapter to summarize all the factors that have been investigated in relation to the N400. It should suffice to say that it varies reliably as a function of manipulations on the semantic level. However, the endeavor of language research by means of event-related potentials would be meaningless, if it was centered around components rather than processes. In this respect a cross-fertilization from the current theories of psycholinguistics is needed.

Obviously, lexical access, word recognition, and semantic analysis on the word and sentence level are necessary prerequisites for our ability to understand. In addition to these processes, the listener or reader has to perform a syntactic analysis that assigns the exact roles to the different constituents of a sentence. One of the important issues in the current

Cognitive Electrophysiology
H-J. Heinze, T.F. Münte, and G.R. Mangun, editors
© 1994 Birkhäuser Boston

discussions within the field of psycholinguistics is how the syntactic representation of an utterance is computed by the human subject. The major classes of theories differ in the way they account for the stage on which information of the various sources is brought together. Two general classes of models can be distinguished in this respect: In interactive models it is generally assumed that information from all sources, that is, from lexical, semantic, discourse, and syntactic levels, is communicated in an unconstrained fashion and is used interactively at an early level to compute a plausible reading of a sentence (e.g., Marslen-Wilson, 1975; Marslen-Wilson and Tyler, 1980; Taraban and McClelland, 1988). Some of these models, for example, specify that the initial syntactic analysis is guided by subcategorization information that has to be derived from the lexicon (Ford et al., 1982; Mitchell, 1990; Mitchell and Holmes, 1985). This class of models can be contrasted with theories that view the language processing system as being composed of distinct modules, each devoted to the analysis of a specific kind of information. In such models the initial syntactic analysis is computed solely on the basis of syntactic principles. Only if the initial analysis fails or a syntactic ambiguity has to be resolved is the information from other sources consulted at a later stage of processing (Ferreira and Clifton, 1986; Forster, 1979; Frazier, 1987; Frazier and Fodor, 1978). In spite of rigorous research in psycholinguistics, the issue is far from being settled. Given the inconsistencies of reaction time studies, some researchers have introduced new techniques, such as eye fixation monitoring, to the investigation of syntactic processes (Ferreira and Henderson, 1990; Frazier and Rayner, 1982; Rayner and Frazier, 1987).

In this chapter we are concerned with the use of ERPs in syntactic processing. While these share the "on-line" character with eye fixation studies, they offer the additional advantage of having multiple dimensions in that parallel processes could be distinguished by their topographical, amplitude and latency features. Clearly, to make ERPs bear on these issues it would be desirable to pinpoint certain features of the ERP that were sensitive to the syntactic information of a sentence or a word. However, until now only a limited number of studies have addressed questions of syntactic processing by means of ERPs (summarized by Kutas and Kluender; see Chapter 8, this volume). In essence, a number of ERP effects have been described in response to different kinds of syntactic violation but no consistent pattern emerged during these studies that could serve as a counterpart to the N400 observed in semantic tasks. The review by Kutas and Kluender underscores that the best paradigm to investigate syntactic processes by means of ERPs has yet to be defined. Also, some of the studies conducted so far are amenable to criticism because of a relatively low signal to noise ratio. Further experimentation is therefore needed to delineate possible ERP effects sensitive to syntactic processing.

This chapter summarizes experiments that have been carried out in our laboratory to study syntactic information in the German, English, and

Finnish languages. In the first part, we present data regarding the electrophysiological effects obtained in sentence-reading tasks. The second part of the chapter is concerned with effects observed for syntactic mismatches in word-pair tasks. The motivation for the use of word-pair tasks was to isolate the effects of syntactic incongruencies from those effects that may be related to a recomputation of the sentence in response to the encountered error. We showed that there are certain similarities between the effects that can be obtained to syntactic violations in word-pair experiments and the more natural sentence-reading tasks. The data presented in this chapter in conjunction with the results summarized by Kutas and Kluender (Chapter 8, this volume) should enable us to investigate syntactic processes in more detail in the near future.

Syntactic Incongruencies in Sentence-Reading Tasks

The starting point for our own observations was an experiment carried out by Kutas and Hillyard (1983) in which the subjects were presented with semantic and syntactic incongruencies (verb tense, noun number) in a rapid serial visual presentation paradigm. For the semantically inappropriate words, a typical centroparietal negativity was obtained with a peak latency of 400 msec, thus qualifying as an N400 component. In contrast, the syntactic errors were associated with only a small negativity that showed a somewhat more frontal distribution. The negativity in response to the syntactic errors was followed by a rather broad positivity. There have been a number of replications of this original study using Spanish (Kutas et al., in manuscript) and Dutch (Hagoort, Brown, and Groothusen, in press). In the Spanish experiment, Kutas and coworkers obtained a rather complicated pattern of results in that a group of monolingual speakers of Spanish showed both a low-amplitude negativity and a late positivity for the syntactic incongruencies, while the same verb tense violations led to only a late positivity in a group of bilinguals. Hagoort, Brown, and Groothusen, in press, obtained positive shifts with an onset latency of about 500 msec in response to number disagreements as well as other syntactic violations. In the latter study, however, a low-amplitude negativity preceding the positive shift might have gone undetected because of a rather low signal-to-noise ratio. No semantic errors had been included in the Hagoort experiment. While the previous studies sought for a single ERP response that could serve as a marker for syntactic processing, Neville et al. (1991) applied a different logic envisioning separate processing devices for different types of syntactic information. Therefore, Neville et al. contrasted several different syntactic violations and described a negativity with left frontal maximum and an onset latency of about 300 msec for phrase-structure violations; for other violations, different ERP effects were found.

   Taken together these studies suggested that while a late positive response can be reliably obtained for syntactic errors this effect can in some instances be preceded by a rather low-amplitude negativity with a frontal distribution. Several reasons led us to replicate and extend the earlier sentence reading studies. First, languages differ in the means of communicating syntactic information (Bach, Brown, and Marslen-Wilson, 1986; Bates et al., 1984; Cuetos and Mitchell, 1988). For example, English as the language that has been mostly used in the above-mentioned studies relies on word order as one of the primary sources for syntactic processing, while highly inflected languages (e.g., Serbo-Croatian, Turkish, Finnish) use morphosyntactic features (such as suffixes) to convey syntactic information. Thus it is conceivable that certain languages, in our case German, might be better suited than English to delineate ERP effects to syntactic violations. Another reason for doing another sentence-reading experiment was that none of these studies attempted to equalize the significance of semantic and syntactic errors for the subjects. It could well be argued that the reading of a sentence is much more disrupted by encountering a semantic anomaly than by seeing a morphosyntactic incongruent word.

Sentence Verification

In our first experiment we sought to maximize the chances of obtaining a negativity to syntactic mismatches by changing the original Kutas and Hillyard experiment (1983) in two ways. First, the syntactic errors were always presented at the end of a sentence, as were the semantic incongruent words. Second, the subjects were asked to verify each sentence by pressing one of two buttons. The verification procedure was introduced to equalize task significance for semantic and syntactic errors.

To produce a total of 360 sentences, six versions of each of 60 basic sentence frames were generated by changing the final word. Three of these versions were semantically and syntactically correct. In one of the versions a semantically incongruous final word was used. In addition, a version using a final word that was semantically appropriate but syntactically inappropriate was presented. Only nouns were presented in the final position, and case inflection errors were used for the syntactically inappropriate stimuli. Finally, in the sixth version of each sentence frame a closed-class word instead of the appropriate open-class final word was presented. The results for these latter sentences are not presented here.

EXAMPLES FOR SENTENCES

Correct versions:
> (1)   Der Zollbeamte kontrolliert den Lastwagen.
>   The customs officer controls the truck.
> (2)   Der Zollbeamte kontrolliert den Koffer.
>   The customs officer controls the suitcase.

(3)   Der Zollbeamte kontrolliert den <u>Reisepass</u>.
      The customs officer controls <u>the passport</u>.

Semantically incorrect:
(4)   Der Zollbeamte kontrolliert den <u>Spaß</u>.
      The customs officer controls the <u>joke</u>.

Syntactically incorrect:
(5)   Der Zollbeamte kontrolliert den <u>Koffers</u>.
      The customs officer controls the <u>suitcase's</u>.

Closed-class condition:
(6)   Der Zollbeamte kontrolliert den <u>obwohl</u>.
      The customs officer controls the <u>though</u>.

The sentences were presented one word at a time (duration, 200 msec; stimulus-onset asynchrony, 600 msec) within a dot matrix (15×36 dots) that subtended 1.14 degrees in height and 5.7 degrees in width. The interval between two sentences was 5 sec.

The task of the subjects ($n = 12$) was to judge, as fast as possible, whether or not a given sentence was correct. Whenever they observed an error, semantic or syntactic, they had to press the "no" button, while for the correct sentences a "yes" button had to be pressed. Assignment of buttons to the left and right hand was counterbalanced across subjects. The EEG was recorded with time constants of 10 sec from seven scalp sites against the left mastoid process. The data processing procedures included artifact rejection and digital filtering before the conventional averaging procedures.

Figure 1 shows the grand average of the ERPs to the final words. The typical ensemble of peaks can be observed with an early negativity at about 100 msec followed by a positivity at a latency of 200 msec. The waveforms for both semantically and syntactically incorrect words separate from the waveforms to the correct endings beginning at approximately 250 msec. The semantically incorrect as well as the syntactically incorrect terminal words were associated with more negative ERPs in the 250- to 550-msec range. The differences between the ERPs to the two error classes and the correct final words were of comparable magnitude. The ERPs were quantified by mean amplitude measures within successive time windows relative to baseline. Statistical analyses revealed significant main effects of stimulus class for the area measures in the time windows 200–400 msec, 300–500 msec, and 400–600 msec. The pairwise comparison between the ERPs to semantic and syntactic errors did not show any significant differences.

The results for syntactically inappropriate words in this experiment are clearly different from those obtained previously in similar paradigms (Hagoort, Brown, and Groothusen, in press; Kutas and Hillyard, 1983; Kutas et al., in manuscript). Syntactically incongruent terminal words gave rise to a negativity with about the same magnitude as the semantically terminal incongruent words. The current experiment was different from the other

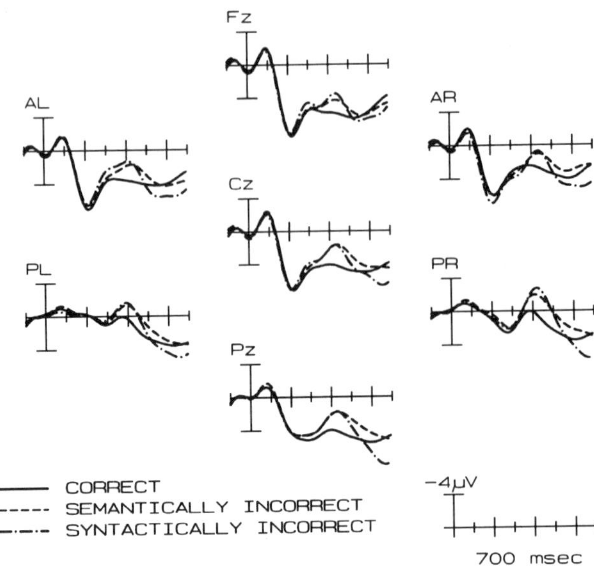

Figure 1.   Grand average ERPs ($n = 12$) to words presented visually at sentence terminal positions. A verification task was given to subjects. ERPs to semantically incongruent words were associated with increased negativity beginning at about 250 msec. A similar negativity was also observed for words that were semantically appropriate but had wrong case-inflection (syntactically incongruent).

studies in several ways. First, the syntactic error was always presented for the sentence terminal word, which made the sentences highly predictable. Second, each of the sentence frames was repeated six times, further adding to the predictability of the sentences. This structure might have led the subjects to fill in the candidate words for the sentence terminal words such that any case of mismatch induced a negative component. Further, it might be argued that the verification task possibly gave rise to negative ERP components that are specific to the task but resulting from the syntactic processing itself. Rösler et al. (in press) presented sentences with semantically and syntactically anomalous words (past participles of verbs) at the sentence terminal position to subjects whose task was a lexical decision for the terminal word. Under these conditions a smaller and more anterior negativity was found for the syntactic errors than for the semantic incongruencies, which might be taken as support that the pattern of results observed in our study is at least partly caused by the verification task.

Syntactic, Semantic and Orthographic Errors Within Stories

To address these questions, for a second experiment it was decided to present syntactic errors embedded within sentences forming short stories. In addition to syntactic errors, semantic and orthographic errors were included to test the specificity of (1) the earlier negativity and (2) the late positive shift in response to the syntactic errors. The subjects' task was to silently read the stories and to answer short questionnaires given during breaks between stories. The subjects were informed that occasionally there would be a "wrong" word, but that these words were irrelevant for answering the questions.

Three short tales with a total of 4187 words served as the basis for the generation of stimuli. From these, 384 nouns with intermediate positions within a sentence were selected and designated the 'critical words.' Four sets of stories (scenarios) were prepared such that at <u>each</u> critical position each of the following conditions was represented <u>once</u>:

(1)   True (96/384 critical words):
         E.g., Die Hexe benutzte ihren <u>Besen</u>, um zum Wald zu fliegen.
      (The witch used her <u>broom</u> to fly to the forest.)
(2)   Semantically incorrect (96/384 critical words):
         e.g., Die Hexe benutzte ihren <u>Löffel</u>, um zum Wald zu fliegen.
      The witch used her <u>spoon</u> to fly to the forest.
(3)   Syntactically incorrect (96/384 critical words):
         e.g., Die Hexe benutzte ihren <u>Besens</u>, um zum Wald zu fliegen.
      (The witch used her <u>broom's</u> to fly to the forest.)
(4)   Orthographically incorrect (but phonologically correct, 96/384 critical words):
         e.g., Die Hexe benutzte ihren <u>Behsen</u>, um zum Wald zu fliegen.
      (The witch used her <u>broome</u> to fly to the forest.)

The syntactically incorrect words were generated by using the semantically appropriate word with a wrong case inflection. In the case of the orthographically incorrect words, application of the German grapheme-to-phoneme conversion rules yielded the pronunciation of the correct word. The semantic violations were obtained by exchanging a particular critical word with another critical word such that the physical stimuli contributing to the "semantic" ERP were identical to those of the control condition. Each subject saw only one scenario, and thus each condition (true, semantically incorrect, syntactically incorrect, orthographically incorrect) was presented 96 times. The mode of presentation and recording conditions were similar to the previous experiment except for the slightly longer (SOA) of 800 msec between words and the use of four additional electrode sites.

In presenting the results, we first focus on the earlier effects in the 200- to 600-msec range, which take the form of negativities for the semantic and syntactic errors. The grand average ($n = 12$) ERPs to the critical words are depicted in Figure 2. The waveforms to the correct words are similar to the ones observed in the previous experiment. With an onset latency of 250 msec, the semantically incorrect words are characterized by a prominent negativity that has a centroparietal maximum. After 550 msec, the waveform to the

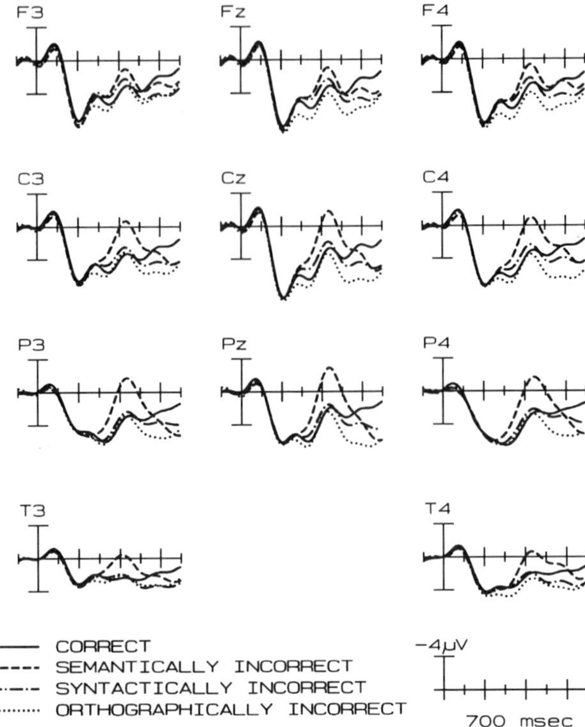

Figure 2.   Grand average ERPs ($n = 12$) to words presented at intermediate positions within sentences forming short stories. Semantically inappropriate words were associated with prominent N400 component. Syntactic errors elicited low-amplitude negativity with frontal maximum followed by positive component. Orthographic errors produced more positive-going waveform beginning at approximately 250 msec.

semantically incorrect word takes a more positive course. As in the previous experiment, the syntactically incorrect words were associated with a more negative ERP than the correct words beginning at about 250 msec. However, this effect was much less pronounced than that seen for the semantically incorrect stimuli and showed a more anterior distribution. The ERPs to the orthographically incorrect words do not show an enhanced negativity but rather are characterized by a more positive waveform beginning at 250 msec.

The difference waves obtained by subtracting the ERPs to the (correct) control sentences from the waveforms to the different error classes are shown in Figure 3. The "semantic" difference wave takes the form of a large amplitude monophasic negativity with a peak latency of 400 msec. In contrast, the negativity in the "syntactic" difference waves had a smaller amplitude and a relative maximum over the anterior part of the scalp. An

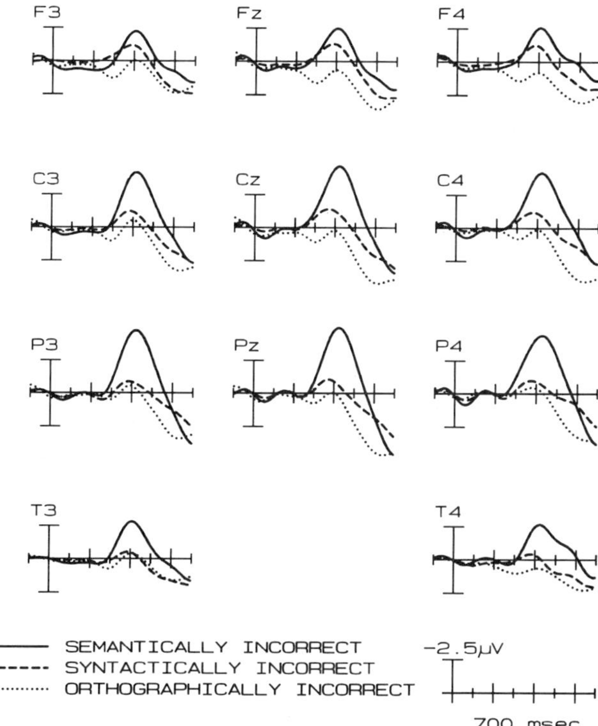

Figure 3.   Difference waves (grand averages) obtained by subtracting ERPs to control words from ERPs to different error classes. By this, the differential brain activity associated with processing of different types of errors is rendered visible. Responses to syntactic errors are characterized by negativity with frontal maximum.

extended positivity with an onset latency of 200 msec was the only effect in the "orthographic" difference wave. To further illustrate the topographical difference between the negativities recorded in response to semantic and syntactic errors, the normalized area measures (time window, 250–450 msec) are plotted as bars at positions roughly corresponding to the scalp sites used (Fig. 4). An in depth account of the statistical results can be found elsewhere (Münte, Schuchardt, and Heinze, in press). Of particular importance for the present question regarding the specificity of ERP effects to syntactic and semantic errors is the fact that analyses of variance on both original and rescaled area measures (according to the suggestion by McCarthy and Wood, 1985) obtained for the difference waves revealed significant condition (semantic vs. syntactic) × topography interactions (time windows, 200–400 msec and 300–500 msec; all, $p < .005$).

Unlike the first experiment, which showed negativities of comparable amplitude and distribution to syntactic and semantic errors, this study revealed clearly different negativities for the two error conditions (Fig. 4). In this respect, the results are similar to those obtained by Kutas and Hillyard

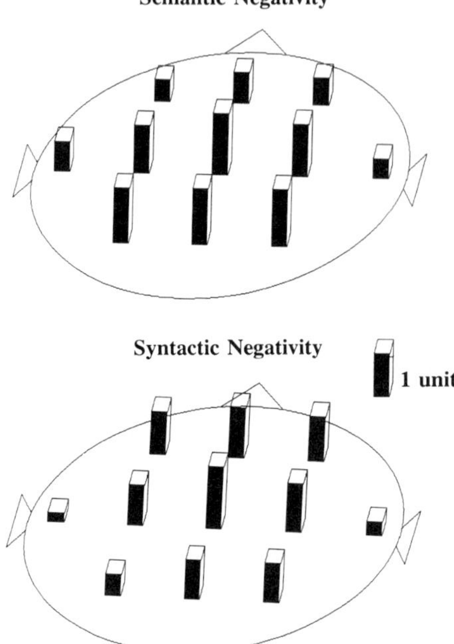

Figure 4. Story-reading experiment. Mean amplitude (time window, 250–450 msec) of difference waves for semantic and syntactic errors is plotted as bar graph at positions roughly corresponding to recording positions. To illustrate relative maxima, normalized amplitudes were used. For semantic errors, a centro-parietal maximum was found corresponding to the distribution usually found for N400; for syntactic errors, an anterior maximum was observed.

(1983), Kutas et al. (in manuscript), and Neville et al. (1991; phrase-structure violation) and Rösler et al. (in press). Differences in relative distribution of ERP effects have served as an argument for the separation and definition of components (Holdstock and Rugg, 1993; Ruchkin et al., 1988). The very least that can be stated about the negativities to the semantic and syntactic errors is that (partially) different neuronal populations must be active. The anterior negativity in response to syntactic errors therefore appears to be the prime candidate for an ERP effect specific to syntactic processes and should be targeted in more detail in future experiments.

We now briefly discuss whether similar arguments can also be made for effects to the different error classes observed later in the recording epoch. Figure 5A shows the ERPs for the Cz derivation encompassing the responses to critical words and the two following words. All three error classes were associated with a long-lasting positive shift. The onset latency of the positive shift differs between the three error classes, most likely resulting from the overlap of the preceding negativities in the semantic and syntactic conditions. In Figure 5B, the difference waves are depicted for the three midline derivations. Statistically, highly significant effects were found comparing area measures from the successive time windows in the 600- to 1200-msec range for the three error classes versus the control sentences. The positive shift showed maxima at Pz for all three difference waves. The difference waves suggest a differential anterior-posterior distribution of the positive shift for the three error classes. Whether or not this difference holds in future studies

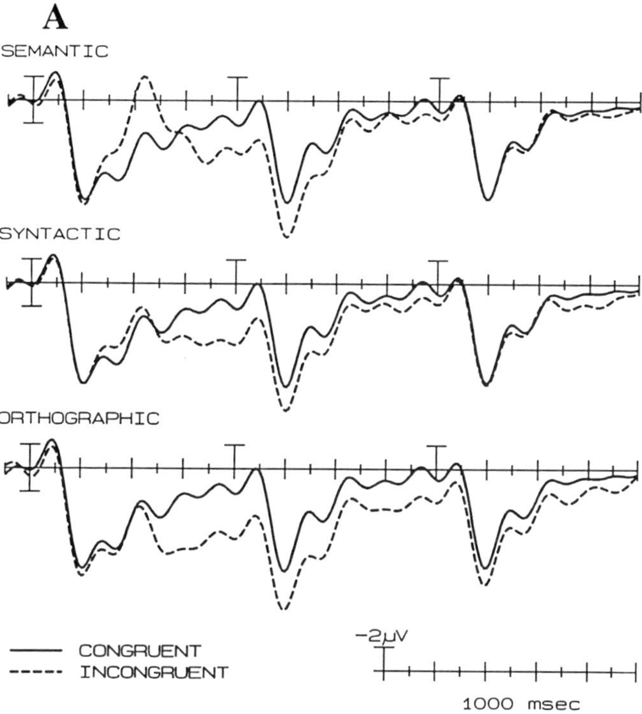

Figure 5.   A. Story-reading experiment. To illustrate later effects associated with different types of errors, ERPs to semantically, syntactically, and orthographically incongruent words and two subsequent words are plotted together with ERPs from control sentences (Cz electrode). Negativities found for semantic and syntactic conditions were both followed by long-lasting positive shift. **B.** Difference waves for midline-electrodes plotted on extended time scale.

remains to be seen. The primary result of this study, however, is that all error classes are associated with a positive shift of similar appearance leading to reservations as to the specificity of this ERP phenomenon.

A tentative interpretation of the positive shifts is that they reflect in some way the efforts of the sentence comprehension system to restore the meaning of the sentence in spite of the violation encountered. In this respect, the questionnaire task used in the present study might have forced the subjects to attempt a reconstruction of the meaning more than in previous studies. This might explain that in the present study positive shifts to semantic incongruencies were found to be quite prominent. A similar interpretation of their late positivity to subjacency and phrase structure violations has been advanced by Neville et al. (1991).

To summarize, we obtained an anteriorly distributed negativity to syntactic violations that contrasted with the N400 to semantic violations. In addition, a late positivity was observed in response to all three types of violations that appears to be nonspecific.

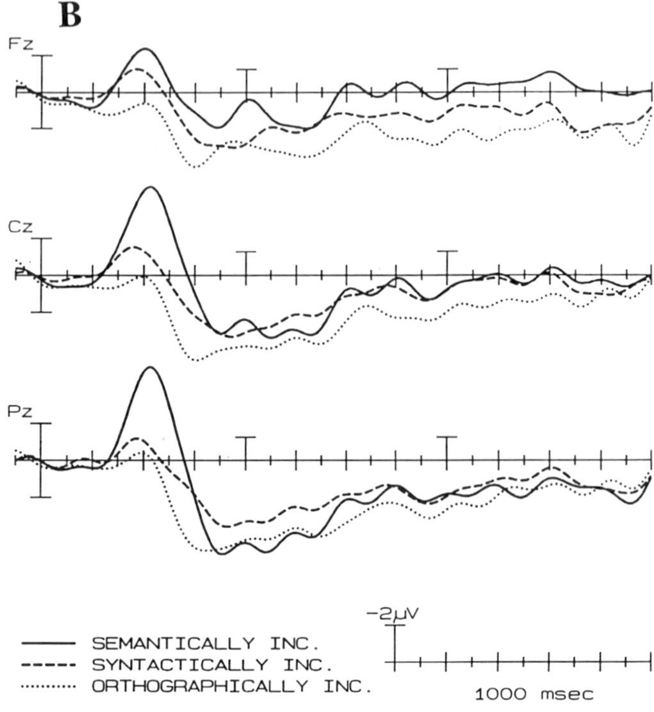

Figure 5.   (*continued*)

Syntactic Violations in Word-Pair Paradigms

As has been illustrated previously in this chapter, one of the problems in using sentence-reading paradigms for the investigation of ERP effects of syntactic processing is the difficulty of deciding whether the observed effects result from the processing of the syntactically incongruous word per se or rather from recomputing efforts by the sentence processing system. At least for the late positive shift seen for the syntactic, semantic, and orthographic errors in the story-reading experiment, the latter explanation might apply. In addition, one is faced with the effects of overlap between negative and positive components in that both were elicited in response to syntactic errors. We therefore sought for a paradigm that more easily would allow a delineation of the electrophysiological effects of syntactic incongruency. One of the most basic processes in the computation of the syntactic representation of an utterance must be the conjunction of single words to phrases. The following experiments address this function by presenting words that may or may not be conjoined to phrases.

The basic design is as follows. In the first position of a word pair, an article or pronoun is presented followed in the second position by either a verb or a noun. These are combined in such a way as to match [e.g., Das–Haus; The (neutral)–House] or to not match [Der–Haus; The (masculine)–House]. The experiments here are similar to a number of behavioral experiments that have addressed the issue of how word recognition is influenced by syntactically appropriate or inappropriate primes using lexical decision and naming tasks (Gurjanov et al., 1985; Katz et al., 1987; Lukatela et al., 1982, 1983; Seidenberg et al., 1982, 1984). Of particular importance in this line of research are those experiments that have used the Serbo-Croatian language (e.g., Gurjanov et al., 1985; Katz et al., 1987; Lukatela et al., 1982, 1983) because in this highly inflected language the case role of a word in a sentence is determined by morphological information contained in a suffix. In essence, these experiments provided evidence for facilitation of processing of the second word of a pair when it had been preceded by a syntactically correct first word in lexical decision but not in naming tasks. This pattern of results was interpreted as being consistent with the view that syntactic information processing is purely postlexical.

At this stage of our research program, the primary question addressed by the word-pair paradigms discussed next, however, was not to decide whether or not syntactic information processing is pre- or postlexical. Rather, the aim of these experiments was to define the ERP effects to syntactic incongruencies in word pairs and to compare them to the effects in sentence-reading tasks.

Word-Pair Paradigm with Syntactic Judgment Task

For the first experiment in this series it was decided to use a syntactic judgment task, that is, the subjects had to judge whether or not the preceding pair of words formed a syntactically possible combination. To produce the stimulus lists, 216 German verbs and 216 German nouns were selected. Care was taken to exclude words that could be used as verbs or nouns (e.g., RENNEN, English RUN) were ambiguous as to the word class. Inflected verbs and nouns were used. The 216 verbs and 216 nouns were combined with articles and pronouns to yield the following six classes of word pairs:

(1) Verb, correct (72 pairs):
     ICH–VERTEIDIGE (I–DEFEND)
(2) Verb, inflection error (72 pairs):
     ER–BEWUNDERE (HE–ADMIRE)
(3) Verb, class error (72 pairs):
     DER–VERTEILE (THE–DISTRIBUTE)
(4) Noun, correct (72 pairs):
     DER–WURM [THE (masculine)–WORM]
(5) Noun, inflection error (72 pairs):
     DER–HAUS [THE (masculine)–HOUSE]
(6) Noun, class error (72 pairs):
     DU–PARLAMENT (YOU–PARLIAMENT)

Each of the critical words (nouns, verbs) was seen just once by each subject. To ensure that all words were presented in each condition equally often, three different scenarios were constructed, each of which was given to five subjects. Before each trial a dot matrix appeared within which the first word of a pair appeared for 200 msec followed by (SOA, 800 msec) the second word, which was again presented for 200 msec. After an additional 1200 msec a delayed response (correct? wrong class? wrong inflection?) was required. The responses were nearly perfect and are not reported here.

The ERPs to the first words did not differ for words that predicted nouns (e.g., *der, mein, die, unser*) and words that predicted verbs (e.g., *ich, du, er*; see Fig. 6B). Clearcut ERP effects related to syntactic incongruity were seen for the words in the second position as exemplified for the nouns in Figure 6A. Both inflection errors and class errors were associated with a more negative-going ERP beginning approximately 300 msec post stimulus. This difference takes the form of a monophasic negativity being maximal over the anterior scalp, as illustrated by the difference waves (Fig. 7). Class and

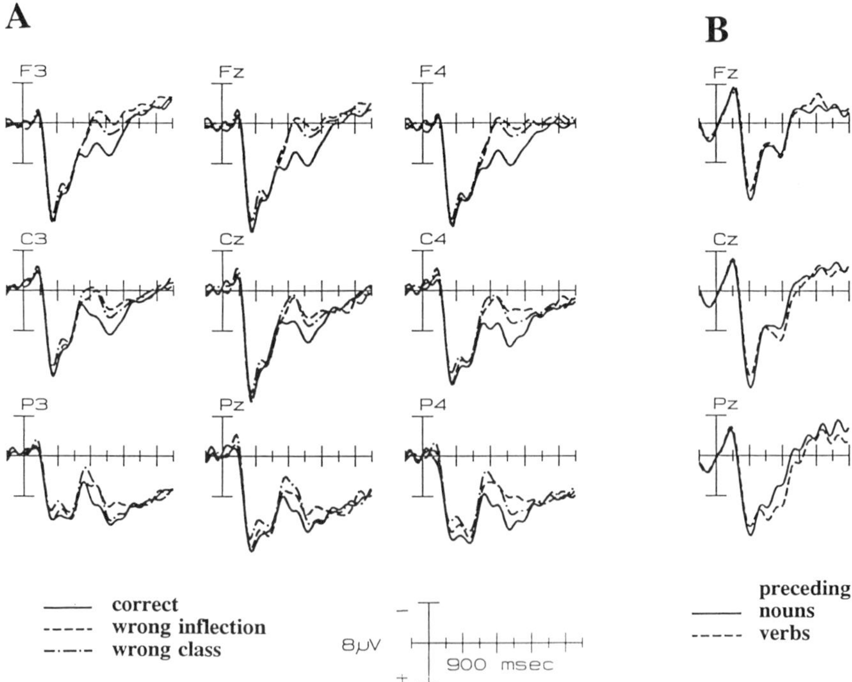

Figure 6A. Grand average ERPs (*n* = 15) for the second words of word-pair in a syntactic judgment task. Words having wrong inflection or wrong class with regard to prime stimulus that had appeared in the first position of a pair were associated with a more negative ERP beginning at approximately 300 msec. **B**. ERPs to first words of word pairs did not show strong differences between primes that predicted verbs and primes that predicted noun.

inflection errors were similarly associated with a negativity. The statistical analyses corroborated the visual inspection in that for successive time windows from 300 msec onward highly significant main effects for (condition: correct vs. wrong inflection vs. wrong class) were found.

The major finding of the current study is thus a monophasic negativity for syntactically incongruous words in the 300- to 600-msec range. While there is a superficial similarity to the N400 component that has been found in semantic tasks, the distribution of the negativity argues against an interpretation of the current effects in terms of an N400. In contrast, there is a remarkable parallel to the distribution of the anterior negativity in the story-reading task described earlier. One question arising in the interpretation of this result is to what extent the negativity is related to the task requirements of making a syntactic judgment. As was illustrated for the sentence-reading experiments, the introduction of a sentence verification task

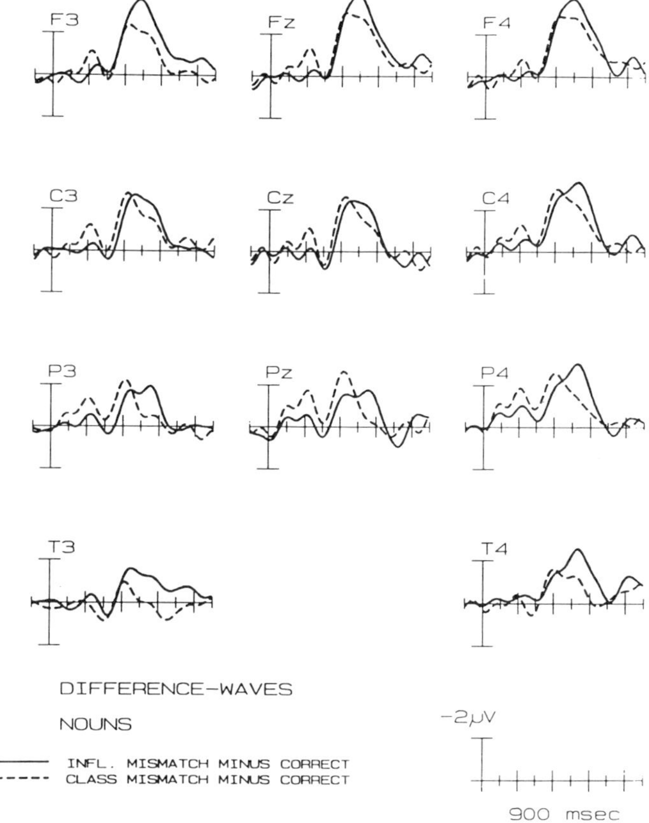

Figure 7.   Difference waves for second words of syntactic judgment experiment. Difference waves were computed by subtracting ERPs to correct words from ERPs to those words that were of wrong inflection or wrong class. Monophasic negativity was observed in both cases with frontal maximum.

might have induced an amplitude increase of a negative component obtained for syntactically incongruous words. Thus, further experiments were conducted to check whether the negativity to syntactic incongruencies could also be obtained in absence of an overt syntactic judgment.

Word-Pair Paradigm with Lexical Decision Task

The stimuli of the previous experiment were used again. In addition to the six word-pair classes of the preceding study (see previous), two additional

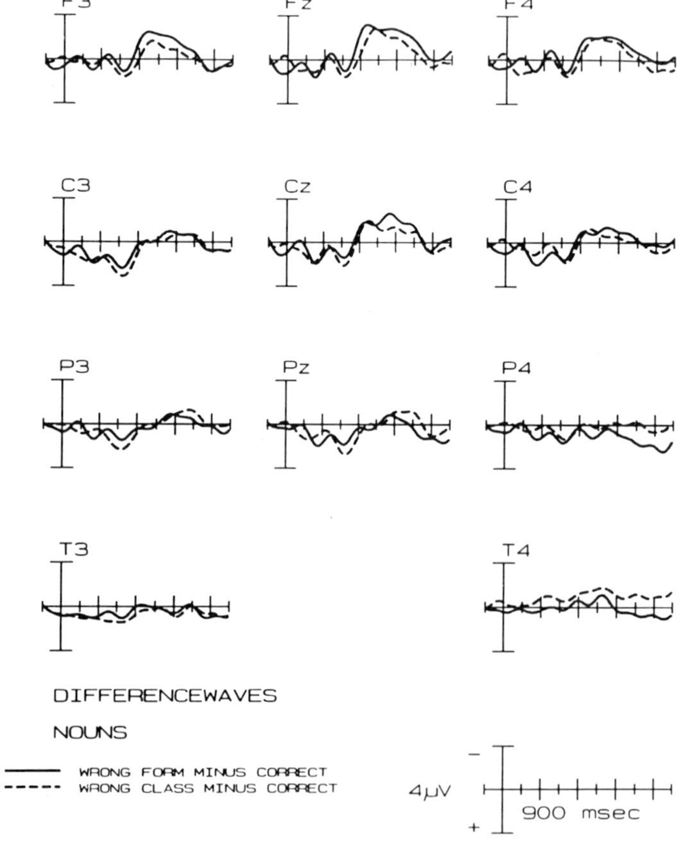

Figure 8.   Difference waves ($n = 2$) for second words of a pair. In this experiment, a modified lexical decision task was used (see text). Again, a monophasic negativity was revealed for words having wrong inflection or wrong case. Negativity was of smaller amplitude and has less widespread distribution, as for syntactic judgment task (see Fig. 7).

types of word pairs were constructed:

(7)   nonword (60 pairs):
    e.g., DU–NARG (YOU–NARG)
(8)   repetition (60 pairs):
    e.g., MEIN–MEIN (MY–MY)

The task of the subjects ($n = 12$) was to press a button held in the right hand whenever a nonword was present in the second position of a word pair, and to press a button held in the left hand when a repetition was encountered. In this way, we ensured that each word had to be read by the subjects. The subjects were not informed that the processing of the syntactic information was the primary purpose of the study. None of the subjects had participated in other syntactic experiments. The nonword and repetition pairs were presented randomly intermixed with the other word-pairs.

The difference waves for this study were computed as for the previous one (Fig. 8). A negativity was again observed for both inflection and class errors. In this task, a clearly more frontal distribution was found. This was corroborated by the statistical analyses that revealed a significant effect for the validity factor only for the frontal electrode pair (area measures: 300–500 msec $p < .04$; 400–600 msec $p < .03$, 500–700 msec $p < .02$) and the midline electrodes (400–600 msec $p < .03$; 500–700 msec $p < .05$).

As the lexical decision task gave rise to an ERP effect related to syntactic incongruency similar to that of the syntactic judgment, overt attention to the syntactic relation is not a necessary prerequisite for the generation of the negativity to syntactic errors. This view is further supported by an additional experiment in which the subjects' task was to decide whether or not a letter presented 1200 msec after the word pairs was contained in one of the words of the pair. In this paradigm, we could also observe a monophasic negativity to syntactic incongruent second words (Münte, 1992).

Comparison of Semantic and Syntactic Judgment Tasks:
Topographical Differences

The previous experiments showed that it is possible to delineate a reliable ERP response to syntactic incongruencies in word-pair paradigms. While the inspection of the waveforms suggested that the obtained negativity is more anteriorly distributed as compared to the N400 typically observed in semantic tasks, a direct comparison of the effects of syntactic and semantic incongruencies would be needed to make a more definite statement. To this end recordings from an extended array of electrodes were made. Moreover, to assess whether the negativity can be found in other languages besides German, we recorded from an English-speaking subject group at Dartmouth Medical School (New Hampshire, U.S.A.) (see Münte, Heinze, and Mangun, 1993 for detailed results).

Two hundred pairs of English verbs that could be used interchangeably at least in some circumstances, and two hundred pairs of synonymous nouns formed the basis of the stimulus lists. No words were included that could be used as either verbs or nouns. For the construction of the syntactic condition personal (I, you, we) and possessive pronouns (your, my, our) were used as well. These stimuli were grouped to yield the following word-pair classes:

(a) semantic judgment task (SJT)
   (1) verb, valid: spend–squander
   (2) verb, invalid: excavate–migrate
   (3) noun, valid: gangster–robber
   (4) noun, invalid: parliament–cube
(b) syntactic judgment task (GJT)
   (5) verb, valid: you–squander
   (6) verb, invalid: your–migrate
   (7) noun, valid: our–robber
   (8) noun, invalid: I–cube

None of the critical words were repeated. Several scenarios were computed to ensure that all critical words were presented in both tasks and in valid and invalid pairs. The timing of the presentation was similar to the previous experiments. The conditions were presented blockwise with the instructions for the SJT being to decide as quickly as possible by pressing one of two buttons whether the second word of a pair could be used interchangeably with the first or not. For the GJT, a syntactic judgment was required; native speakers of English served as subjects. The EEG was recorded with a 32-channel system using a balanced sternovertebral reference.

Figure 9 shows the grand averages of selected midline electrodes for the

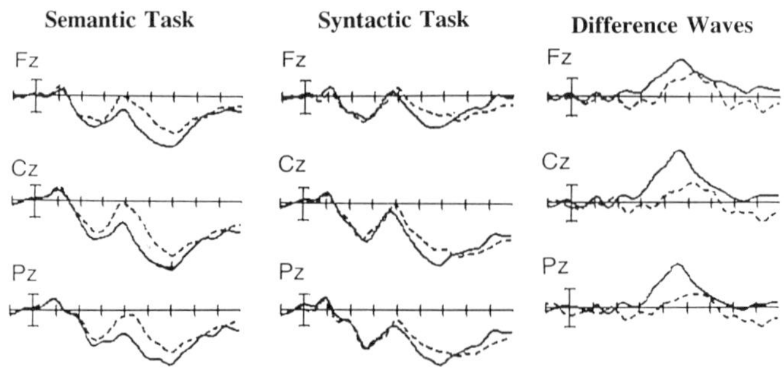

Figure 9.   Comparison of semantic and syntactic judgment tasks in the English language. In left column, ERPs for semantic judgment task ($n = 11$) are depicted for midline electrodes. Semantically inappropriate second words were associated with a more negative waveform beginning at about 250 msec. In middle column, analogous ERPs from syntactic judgment condition are shown. Again, a more negative waveform was found for syntactically inappropriate words. Corresponding difference waves are shown in right column: Syntactic negativity has a more frontal distribution and a later onset latency.

semantic and syntactic judgment tasks as well as the corresponding difference waves. As expected, second words in an invalid relationship gave rise to a more negative waveform as compared to the ERPs to the valid second words. The difference waves illustrate two major features that distinguish between the negativities obtained in the two tasks. First, the relative distribution of the negativity in the syntactic task showed a frontal maximum while for the semantic task a centroparietal maximum was seen. Second, the onset of the syntactic negativity was delayed by approximately 80 msec compared to the semantic negativity. This is better illustrated by isovoltage maps using spline interpolations. The maps are based on area measures for 100 msec time windows centered around the maxima of the difference waves. In the left column of Figure 10 are displayed the results from the English semantic and syntactic tasks. A centroparietal maximum was obtained for the negativity in the semantic task, but the syntactic negativity showed was maximal at frontopolar sites. It should be pointed out that these sites had been included neither in our earlier study on syntactic negativities nor in the studies from other laboratories. Statistically, there was a main effect for validity (valid vs. invalid) in the time windows 300–500 msec, 400–600 msec, and 500–700 msec. The different distribution of the negativities in the two tasks resulted in significant condition × validity × electrode site effects for several scalp sites. For a full account of the statistical results, see Münte, Heinze, and Mangun (1993).

A similar study adding a phonological judgment task was recently conducted in our laboratory (Münte, Kossow, and Heinze, in manuscript) using German stimuli. Only nouns were presented in the second positions in this experiment. For the syntactic tasks these were combined with articles that either did or did not match the nouns for sex (e.g., valid: Das–Haus; the ⟨neutral⟩ house). Thus, the following pairs were obtained for the different tasks:

(a) Semantic task
  (1) valid: Wolkenkratzer–Haus (skyscraper–house)
  (2) invalid: Wurm–Haus (worm–house)
(b) Syntactic task
  (3) valid: Das–Haus (the ⟨neutral⟩–house)
  (4) invalid: Der–Haus (the ⟨masculine⟩–house)
(c) Phonologic task
  (5) valid: Maus–Haus (mouse–house)
  (6) invalid: Kamel–Haus (camel–house)

Different scenarios were again constructed such that each critical stimulus appeared equally often in each of the six conditions but was seen only once by each subject. Twelve native speakers of German participated. A 32-channel recording system was used with the scalp sites referenced to the right mastoid process. The grand average ERPs and the difference waves were similar to the ones seen in the English experiment just described and

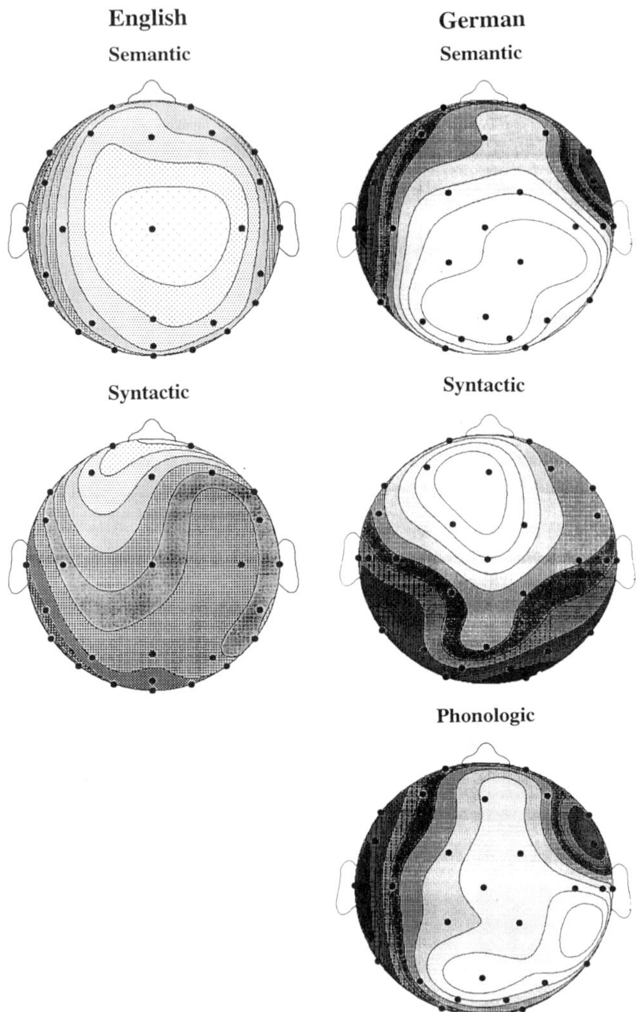

Figure 10.   Topographical maps obtained by spline-interpolation. Left column: Maps computed for difference waves from semantic and syntactic judgment tasks (after data from Münte, Heinze, and Mangun, 1993) use relative scaling to show differences in distribution of negativities in two tasks. Lighter grey values correspond to more negative amplitudes. Mean amplitude measures in 100-msec time windows centered around maximal amplitude of negativities used. Negativity from the semantic task had centroparietal maximum; negativity from the syntactic task had frontopolar maximum. Right column: Maps from similar experiment (after data of Münte, Kossow, and Heinze, in manuscript) in the German language. For semantic and syntactic tasks, findings were similar to English language experiment. A phonological task (rhyme judgment) led to negativity with strong right-hemispheric preponderance.

only the isovoltage maps are shown here (see right column of Fig. 10); these were obtained in a fashion similar to that of the English experiment. The negativity in the semantic difference wave again showed a maximum over centroparietal areas of the scalp; the scalp fields for the syntactic judgment task showed a left frontal maximum. Thus, the general pattern of results was replicated for English and German materials. At present, we do not have an explanation for the slightly more posterior maxima in the distributions of both negativities in the German study, but one possibility is the use of a noncephalic reference in the American and a mastoid reference in the German study.

For the phonological task, the negativity for the nonrhyming stimuli had its maximum over the right parietotemporal area, which corresponds nicely with earlier results (Rugg and Barrett, 1987) obtained in English with a similar task. These researchers similarly described a more pronounced right-hemispheric preponderance of the negativity in a rhyme-matching task as compared to a semantic N400.

Morphosyntactic Incongruency Effects in the Absence of
Semantic Information: Evidence from Finnish Pseudowords

As was stated in the introduction, a major dispute in psycholinguistics concerns the independence of language modules, with one position being that the individual modules subserve their processing function in an encapsulated fashion without being informed by other sources. The extreme counterpart of this position assumes a processing network with the extraction of different level information going on in parallel and with no defined special processing modules.

As it appears from this argument and from word-pair tasks that a negative component can be elicited in response to syntactically incongruous words, we have been searching for a way to isolate semantic and syntactic levels of language processing such that only syntactic information is available to the subject. We hypothesized that if the syntactic processor extracted information from the incoming string of words without taking other information into account, it should do so even under circumstances that do not provide semantic information. We further hypothesized that we should be able to record a negativity to syntactically incongruous as compared to the congruous stimuli if in fact the negativity was in some way related to the activity of a syntactic processing module.

Gurjanov et al. (1985) and Katz et al. (1987) reported auditory and visual lexical decision experiments in which they tested whether or not inflected Serbo-Croatian pseudoadjectives influenced decision times for subsequently presented nouns. As with real adjectives as primes, lexical decision times to nouns were faster when they were preceded by a pseudoadjective that matched the noun in case.

In this study we used pseudowords that were derived from real Finnish words. Finnish and Serbo-Croatian have in common that both languages are highly inflected and that the case role of a word is largely assigned on the basis of inflectional morphology. However, while for Serbo-Croatian the specific case denoted by an inflectional suffix can only be retrieved when the gender of the word stem is known, in Finnish all case inflection suffixes are unambiguous. We used a word-detection task with a delayed response to ensure that subjects had to read each of the stimuli. The goal of this pilot study was to obtain evidence for the action of the syntactic processor in absence of semantic content. Sixteen subjects (12 women, 2 left-handed) reporting Finnish as their mother tongue who had lived between 9 months and 22 years in Germany and spoke German as their second language, participated in the study. The recordings were made at the neurophysiological laboratory at Hannover Medical School.

A total of 240 Finnish pseudowords were generated by changing one or two letters of a real word. A suffix marking the case inflection was added to each of the pseudoword stems using 11 of the total of 15 cases of the Finnish language. Care was taken that the phoneme structure of the pseudowords was comparable to real Finnish words. The 240 stimuli were presented pairwise with a display similar to the study described earlier except that each of the stimuli was presented for 400 msec instead of 200 msec. Of the resulting 120 pairs, 60 were morphosyntactically congruent; that is, both pseudowords were of the same case (e.g. tuus**alla**–heikä**ällä**; **with** tuusa–**with** heikää; morphosyntactically congruent, MS +); the other 60 pairs contained words with different cases (e.g., meko**selle**–turl**ina**; **to the** mekonen–**as a** turli; morphosyntactically incongruent, MS−). Each of the 11 cases was used between 9 and 19 times in the stimulus lists. Seven of these cases have more than 1 case inflection marker. For these cases the particular suffix was determined by the vowels of the (pseudo)word stems according to the rules of vowel harmony of the Finnish language. None of the stimuli was repeated. Two scenarios were constructed to ensure that each critical stimulus was presented once in a (MS +) pair and once in a (MS−) pair. In addition to the pseudoword pairs, 20 pairs that contained a real Finnish word in one of the positions were presented randomly intermixed with the (MS +) and (MS−) pairs. At 1600 msec, after the onset of the second stimulus of the pair, the question **sanako?** (was there a word?) appeared on the videoscreen, and the subjects had to make a response by pressing one of two buttons held in their hands.

The ERPs were recorded from 11 scalp sites and biosignals were processed as described for the word-pair syntactic judgment task. The resulting ERPs for the second stimuli of the pairs are depicted in Figure 11. A major feature distinguishing the ERPs to the pseudowords in this experiment from the ERPs to word stimuli in the previous studies is a more negative waveform beginning at about 250 to 600 msec. This negativity most likely represents activity similar to the semantic N400 component, because large N400 components have been found for pseudowords in a large number of lexical decision experiments (e.g., Bentin et al., 1985; Münte, Künkel, and

Heinze, 1989). The words that were morphosyntactically incongruent with respect to the preceding word, however, were characterized by an additional rather widespread negativity beginning approximately 350 msec after stimulus onset. Its amplitude was smaller than the effects observed for syntactically incongruent words in the experiments using a syntactic judgment task but similar to the negativities seen for lexical decision and letter monitoring tasks. The statistical analyses revealed a significant main effect of morphosyntactical incongruency (MS + vs. MS−) for the time windows 400–600 msec ($p < .04$) and 500–700 msec ($p < .02$).

The salient point in the ERP data from Finnish pseudowords is that apparently a negativity related to morphosyntactical incongruency can be obtained in absence of any semantic meaning. Such an effect would be of enormous importance in two respects. On the one hand, this would indicate

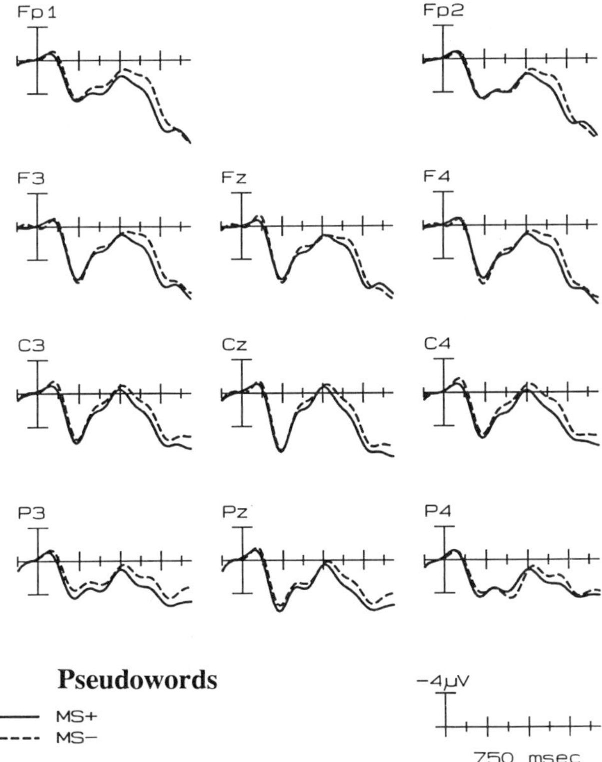

Figure 11.   Grand average ERPs ($n = 16$) from a Finnish-speaking subject group. Shown are responses to second stimuli of stimulus pairs consisting of Finnish pseudowords. Pseudowords were morphosyntactically incongruent (MS−) with respect to preceding word were associated with more negative waveform compared to morphosyntactically congruent (MS +) pseudowords.

that the effects that we have observed in our series of word-pair tasks are in fact related to syntactic processing. This can serve as an argument in support of using the ERP technique in the investigation of syntactic processing.

On the other hand, and more important, an effect of syntactic incongruency in the absence of semantic content could be taken as evidence in favor of theories that assume a modular architecture of the language processing system with a processing module specialized in syntactic analysis (Forster, 1979). This interpretation is critically dependent on the question whether the observed effects can be attributed solely to the morphosyntactic incongruency or whether other alternative explanations can be given. In addition to morphosyntactic congruency, the MS+ and MS− pseudoword pairs also differed in suffix identity. While for the MS− pairs, all case-inflection suffixes were different, this was the case for only 40% of the MS+ pairs, that were derived from those cases that can have more than one suffix.

In a related experiment (Münte and Heinze, in manuscript), we used Finnish adjective–noun word pairs in a syntactic judgment task. The stimulus pairs of this experiment had been specifically composed to test for the effects of suffix identity in relationship to syntactic congruity. In this experiment, ERPs to syntactically incongruent nouns were more negative than syntactically congruent nouns regardless of whether the congruent words had the same or a different suffix as the preceding adjectives. However, there was a lower amplitude negativity with an onset of 250 msec that differentiated the congruent nouns with different suffixes from those with the same suffixes as the verbs. On the basis of the onset latency of this suffix-identity effect, it would seem unlikely that the negativity observed in the current pseudoword experiment can be attributed to suffix identity rather than morphosyntactic incongruity. While additional experimentation is needed to further clarify this issue, the use of pseudowords from highly inflected languages appears to be very promising for the delineation of "pure" syntactic incongruency effects. Such data can in turn be used to qualify the effects of syntactic congruency from experiments with real words.

ERP Effects to Syntactic Incongruencies and Linguistic Theory

Why should psycholinguists pay attention to ERP results? The typical answer of an ERP researcher would be that event-related potentials can serve as an online measure to investigate syntactic processing. Thus, ERPs, in principle, provide a means to dissociate different processing stages in that we might be able to describe specific ERP components tied to these stages. Until now, however, ERP results in the syntactic domain fell considerably short of this premise. As illustrated by the chapter by Kutas and Kluender (Chapter 8, this volume), a variety of different ERP effects to an equally large variety of syntactic violations have been described, casting doubts on the utility of ERPs in this domain.

One of the tasks to be tackled during the next few years is to distinguish ERP responses that are specific to syntactic processing from those that occur unspecifically as a consequence of having encountered an error during reading or listening. In this way, the number of candidates for an ERP component specific to syntactic processing should be reduced.[1]

In this chapter, we have tried to make a contribution toward this end in that several different paradigms were used to test the specificity and generality of certain ERP effects. The most consistent effect observed in these tasks was a negativity with an anterior maximum that could be contrasted with the centroparietal N400 to semantic incongruencies in both word-pair and sentence-reading tasks. Logically, the different distributions of these negativities must be the consequence of different brain systems active during semantic and syntactic processing. The use of both overt syntactic judgment tasks and tasks that covered the main purpose of the experiments (such as lexical decision, word detection, letter monitoring) demonstrated that this anterior negativity is largely independent of task.

The finding that different brain systems subserve the analysis of semantic and syntactic information bears a strong resemblance to the notion of modules that are specialized for certain aspects of language processing, as is proposed by some of the current theories of language processing (e.g., Forster, 1979). It is intriguing to speculate that the negativities observed in the experiments mentioned previously as well as in a variety of other studies represent the physiological correlates of the action of the processing modules. One of the critical issues in this respect has been raised by Kutas and Kluender (Chapter 8, this volume) in that using violations of a certain type (e.g., "syntactic") in no way guarantees that we "know exactly what is 'expected' and therefore what is 'violated'." This criticism might apply to the syntactic incongruencies in the word-pair and sentence-reading tasks that we have used. One could argue that a syntactic violation within a sentence or a word-pair like "We–Dog" simply does not make sense, and any ERP effects observed in response to these violations are the result of the difficulty of computing a meaning for the observed string of words rather than the consequence of the syntactic violation per se. The experiment using Finnish pseudowords represents an important control in this respect because in that study no meaning could be extracted from the word stimuli and the task was not geared toward the evaluation of the syntactic relation between the pseudowords. From these results, it appears that a syntactic violation presented in isolation is capable of eliciting a brain response, thus lending support to theories that propose a specialized syntactic processing system that performs its analysis without being informed by semantic or other level

---

[1] This is not to say that there could only be one single ERP effect related to syntactic processing. In fact, one reason for the relative uniformity of our results is probably the use of simple (morpho)syntactic errors such as case-, gender-, and word-class disagreement.

information. Additional experiments using sentences composed of German pseudowords are currently under way to follow up on this finding and to rule out the possible confounds just discussed. We should thus shortly be able to use the ERP responses described in this chapter to test predictions derived from theories of grammar. A promising area would be the investigation of (possible) interactions or hierarchical dependencies between different processing stages.

Future Developments

Most of the research using ERPs in language processing has used violations as vehicles for the investigation of target processes. This chapter exemplifies this approach, the advantage being that the time of onset of the violating stimulus can serve to precisely time-lock the event-related brain activity. However, in future experiments the use of language material without violations should be used to derive measures for syntactic complexity and parsing strategies. The effects obtained for syntactic incongruencies will nevertheless be useful in the future as they can be used to probe the syntactic processing system in different states, that is, by investigating the interactions between ERP effects of violations and different levels of syntactic complexity. Experiments using highly inflected languages have some promise for the future as the syntactic information is more densely packed in these languages. Extended use of such languages will allow us to more easily distinguish between ERP effects that are directly related to syntactic violations and those electrophysiological effects which are indirectly related to the occurrence of a violation. Another important step in the further characterization of ERP responses to syntactic violations will be the use of auditory stimulation to determine whether these effects are modality independent.

Acknowledgments. We are indebted to G. R. Mangun for his collaboration in the American study. K. Kossow, R. Reekers, S. Rohs, M. B. Scholz, S. Schuchardt, and K. Wiese were involved in the various studies contained in this paper. Sinikka Münte prepared the stimuli for the Finnish experiment.

References

Bach E, Brown C, Marslen-Wilson WD (1986): Crossed and nested dependencies in German and Dutch: A psycholinguistic study. *Lang Cognit Proc 1*:249–262.
Bates E, MacWhinney B, Caselli C, Debescovi A, Natale F, Venza V (1984): A cross-linguistic study of the development of sentence interpretation strategy. *Child Develop* 55:341–354.
Bentin S, McCarthy G, Wood CC (1985): Event-related potentials associated with semantic priming. *Electroencephalogr Clin Neurophysiol* 60:343–355.
Cuetos F, Mitchell DG (1988): Cross linguistic differences in parsing: Restrictions on the use of the late closure strategy in Spanish. *Cognition* 30:73–105.

Ferreira F, Clifton C (1986): The independence of syntactic processing. *J Mem Lang* 25:348–368.

Ferreira F, Henderson JM (1990): Use of verb information in syntactic parsing: Evidence from eye movements and word-by-word self paced reading. *J Exp Psychol Learn Mem Cognit* 16:555–568.

Fischler I (1990): Comprehending language with event-related potentials. In: *Event-Related brain potentials: Basic Issues and Applications*, Rohrbaugh JW, Parasuraman R, Johnson R, eds. New York: Oxford University Press.

Ford M, Bresnan J, Kaplan RM (1982): A competence-based theory of syntactic closure. In: *The Mental Representation of Grammatical Relations*, Bresnan J, ed. Cambridge: MIT Press.

Forster KI (1979): Levels of processing and the structure of the language processor. In: *Sentence Processing: Psycholinguistic Studies Presented to Merril Garrett*, Cooper WE, Walker ECT, eds. Hillsdale, NJ: Erlbaum.

Frazier L (1987): Sentence processing: A tutorial review. In: *Attention and Performance*, Coltheart M, ed. Hillsdale, NJ: Erlbaum.

Frazier L, Fodor JD (1978): The sausage machine: The new two-stage parsing model. *Cognition* 6:291–325.

Frazier L, Rayner K (1982): Making and correcting errors during sentence comprehension: Eye movements in the analysis of structurally ambiguous sentences. *Cognit Psychol* 14:178–210.

Gurjanov M, Lukatela G, Moskovljevic J, Savic M (1985): Grammatical priming of inflected nouns by inflected adjectives. *Cognition* 19:55–71.

Hagoort P, Brown C, Groothusen J: The syntactic positive shift as an ERP measure of syntactic processing. *Lang Cognit Proc* (in press).

Halgren E (1990): Insights from evoked potentials into the neuropsychological mechanisms of reading. In: *Neurobiology of Higher Cognitive Function*, Scheibel AB, Wechsler AF, eds., New York: Guilford Press.

Holdstock JS, Rugg MD (1993): Dissociation of auditory P300, differential brain response to target and novel non-target stimuli. In: *New Developments in Event-Related Potentials*, Heinze HJ, Münte TF, Mangun GR, eds. Boston: Birkhäuser.

Katz L, Boyce S, Goldstein L, Lukatela G (1987): Grammatical information effects in auditory word recognition. *Cognition* 25:235–263.

Kutas M, Hillyard SA (1980): Reading senseless sentences: Brain potentials reflect semantic incongruity. *Science* 207:203–205.

Kutas M, Hillyard SA (1983): Event-related brain potentials to grammatical errors and semantic anomalies. *Mem Cognit* 11:539–550.

Kutas M, Van Petten C (1988): Event-related brain potential studies of language. In: *Advances in Psychophysiology 3*, Ackles PK, Jennings JR, Coles MGH, eds. Greenwich: JAI Press.

Kutas M, Bates E, Kluender R, Van Petten C, Blesch F: What's critical about the critical period? Effects of early experience in language processing in bilinguals (in manuscript).

Lukatela G, Kostic A, Feldman L, Turvey M (1983): Grammatical priming of inflected nouns. *Mem Cognit* 11:59–63.

Lukatela G, Moraca J, Stojnov D, Savic M, Katz L, Turvey MT (1982): Grammatical priming effects between pronouns and inflected verb forms. *Psychol Res* 44:297–311.

Marslen-Wilson WD (1975): Sentence perception as an interactive parallel process. *Science* 189:226–228.

Marslen-Wilson WD, Tyler LK (1980): The temporal structure of spoken language understanding. *Cognition* 8:1–71.

McCarthy G, Wood CC (1985): Scalp distributions of event-related potentials: an ambiguity associated with analysis of variance models. *Electroencephalogr Clin Neurophysiol* 62:203–208.

Mitchell DC (1990): Verb-guidance and other lexical effects of parsing. *Lang Cognit Proc* 4:123–154.

Mitchell DC, Holmes VM (1985): The role of specific information about the verb in parsing sentences with local structural ambiguity. *J Mem Lang* 24:542–559.

Münte TF (1992): *Hirnelektrische Korrelate von Sprache.* Habilitation-Thesis, Medizinische Hochschule, Hannover.

Münte TF, Heinze HJ: Brain potential correlates of morphosyntactic processing in the Finnish language (in manuscript).

Münte TF, Heinze HJ, Mangun GR (1993): Dissociation of brain activity related to syntactic and semantic aspects of language. *J Cognit Neurosci* 5:343–352.

Münte TF, Kossow K, Heinze HJ: Brain activity to words judged for phonologic, semantic, or syntactic properties: Topographical differences (in manuscript).

Münte TF, Künkel H, Heinze HJ (1989): Semantic distance and the electrophysiological priming effect. In: *Brain Dynamics*, Vol. II, Basar E, Bullock TH, eds., New York: Springer.

Münte TF, Schuchardt S, Heinze HJ: ERP effects specific to semantic, syntactic and orthographic errors in a story reading task *Biological Psychology* (1993, in press).

Neville H, Nicol JL, Barss A, Forster KI, Garrett MF (1991): Syntactically based sentence processing classes: Evidence from event-related brain potentials. *J. Cognit Neurosci* 3:151–165.

Rayner K, Frazier L (1987): Parsing temporarily ambiguous complements. *Q J Exp Psychol* 39A:657–673.

Rösler F, Friederici A, Pütz P, Hahne A: Event-related potentials while encountering semantic and syntactic constraint violations. *J Cognit Neurosci* (in press).

Ruchkin DS, Johnson R, Jr., Mahaffey D, Sutton S (1988): Towards a functional categorization of slow waves. *Psychophysiology* 20:339–353.

Rugg MD, Barrett SE (1987): Event-related potentials and the interaction between orthographic and phonological information in a rhyme judgment task. *Brain Lang* 32:336–361.

Seidenberg MS, Tannenhaus MK, Leiman JL, Bienkowski M (1982): Automatic access of the meanings of ambiguous words in context: Some limitations of knowledge-based processing. *Cognit Psychol* 14:489–537.

Seidenberg MS, Waters G, Sanders M, Langer P (1984): Pre- and post-lexical loci of contextual effects on word recognition. *Mem Cognit* 12:315–328.

Taraban R, McClelland JL (1988): Constituent attachment and thematic role assignment in sentence processing: Influences of content-based expectations. *J Mem Lang* 27:597–632.

Chapter 10

ERP Mapping: A Tool for Assessing Language Disorders?

DANIEL BRANDEIS AND DIETRICH LEHMANN

The topography of event-related potential (ERP) maps reflects the intracranial distribution of neural activity. Map topography can be quantified by spatial features, such as the locations of map maximum and minimum potentials, the locations of the centroids of the positive and the negative scalp sites versus the average reference, or the parameters of equivalent dipoles. Such spatial features usually offer significant data reduction compared to the original maps and retain their robust topographic characteristics. Space-oriented ERP map analysis (Lehmann and Skrandies, 1984) is based on the extraction of such spatial map descriptors. The method can provide critical evidence about the nature of the ERP changes, and can distinguish whether brain activity patterns are altered, attenuated, or delayed (e.g., Brandeis and Lehmann, 1989; Brandeis, Naylor, Halliday, Callaway, and Naylor, 1992).

The ERP topography can reflect how a given word is understood. Brown and Lehmann (1979) recorded ERP maps to homophone words that were understood as nouns or as verbs, depending on the preceding sentence. Noun and verb meanings elicited different ERP topographies, as shown with spatial analysis methods. These different topographies were consistent across different English and Swiss German homophones. ERP topography thus reveals a general principle of brain organization by connotative meanings or syntactic classes. ERP topography also discriminates between pattern-

Cognitive Electrophysiology
H-J. Heinze, T.F. Münte, and G.R. Mangun, editors
© 1994 Birkhäuser Boston

masked words and pseudoletter strings that subjects cannot distinguish at better-than-chance accuracy. Brandeis and Lehmann (1986) found that ERP map maximum locations were more left lateralized for such words than for nonwords at latencies greater than 300 msec, suggesting that discriminative lateralized processing of words versus nonwords can continue without leading to conscious discrimination.

The discovery of the N400 component to semantically incongruous sentence endings by Kutas and Hillyard (1980), and subsequent work focussing on N400 linguistic determinants have further established that ERPs are useful online measures of language processing (as reviewed by Kutas and Van Petten, 1988, for example). The N400 is small or absent for predictable congruous endings and is reduced for incongruent endings that are semantically related to the predictable ending (Kutas, Lindamood, and Hillyard, 1984), suggesting that some form of semantic priming of the ending contributes to N400 decreases. Anomalous syntactic word forms, such as errors in case or numerous, can also produce a N400-like component, at least for German sentence endings (Münte, Heinze, and Prevedel, 1990). For words outside a sentence context, N400-like components are attenuated by word repetition and by semantic priming (Bentin, McCarthy, and Wood, 1985; Rugg, 1985).

Space-oriented ERP analysis may be useful to clarify whether N400-like components in different conditions, modalities, or patient groups arise from the same neural generators. This may be particularly important in studying language-disturbed patients where language processes may be affected in different ways. Robust topographic measures could help to clarify the language processing patterns of individual patients in clinical applications.

Because adequate spatial sampling (with good spatial resolution and wide coverage) requires a large number of channels, problems of robustness for statistical testing and interpretation may arise without spatial feature extraction. For example, statistics for selected channel pairs may suggest reversed lateral asymmetries of auditory and visual N400 components even if statistics for the complete normalized amplitude distributions reveal no topographic differences (Holcomb and Neville, 1990).

We examined robust topographic map features in the N400 time range during sentence reading, using a modified N400 paradigm with simply structured sentences (Brandeis, Lehmann, and Mingrone, 1990). ERP maps to 4-word sentences (1.8 words per second) with different types of expected and unexpected endings were recorded with a commercially available mapping system (Biologic Brain Atlas) from 20 channels with Electrooculogram (EOG) rejection at 125 Hz (analog bandpass, 0.1–30 Hz, digital highpass filter with a 1-Hz cutoff). Young adults ($n = 17$, mean age 27 years; 10 right-handers with no history of familial sinistrality) served as subjects. Here we focus on topographic ERP results for adjectives used as endings of sentences about color (Karniski, Vanderploeg, and Lease, 1993), such as "the sky is blue," in German. For "color word" blocks, the correct endings were the best color completions (here, "blue"), and the incorrect endings were inappropriate colors (such as "green"). In "other word" blocks, the correct endings were again the best color completions, but the incorrect endings were noncolor adjectives (unrelated to colors; for example, "shy"). Subjects tried to remember the sentences. Response sentences (12%, followed by the question "no–yes?" and requiring a corresponding button press) were excluded from the ERP averages.

The ERP data were reformatted as average reference map series. Technical zero baselines were used throughout the analyses. The grand mean ERP map series to the final item was used for adaptive segmentation (Lehmann and Skrandies, 1980). The peaks of the global dissimilarity time function identified the latencies of maximal topographic change between successive grand mean maps and delimited segments of relatively stable topography (Lehmann and Skrandies, 1984). This procedure indicated the presence of an early (244–364 msec) and a late (364–444 msec) segment in the N400 time range.

For each map, the centroid (center of gravity) locations of the positive and the negative areas were used as the spatial map descriptors. This centroid computation resulted in a left-right and an anterior-posterior coordinate value for the two centroids of each map. Segment means of these coordinate values were analyzed in repeated-measures ANOVAs, as was the segment's average global field power (a global map amplitude measure; Lehmann and Skrandies, 1980).

Different lateral centroid locations in the early and the late N400 segment confirmed the presence of two components in the N400 time range. The time-varying centroid locations were visualized as trajectories of the anterior-posterior and the left-right centroid coordinates (Brandeis and Lehmann, 1986). These centroid trajectories (Fig. 1) confirmed that the two N400 segments covered the times at which topographic ERP differences between correct and incorrect endings were present. The centroid location statistics were consistent with the topographic effects in the N400 time range seen in Figure 1: A massive topographic change occurred with incorrect endings in the "other word" condition in which a posterior negativity and an anterior positivity replaced the posterior positivity and anterior negativity seen for correct endings. This topographic change pattern occurred in each subject and was found in both N400 segments. Similar but smaller topographic changes occurred for incorrect color words in the early N400 segment. In contrast to these topographic measures, the global field power measure showed no systematic effects.

The segment maps in Figure 2 show the full N400 topography and confirm that spatial feature analysis captures the effects on map topography very well. The difference maps (obtained by subtracting correct from incorrect ERP maps) showed clear parietal minima and prefrontal maxima in the late N400 segment. Lateralization differences between the two N400 segments were less pronounced than in the raw maps. Global field power

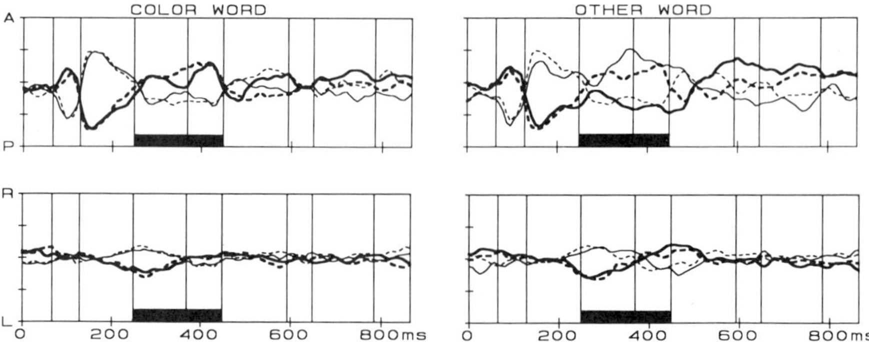

Figure 1   Centroid trajectories for sentence endings from "color word" and "other word" blocks ($n = 17$). Separate trajectories for anterior-posterior (P-A, top) and left-right (L-R, bottom) position of positive (*thin lines*) and negative (*heavy lines*) centroid and for correct (*dashed lines*) and incorrect (*solid lines*) endings. Tic marks on A-P and the L-R axes at one electrode intervals; Cz at center electrode of both axes. *Vertical lines* and *dark horizontal bar* identify early and late N400 segments. Massive topographic change ("reversal") of anterior-posterior centroid locations with incorrect "other word" endings is largely confined to N400 segments. Left-right trajectories show that early and late N400 segments are topographically distinct.

was larger for the "other word" than for the "color word" difference maps, a finding that is consistent with the attenuated N400 found by Kutas, Lindamood, and Hillyard (1984) for semantic associates of the predictable completion.

The map topography to expected, correct sentence endings clearly differed from an attenuated N400. This indicates that different intracranial distributions of neural activity were involved in processing correct and incorrect endings. Amplitude modulation (scaling) or a polarity reversal could not account for these different topographies to correct and incorrect endings.

The same sentences were used to study 10 patients (age range 30–68 years) with left-hemispheric cerebrovascular insult and associated language problems. The residual language problems of these patients were highly variable, as illustrated by their picture-naming scores, which ranged from 0 to 88 (maximum = 90). We examined their N400 segment topography to endings in the "other word" blocks, thereby focusing on the conditions and segments with the most consistent topographic results in the young group. Because of patient heterogeneity, no group comparisons were done. Scatterplots of patient topographic changes (Fig. 3) showed that only a few patients exhibited the typical topographic change pattern in the early N400 segment, but that most patients showed it in the late N400 segment.

The relevant topographic changes in the expected directions were ex-

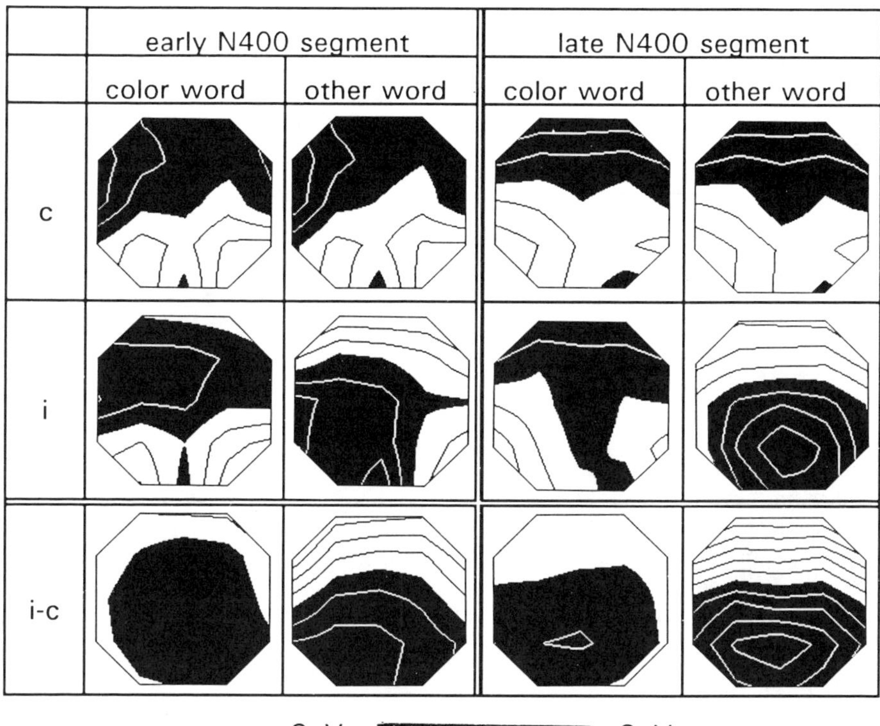

Figure 2   ERP maps for "color word" and "other word" blocks in two N400 segments ($n = 17$). Anterior-negative/posterior-positive topography to correct ("c", top row) color words is replaced by posterior-negative/anterior-positive N400 topography to incorrect ("i", middle row) noncolor words in "other word" blocks. Incorrect color words lead to much smaller, but qualitatively similar, topographic changes. "Incorrect–correct" difference maps ("i–c", bottom row) are strongest in "other word" blocks. Left-right map asymmetries in early N400 segment maps are absent or reversed in late N400 segment maps. Average reference potential maps show head seen from above, nose up.

pressed in a single value, called the topographic N400 change score. This topographic N400 change score from correct to incorrect "other word" ERP maps was computed by adding the posterior shift of the negative centroid and the anterior shift of the positive centroid (i.e., the x and y coordinate values from Fig. 3). For the early N400 segment, this topographic N400 change score correlated with the naming scores of the seven mild patients (naming scores at least 80/90). The mildly affected patients with the poorest naming scores actually had negative topographic N400 change scores (reversed topographic change patterns) that contributed to this correlation. The three severely aphasic patients had naming scores less than 45/90 and

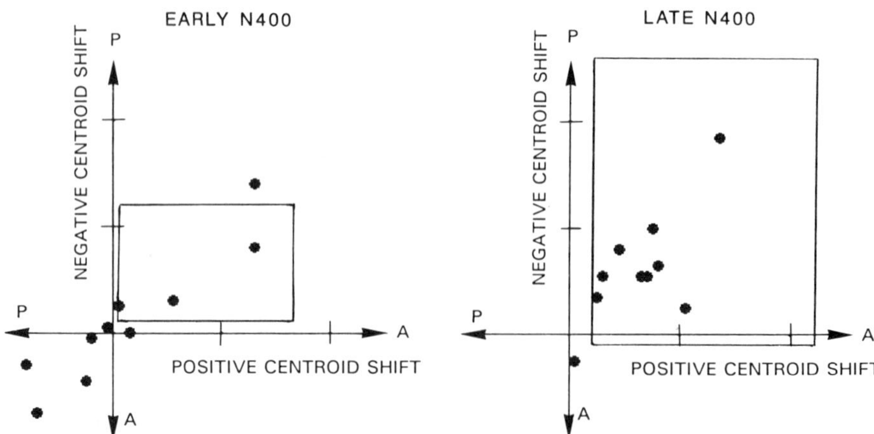

Figure 3   Topographic shifts of map centroid locations from correct to incorrect endings in "other word" blocks for each N400 segment. Each dot represents a left-hemispheric patient. Dot coordinates reflect segment mean of topographic shifts of positive map centroids in anterior, and of negative map centroids in posterior direction. Rectangles enclose range of topographic shifts found in 17 young adults and illustrate consistency of their shift pattern. More patient entries were outside the young group range in early than in late N400 segment. Centroid locations are computed from average reference maps; tic marks indicate one electrode distance.

topographic change scores close to 0. Correlations with patient age, with their sentence comprehension scores, and correlations between language performance measures and total topographic N400 shift in the late N400 segment were all nonsignificant.

First results from healthy elderly subjects ($n = 15$; mean age, 70 years, with no language problems) indicated that their map asymmetries in the early N400 segment were reduced. In the early N400 segment from "other word" blocks, some elderly subjects showed the topographic change pattern found in the young group, while some showed the reversed topographic change pattern seen in some mild patients. In the "color word" blocks, however, elderly subjects showed a consistent normal topographic change pattern in the early N400 segment, arguing against a general delay of N400-like patterns with age.

Our space-oriented analysis of ERP maps during sentence reading resulted in topographic findings that might not have been detected with less extensive spatial sampling. The N400 showed not only the characteristic posterior negativity but also a prominent prefrontal positivity. Segmentation of young adult ERP maps identified two brief segments with language-related ERP topography in the N400 time range. The segments differed in map asymmetry and in their sensitivity to left hemisphere damage and aging, which provided additional support for activation of different language-related processes in the N400 time range.

Highly robust topographic results were found in the "other word" blocks. The massive topographic change from correct color endings to unrelated noncolor endings characterized the N400 and pointed to distinct processes that are reliably invoked by new, unrelated language information in the context of our simple, constrained sentences. The reduced topographic N400 change (and the less pronounced difference maps) between correct and incorrect color words in the "color word" block corresponded well with the finding of a reduced N400 for incongruous words that are semantic associates of the best completion (Kutas, Lindamood, and Hillyard, 1984).

Topographic N400 patterns and particularly the early topographic change pattern to unrelated noncolor words differed for fully recovered, mild, and severe patients with left hemisphere damage. This suggests that such topographic change patterns may be useful for clinical classification of language processing patterns, for measuring subclinical changes in language processes, and for assessing the recovery of language functions. The selective sensitivity of topographic patterns in the early, asymmetric N400 segment to our "mild" patients' naming performance underlines the importance of adequate spatial and temporal resolution in clinical applications. The correlation with naming is consistent with other evidence that the N400 is more closely related to verbal production than to comprehension abilities, even in tasks that require no verbal output. Kutas, Hillyard, and Gazzaniga (1988) reported that among split-brain patients with right hemisphere competence for language comprehension only those with additional right-hemispheric control of speech produced a N400 to incongruent words presented to the right hemisphere.

These findings suggest that both delays (absence of early topographic change) and altered processes (reversed early topographic change patterns; reduced segment map asymmetries) may be involved in language deficits after left-hemispheric cerebrovascular insult and in a subset of healthy elderly subjects. In addition, the findings illustrate the advantages of combining a robust topographic approach to ERP analysis with linguistic manipulations in a reading paradigm. Such a combined approach is necessary to relate language processing deficits to topographic ERP patterns of high temporal and cognitive specificity.

ERP analysis with simple spatial map descriptors utilizes topographic information in a robust manner. It is not confined to the few instances in which simple models of the head and the intracranial source constellation (Kavanagh et al., 1978) are adequate. Topographic ERP descriptors in the N400 time range offer a unique, space-oriented view on brain electrical activity during reading, and topographic approaches will be essential to compare and understand N400-like patterns produced by different patients and by different experimental manipulations. Still, it is encouraging that several of our topographic findings converge with results from previous, waveshape-oriented N400 studies.

Acknowledgments. Work supported by grant 31-9395.88 from the Swiss National Science Foundation. We are grateful to Walter Mingrone and Dr. Michaela Krause for help with subject testing and data analysis, to Dr. Walt Karniski for the idea of using color endings, and to Dr. Ralph Baumgartner and Professor Walter Waespe for patient referral.

References

Bentin S, McCarthy G, Wood CC (1985): Event-related potentials, lexical decision and semantic priming. *Electroencephalogr Clin Neurophysiol* 60:343–355.

Brandeis D, Lehmann D (1986): Event-related potentials of the brain and cognitive processes: Approaches and applications. *Neuropsychologia* 24:151–168.

Brandeis D, Lehmann D (1989): Segments of ERP map series reveal landscape changes with visual attention and subjective contours. *Electroencephalogr Clin Neurophysiol* 73:507–519.

Brandeis D, Lehmann D, Mingrone W (1990): N400 maps in sentence reading: Robust topographic changes and priming. *Brain Topogr* 3:247–248.

Brandeis D, Naylor H, Halliday R, Callaway E, Yano L (1992): Scopolamine effects on visual information processing, attention and event-related potential map latencies. *Psychophysiology* 29:315–336.

Brown WS, Lehmann D (1979): Verb and noun meaning of homophone words activate different cortical generators: A topographical study of evoked potential fields. *Exp Brain Res (Suppl.)* 2:159–168.

Holcomb PJ, Neville HJ (1990): Auditory and visual semantic priming in lexical decision: A comparison using event-related potentials. *Lang Cognit Process* 5:281–312.

Karniski W, Vanderploeg R, Lease L (1993): Virtual N400 and slow wave topography to auditory sentence incongruity. *Brain Lang* 44:58–79.

Kavanagh RN, Darcey TM, Lehmann D, Fender DH (1978): Evaluation of methods for three dimensional localization of electrical sources in the human brain. *IEEE Trans Biomed Eng BME* 25:421–429.

Kutas M, Hillyard SA (1980): Reading senseless sentences: Brain potentials reflect semantic incongruity. *Science* 207:203–205.

Kutas M, Van Petten C (1988): Event-related brain potential studies of language. In: *Advances in Psychophysiology*, Ackles PK, Jennings JR, Coles MH, eds., pp. 139–187. Greenwich, CT: JAI Press.

Kutas M, Hillyard SA, Gazzaniga MS (1988): Processing semantic anomaly by right and left hemispheres of commissurotomy patients: Evidence from event-related brain potentials. *Brain* 11:553–576.

Kutas M, Lindamood TE, Hillyard SA (1984): Word expectancy and event-related brain potentials during sentence processing. In: *Preparatory States and Processes*, Kornblum S, Requin J, eds., pp. 217–238. Hillsdale, NJ: Erlbaum.

Lehmann D, Skrandies W (1980): Reference-free identification of components of checkerboard-evoked multichannel potential fields. *Electroencephalogr Clin Neurophysiol* 48:609–621.

Lehmann D, Skrandies W (1984): Spatial analysis of evoked potentials in man—a review. *Prog Neurobiol* 23:227–250.

Münte TF, Heinze H-J, Prevedel H (1990): Ereigniskorrelierte Hirnpotentiale reflektieren semantische und syntaktische Fehler bei der Sprachverarbeitung. *Z EEG-EMG* 21:75–81.

Rugg MD (1985): The effects of semantic priming and word repetition on event-related potentials. *Psychophysiology* 22:642–647.

Chapter 11

Threshold Variations in Cortical Cell Assemblies and Behavior

N. Birbaumer, W. Lutzenberger, T. Elbert, and T. Trevorrow

There is nothing new under the sun, which is especially true for the neurosciences[1]. Our present situation is best described by the Italian poet Eugenio Montale:

> Codesto solo oggi possiamo dirti,
> ciò che non siamo, ciò che non vogliamo.
>
> Was heut' wir sagen können, ist nur das,
> was wir *nicht* sind, was wir *nicht* wollen.
>
> Today we can only say
> what we are <u>not</u>, what we <u>don't</u> want.

---

[1] Our theoretical position on electrocortical mechanisms (described in detail in Birbaumer et al., 1990) is based on Hebb's (1949, 1961) concept of cell assemblies. The neuroanatomy of cell assemblies is described by V. Braitenberg and his group (Braitenberg, 1978, 1984; Braitenberg and Schüz, 1991; Schüz and Palm, 1989), and the mathematical modeling rests on algorithms and ideas applied by Palm (1982) and Elbert (1987; Elbert and Rockstroh, 1987). Mayer-Kress et al. (1988), Freeman (1991), and Basar, Basar-Eroglu, and Schult (1989) were forerunners in the use of deterministic chaos for the analysis of fast EEG changes to describe the status of cell assemblies.

Cognitive Electrophysiology
H-J. Heinze, T.F. Münte, and G.R. Mangun, editors
© 1994 Birkhäuser Boston

Before we concentrate on the behavior-modifying properties of cortical potentials, let me clarify our position about the meaning of brain electrical activities in more general terms, because it deviates from the now-prevailing view of cognitive neuroscience or what we call *neuropsychoanalysis*: event-related potential components interpreted as psychoanalysts interpret dreams. By using this term we want to indicate that cognitive psychophysiologists tend to interpret brain electrical activities by inventing catchy names such as in the case of the N400: "semantic mismatch wave" (Kutas and Hillyard, 1980), and by doing so the inferred information processing stage seems to become real; at least there seems to be a corresponding brain processes. In reality, in the case of the N400, for example, incongruent words or sentences, like any other type of incongruity or inconsistency, can elicit a rather unspecific cortical mobilization. We described such phenomena in the early 1970s, and maintain that little is gained if we give them catchy names. It has been obvious for well over a decade that large negativities can be generated in many parts of the brain if priming or mobilization of cell assemblies is called for or activated by external stimulation (Elbert et al., 1982; Rockstroh et al., 1979). Figure 1 represents the result of an early experiment in which the experimental situation becomes ambiguous, or when "control" over

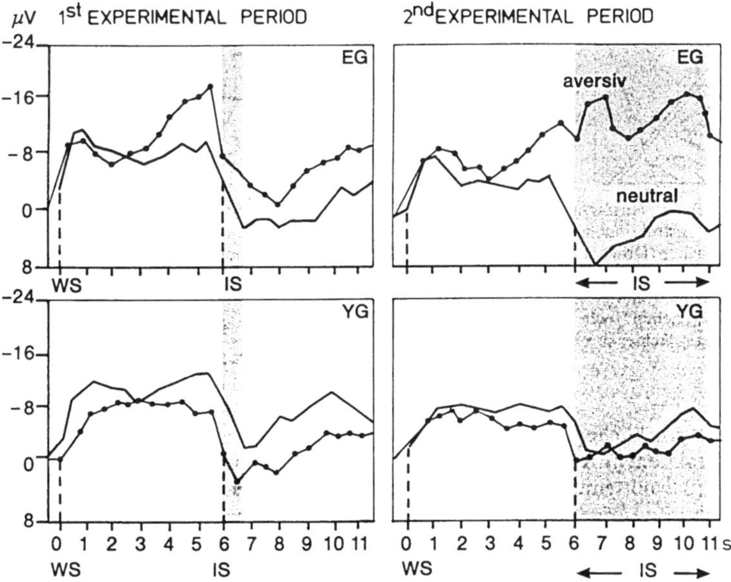

Figure 1.   Large negative postimperative variation (PINV) after loss of control in experimental group (EG, right column) for aversive stimulation only. Neutral imperative stimuli (IS, *hatched line*) do not lead to PINV if the behavioral control (left column, first experimental period) is lost (right column). Control group had no control over IS duration throughout experiment. (from Rockstroh et al., 1979.)

experimental stimuli is withdrawn from the subject. If the subject's expectancy is violated, a poststimulus negative wave develops. Situations with semantic incongruities are but a special case of general priming in arousing conditions.

Cognitive psychophysiologists are tireless inventors of event-related EEG components. Perhaps they hope that their so-called "information processing stages" may ultimately be accepted by the brain, which eventually succumbs under those cognitions and does what it is supposed to do. Before relating the components and waves of event-related potentials to some poorly defined psychological concepts, one should consider the physiological basis of those waves in cortical tissue. If physiological reality matches with psychological terminology, the latter may also receive a breeze of physical reality.

Surface Potentials Indicate Neural Excitability

There is solid neurophysiological evidence, particularly from the work of Caspers, Speckman, and Elger, that local negative variations of the upper cortical layer reflect a lowering of excitation thresholds for the respective assembly. This is not clear for positive variations, which may result from dipole inversion accompanying deeper layer firing or a reduction of upper layer firing. Analysis by Mitzdorf (1985) indicated that most positivities we see in the EEG reflect an increase of excitatory thresholds.

Macroscopically measured potentials largely result from excitatory post-synoptic potentials (EPSPs) in the dendritic trees of the cortex. Stellate cells do not produce open extracellular far-fields. What we see on the scalp or the surface of the cortex is primarily the summation of postsynaptic potentials at pyramidal cells (see also Mitzdorf, 1985). Consequently, a surface-negative potential signifies depolarization of the dendritic tree in vast networks of pyramidal cells, hence indicating an increment in excitability of the underlying cortical tissue.

This suggests that the apical dendritic tree is a candidate for the regulation of cortical excitability thresholds. Depolarization of the apical dendrites causes an efflux of negative charges into extracellular space and back into the neuron in the deeper layers. This results in a polarization of the cortex with the negative pole near the surface. A lowering of thresholds for cortical excitability will result in an increase in surface negativity, while a positive wave is generated when thresholds are set high; therefore, a positive wave, such as the P3, corresponds to an interruption of ongoing activity and to a dysfacilitation of widespread neural activity.

There is also experimental evidence from macroscopic recordings supporting these statements. For instance, drugs that decrease cortical excitability proportionately decrease slow surface negativity. While various anticonvulsant drugs may exhibit quite different effects on earlier components of visual and auditory evoked potentials, as well as on spontaneous EEG activity (Rockstroh et al., 1987, 1991), they all lower CNV (contingent negative variation) amplitude and also the negative DC (direct current) shifts obtained during hyperventilation (Rockstroh, 1990; von Bülow et al., 1989). Certain stimulant drugs may enhance the CNV, although several results have been paradoxical (for review, see McCallum, 1988).

The Concept of Cell Assemblies and Aspects of Memory Storage

The concept of cell assemblies (Hebb, 1949) has become fundamental to models concerned with the functioning of the brain, even though many neuroscientists were reluctant for some 10 or 20 years to accept the lesson they had to learn from a psychologist. Hebb (1949, 1961) postulated that short-term memory is represented in reverberatory circuits, as described earlier by Lorente de No (1943). Once activated, these circuits can maintain excitation as they are formed by a set of neurons, each of which receives excitation from, and gives excitation to, other members of the same set. If a sufficiently large number of neurons in one such cell assembly is activated, the whole set will become active.

The "trick" to memory storage is that the structure of these cell assemblies is flexible and can be changed rapidly. This requires the strengthening of connections between simultaneously active neurons (Hebb's rule), an assumption that has long been considered the physiological basis of learning and memory storage (Hebb, 1949). It is thought that within neural networks the ability of simultaneously active synapses to depolarize the postsynaptic membrane is increased above a certain level of postsynaptic activity, while insufficient activation weakens active synapses. [This is based on Hebb's rule; models of the memory have been suggested by Palm (1982) and others.]

It is generally assumed that plasticity is realized through N-methyl-D-aspartat (NMDA) receptors (for evidence and models of synaptic plasticity, see Bear, Cooper, and Ebner, 1987; Gustafson et al., 1987). Probably all the synapses on spines are subject to modification in their strength, which would mean that three of four cortical synapses are plastic (Braitenberg and Schüz, 1991). Therefore, the buildup and strengthening of a cell assembly requires that a large portion of cells not relevant for the concept of the incoming event be shut off; otherwise, connections would form randomly. Consequently, cortical excitability must be reduced for a fraction of a second or so before a relevant event can be stored in long-term memory.

On the basis of the considerations of the previous section, we have concluded that the widespread reduction in cortical excitability, or inhibition, respectively, will manifest in a widespread positive wave on the scalp (Birbaumer and Elbert, 1988). There is evidence that the P300, and in part the positive slow wave, reflect exactly this process. P300 has been linked to the updating of memory (Donchin and Coles, 1988), and is thought to reflect processes associated with the maintenance of a model of the environment. The larger the information transmission, the more cells should be involved in the network. As updating information must be isolated from background activity, the dysfacilitation must be more widespread, and positivity should increase in expansion and hence in amplitude.

An interesting example of the experimental evidence confirming the inhibitory/disfacilitatory character of the P3 was provided by Rockstroh et al. (1991). Probe clicks to which subjects were asked to respond were delivered during the time course of (an auditory-oddball) P3. Reaction times were significantly slowed, and probe-elicited ERPs were reduced at times when a P3 was present. A comparable RT slowing was not observed in subjects who responded with a pronounced slow negative wave to the targets. These findings are consistent with the proposed dysfacilitatory process.

Brief Outline of the Threshold Regulation Model

To summarize the consequence of such a Hebbian view for the interpretation of EEG and ERPs:

1. The development of cell assemblies depends on *plastic* ("Hebbian") *excitatory* cell systems with a rapid rise time for their construction. The system ideally suited for this purpose are the apical pyramidal dendritic trees of the upper neocortical layer.
2. A cell assembly includes sometimes widespread cortical neurons including sensory, cognitive (meaning), and motor functions. Any restrictive separation into highly specialized "modules," as is fashionable in present-day neuropsychology, is obsolete, vis-a-vis the fact that every sufficiently large pool of neurones of the cortex is connected to every other neuronal pool, forming the anatomic basis of our illusion of a unified consciousness. The meaning and qualitative nature of an event, an idea, an emotion, or a percept, is reflected in the local topography, the topographical "gestalt" of an assembly not in the properties of its parts, the cells or its transmitters.
3. This specificity of an assembly is best reflected in the spatial distribution and frequency of fast-changing electrical activities, such as the EEG and ERP components. It has to be fast because assemblies must have the ability to *ignite explosively* as a whole: a whisper can turn on a full-blown paranoid delusion within the fraction of a second, including all, or nearly all, sensory, motor, and meaning aspects of that delusion.
4. Cell assemblies, and therefore the EEG and ERP, should have properties of *deterministic chaos*: They cannot be totally random, as it would be impossible to create new ideas and percepts if assemblies did not generate novel activity patterns within the fraction of a thought. Rapid-state changes

and bifurcations are characteristics of cell assemblies, which are sensitive to very weak initial conditions that lead instantly to widespread changes in the whole system. These two elements also characterize chaotic systems. Freeman (Freeman, 1991; Skarda and Freeman, 1987) has shown, in the olfactory system of the rabbit, that chaos becomes more prevalent when there is *competition between parts* of assemblies or between several assemblies. We believe that we have hints for a similar mechanism for cognitive processes in the human neocortex (see following).

5.   Whether a given assembly will be ignited and the "contrasts" between assemblies (their foreground-background Gestalt) depends on the *threshold* of an assembly. Usually assemblies are excited by external or internal stimuli; more frequently, they are *primed* by conditioned stimuli occurring before a stimulus that ultimately ignites the assembly. In addition to externally controlled threshold regulation, cell assemblies have their own automatic threshold control, so as to, as Braitenberg poetically states, "discover and isolate ideas . . ." and to reinforce ideas "and keep them separately" (Braitenberg and Schüz, 1991, p. 205) (Fig. 2). Another main function of automatic threshold control is to prevent the transition from an "Einfall to an Anfall" (transition from an idea to a seizure) in an excitatory neuronal network. Inherent to threshold control is its nonlinear transformation of information. Therefore, we should apply principles from nonlinear systems theory (deterministic chaos) to understand the dynamic patterns of threshold regulation. As illustrated in the following section, *slow cortical potentials* (SCP) seem to be a rather useful tool to measure stimulus-response or "idea"-organization. Local negativity represent lowering of thresholds; local positivities represent augmented thresholds for assembly ignition. Because of the inhibitory nature of threshold control, and the lack of much inhibition in the neocortical architecture, subcortical routes are taken that need more time than for the excitatory ignition of assemblies.

We present here some recent data from our laboratory illustrating the points we have made on the *chaotic properties* of fast EEG and ERP components.

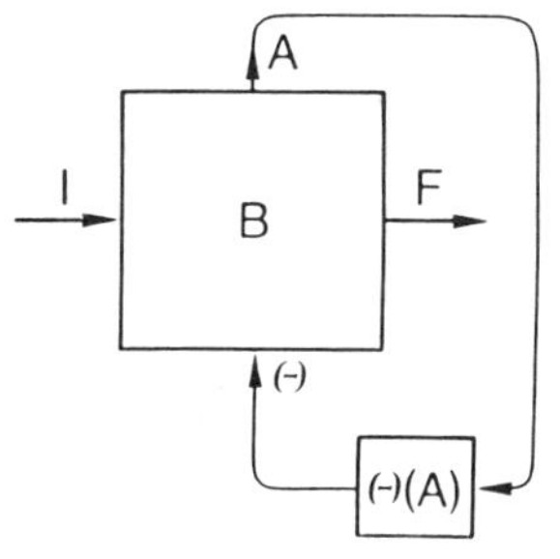

Figure 2.   Threshold control. Brain *B* with input *I* and functional output *F* has in addition output *A* that signals total activity; this is fed into a mechanism of threshold control that sets thresholds of all components at certain level Θ. Form of dependence Θ(A) may vary. [From Braitenberg V, & Schüz A (1991): *Anatomy of the Cortex. Statistics and Geometry*, Berlin: Springer-Verlag.]

Deterministic Chaos in Neural Cell Assemblies

Why introducing deterministic chaos in connectionistic models of brain function and cognition? We (Birbaumer et al., 1990; Lutzenberger et al., 1992), and others before us (Basar, et al., 1989; Freeman and Skarda, 1987) have pointed out that without the presence of chaotic electrical processes the formation of new, unlearned percepts (such as odors) and sensory-motor-patterns cannot be conceptualized or simulated. The chaotic activity has to be deterministic and nonrandom because living biological systems can only function if held within *certain* activity limits through feedback of the ongoing activity. On the other hand, nonlinear changes in the state of neural cell assemblies (NCAs) is a prerequisite of any living system to generate new sensory-motor patterns that control the environment in a dynamic, ever-changing way, patterns that depend on slightly different initial conditions and consequences.

As Freeman (Skarda and Freeman, 1987) has put it, chaos provides the NCAs with a "deterministic 'I don't know' ": Without it, the patterned activity would always return to old, already-formed NCAs leading to exhausting repetition. However, any group of NCAs can be pressed into linear deterministic activity for a certain time period by highly structured external or internal variations. Some pathological conditions, and repetitive behavior without competitive activity, may fall into this category.

Classical behavioral psychology and neuroscience have for a long time exclusively studied linear processes such as habituation or sensitization to simple stimulus configurations. With regard to cognitive activity in the brain ("thinking"), we hypothesize that with increasing competition between NCAs of different cortical localization, the phase space in which a corresponding EEG activity varies becomes multidimensional. On a psychological level, this may correspond to an increasing "dissipation" of the attentional focus. In new situations with multimodal memory demands, the dimensionality ("complexity") of the EEG should be higher. This is similar to the Basar et al. (1989) description of increased coherence of the EEG phase space with increased probability of stimulus occurrence in simple cognitive tasks, and decreased coherence with increasing difficulty to predict stimulus occurrence.

Intelligence and Brain Chaos

If our hypothesis is correct, the dimensionality of the human EEG should increase with higher processing capacities, as measured by tests of intelligence. Dimensionality in one context is described as the ability of a certain space to contain a set of points from a time series, such as the EEG. Mathematically, we are calculating the fractal dimensions of the EEG trace and the fractal dimension of the EEG attractor. An attractor is the "frozen" dynamics of a system, if left without external perturbations. (For methodological details, see Lutzenberger et al., 1992). In theory, the more dimensions

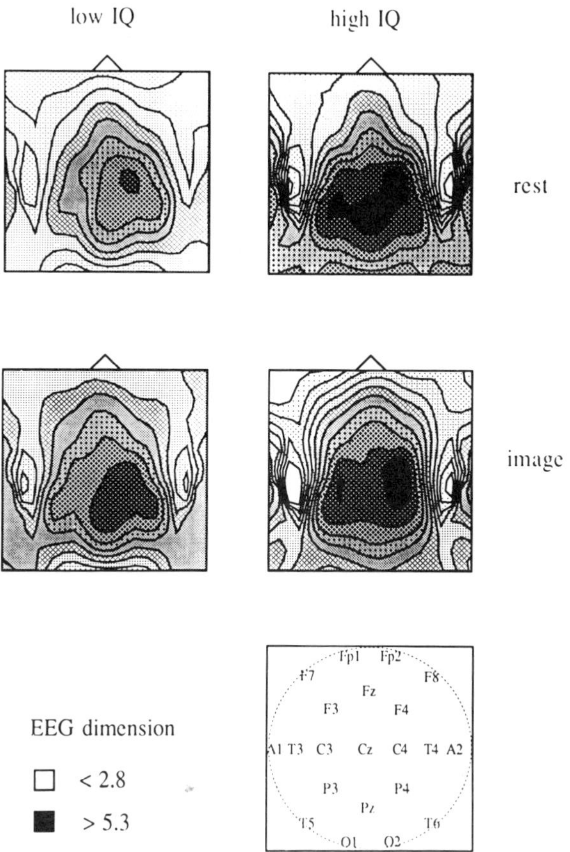

Figure 3.   Chaotic dimensions of 20 EEG channels (lower right) of 10 Ss with high IQs (<110) and 10 Ss with low IQs (>90). Darker color indicates increased dimensions. (From: Lutzenberger W, Birbaumer N, Flor H, Rockstroh B, Elbert T, (1992) Dimensional analysis of the human EEG and intelligence. Neuroscience Letters 143, 10–14. Reprinted with permission.)

of the space state, the more we need to localize all points of our time series within this space, and the more independent processes are building our empirically found time series.

Intelligence may be defined in accordance with Rohracher's subtle description (Rohracher, 1976) of the "performance level of psychological functions for the solving or coping with new situations" ("Der Leistungsgrad psychischer Funktionen bei der Bewältigung neuer Situationen").

Twenty male subjects, (Ss), selected to differ on a broad range in IQ levels (CFT, culture-free intelligence test of Cattell) performed a concentration test (continuous performance test, CPT), engaged in emotional imagery, and participated in eyes-open rest, during which 16 EEG channels were recorded according to the 10–20 system. The chaotic dimensions of the EEG (frequency range, 2–35 Hz) channels were calculated by using singular value decomposition (factor analysis) and the reference point method (Mayer-Kress et al., 1988). Data were collected of groups of 20 resting EEGs (2048 values) and in groups of 20 EEG, during imagining "the most pleasurable" and "the most aversive scene" of the Ss life. EEG α- and β-power were transformed for demonstration purposes constructing "brain chaos maps" (Fig. 3). Figure 3 demonstrates the pronounced and significant [$F(1,18) = 7.3$, $p < .01$] difference between 10 Ss with IQs above average and 10 Ss with IQs below 100 (median split). Higher IQs exhibited approximately five-dimensional phase space during rest; lower IQs exhibited only three to four-dimensional phase space, particularly at central-parietal locations.

Tests of concentration (d2, Bourdon test) and personality questionnaires measuring risk for psychopathology failed to show meaningful differences in EEG dimensionality. Positive and negative emotional content of the image had no influence on the EEG complexity. However, Figure 4 indicates that the EEG dimensionality does not represent a task-independent *trait* variable such as IQ. During imagery, the differences between high and low IQ groups vanishes: the low-IQ Ss EEGs becomes significantly more complex, whereas the high-IQ Ss EEGs become simpler. Figure 4 also indicates that data from an independent sample of medical students (whose average IQ is usually above 100) confirms the positive relationship between IQ and increasing dimensionality.

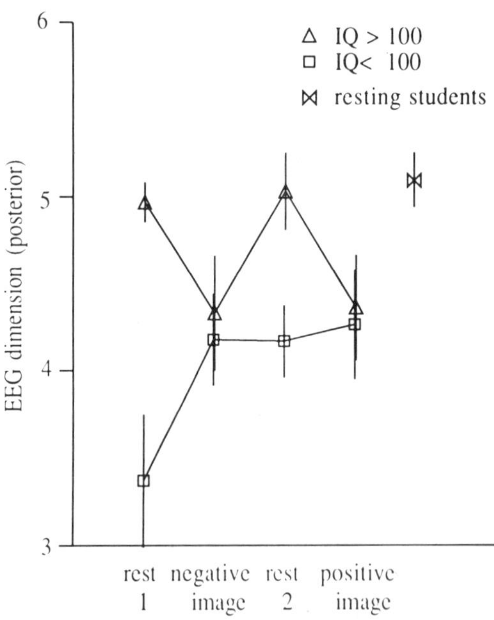

Figure 4.   Chaotic dimensions of parietal EEG during rest condition and during emotionally negative and positive image (see text for explanation).

In an earlier study (Lutzenberger et al., 1992), we used a variety of tasks that cut across sensory modalities including touch, vision, imagery, and verbal processing that reflect neuropsychological processes which differentially engage both the frontal and more posterior areas of the cortex. The outcome shows variations between scalp sites for all measures and also variations between tasks in terms of dimensionality of the EEG attractor as well as of dimensionality of the EEG trace itself. A statistical comparison between the maps generated by means of the various measures shows that different informations are extracted when using the different measures. Results demonstrate that traditional Fourier analyses only extract limited information from the EEG and thus do not present a complete picture.

The methodology of nonlinear dynamics seem to result in higher brain variability for nonprimed situations, with an interaction between different memories of sensory and motor modalities. If task demands are primed by preceding signals, the electric activity turns into a more deterministic mode, represented more clearly, as mentioned earlier, by *slow brain potentials*. Rest, fantasizing, imagery, nonorientated "free-floating" thinking, etc. do not directly determine threshold regulation as does immediate goal-oriented behavior, reflected in SCPs positivities and negativities. For those more spontaneous processes, rhythmicity is mandatory and higher dimensions are expected. Let us return to those situations in which highly deterministic goal-directed behavior and thought are produced by low excitatory thresholds of local NCAs, in particular those of the apical dendrites of the neocortex.

Changes in Slow Cortical Potentials Facilitate or Retain
Behavior and Thought

During the past 15 years we have tried to illustrate the threshold-regulating properties of SCPs with different methodologies. As an example that departs from classical psychophysiological experimentation, we summarize studies utilizing operant conditioning of slow cortical potentials.

In a series of studies (reviewed in Birbaumer et al., 1990; Rockstroh et al., 1989) using healthy Ss and clinical patients, Ss learned to control local negativities and positivities of 6- to 10-sec duration using the biofeedback device that we developed in the late 1970s (Elbert et al., 1980) and the principal operation of which is illustrated in Figure 5.

Ss watched a TV screen that displayed their slow cortical potentials (SCP) from a particular scalp location. SCPs are presented continuously, and are represented by an outlined rocket ship moving back and forth in a horizontal plane for 6–10 seconds per trial (depending on experimental questions). The closer the rocket moves towards two vertical bars on the right, the more negative or positive the particular potential has become. Appearance of the letter A or B on each trial indicates whether positivity or negativity should be produced. Correct responses are usually reinforced with

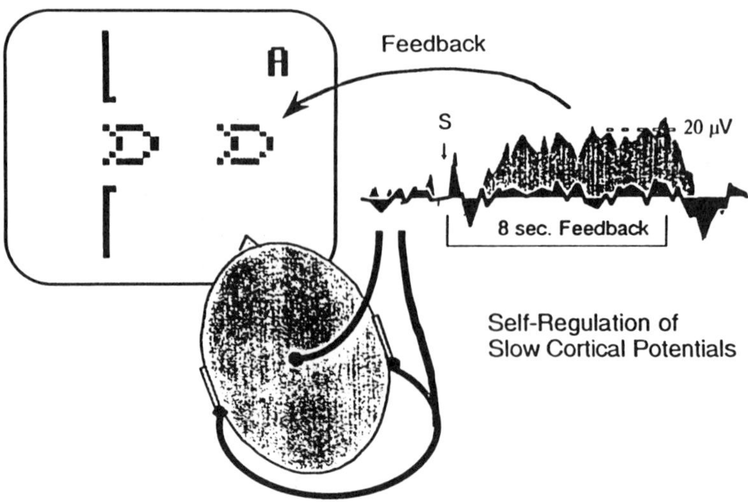

Visual feedback of slow EEG-changes:

Figure 5.   Biofeedback device for instrumental learning of slow brain potentials. Screen with feedback stimulus ("rocket") and indication for cortical negativity ("A") or positivity ("B"). More negativity or positivity moves rocket forward (see text for explanation).

points appearing after each trial on the screen. Points are later exchanged for money. Usually there are 40–60 feedback trials, with a pseudorandom succession of A (i.e., positivity), or B (i.e., negativity). Trials of 'A' or 'B' are presented in *transfer* trials, asking the Ss to produce negativity or positivity without feedback. Different kinds of tasks are presented during transfer trials to test the effects of training on behavior and cognition. Ss are not informed as to whether A or B represent positive or negative SCP shifts. Data published during the past 15 years are highly consistent. (These and other data ruled out the possibility that unspecific activation or inhibition of the whole cortex is responsible for the effects achieved.)

a.   Ss are able to discriminately produce local frontal, central, parietal, right-left hemispheric positive and negative SP (see Fig. 6 for an example from Birbaumer et al., 1992;

b.   Behavioral and cognitive performance becomes highly predictable if the cortical locus and polarity (positive or negative) of the achieved DC changes are known: for example, left precentral negativity augments tactile performance and response speed of the right hand and vice versa. Positivity attenuates performance.

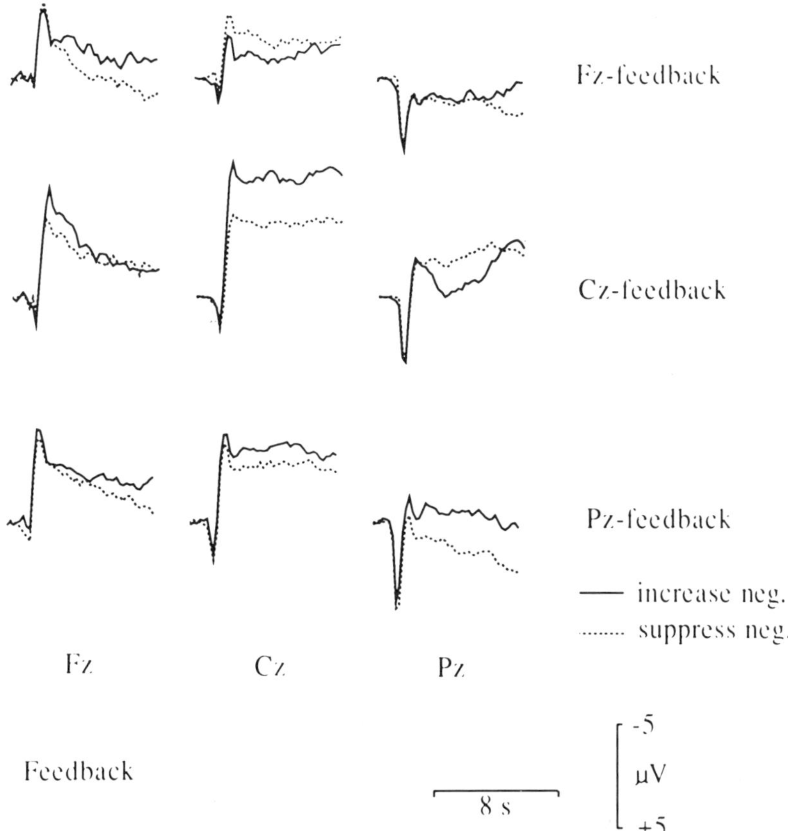

Figure 6.   Slow potential response on negativity and positivity trials in third session of area-specific training (last feedback-only session). Slow potentials recorded from Fz, Cz, and Pz electrodes (abscissa) shown for groups receiving feedback at different mid-sagittal sites (ordinate). Reprinted with permission of Elsevier Science Publishers Ireland, Ltd., from Birbaumer N et al. (1992).

The general relationship between achieved amplitude of central SCP and attentional performance, for example, in a signal detection task (Lutzenberger et al., 1979) follows an invertedly u-shaped function. With more positivity "misses" become more frequent; with extreme negativities, more "false alarms" occur. Different tasks and locations were used, all pointing in the same direction: response speed, vigilance tasks, tactile tasks, motor reflexes (memory tasks led to less clear results) are primed by local negativities and inhibited by local positivities.

Several *clinical groups* with supposed disorders of threshold regulation and attention were investigated: frontal lobe lesions (Lutzenberger et al.,

1980b), psychoses-prone Ss (Elbert et al., 1983), schizophrenics (Schneider et al., 1992), children with attentional problems (Rockstroh et al., 1991), chronic pain patients (Haag et al., 1982), and, recently, drug-resistant epilepsies (Elbert et al., 1991) and patients with psychosomatic disorders (Lutzenberger et al., 1980b).

Patients with functional or neurological disorders of the *prefrontal lobe* such as schizophrenics, Ss with a high risk for schizophrenia, children with attention deficit disorders, and patients with bilateral lesions of the frontal lobes are unable to transfer the learned response and need extended training of as many as 30 sessions to learn SCP regulation. Patients with an increased awareness of internal and visceral bodily processes [such as pain patients or sufferers from migraine (Lutzenberger et al., 1980a) and depressed patients (Schneider et al., 1992)] show performance superior to that of normals.

A double-blind study (Rockstroh et al., 1993) on 24 drug-resistant adult epileptic patients proved the clinical utility of our paradigm and demonstrated, once again, that positivity hightens and negativity lowers excitation threshold. In the epilepsy study, patients had to discriminate central negativities *and* positivities during the first 20 training sessions. They were then trained to produce cortical *positivity only* to suppress seizure activity during the last 8 booster sessions.

Figure 7 shows the slow learning curve of the patients in producing

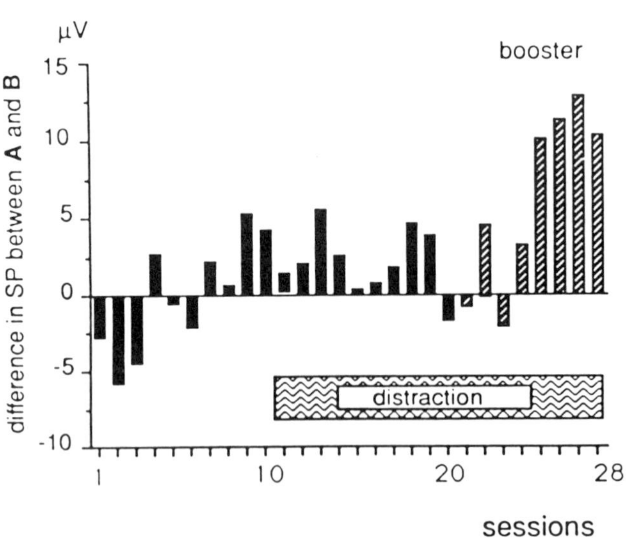

Figure 7.   SP differentiation for the second transfer block (averages across 20 patients).

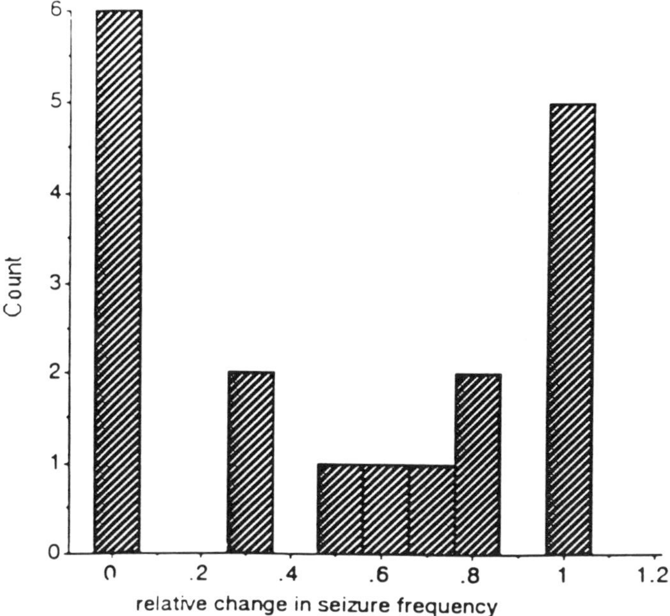

Figure 8.   Frequency distribution of relative changes in seizure frequency, computed as fraction of median seizure incidence during follow-up divided by median during 3-month baseline. Value of 0 on abscissa means that patient has become seizure free during follow-up period; value of 1, seizure incidence has remained unchanged. (From Rockstroh et al. 1993. With kind permission of Elsevier Science Publishers Ireland, Ltd.)

negativity and positivity "on command." The control group had the same conditions but received feedback of its alpha activity. None of the subjects in the control group improved. Figure 8 demonstrates the clinical outcome for the experimental group. With the exception of one patient, only patients who learned to produce negativities and positivities improved or became seizure free. We consider this an impressive confirmation of our hypotheses for "real" clinical life.

Conclusion

Let us conclude by emphasizing that spontaneous waxing and waning of cortical thresholds, in situations of free-floating attention and thought, is represented in spontaneous EEG tracings. These cortical transitions may only become visible if we look at them with a nonlinear, deterministic chaos viewpoint. On the other hand, goal-directed and therefore highly deterministic primed or prepared responses and thought do need local slow brain potentials to occur.

Despite its obvious clinical utility and experimental support, the presented hypothesis retains many doubts and uncertainties. We have not, however, intended to lead the reader into the misery of clarity (E. M. Cioran) or to a special destination. We do not need a special destination; we only wish to have incited your doubts. Let us conclude with a little poetry by the great Italian poet Giorgio Caproni, which characterizes all our scientific efforts more distinctively than we could:

> "Avanti! Ancora avanti"
> urlai.
> Il vetturale si voltó.
> "Signore,"
> mi fecé. "Più avanti
> non ci sone che i campi."

> "Forward, further forward!"
> I howled.
> The coachman turned around.
> "Sir,"
> he said to me, "Forward
> is only the distance."

Acknowledgment. Research was supported by the Deutsche Forschungsgemeinschaft (DFG).

References

Basar E, Basar-Eroglu RJ, Schult J (1989): Chaos and alpha-preparation in brain function. In: *Models of Brain Function*, Corterill RMJ, ed., pp. 365–395. Cambridge: Cambridge University Press.

Bear MF, Cooper LN, Ebner FF (1987): A physiological basis of a theory of synapse modification. *Science* 237: 42–48.

Birbaumer N, Elbert T (1988): P3: By-product of a by-product? *Behav Brain Sci* 11:375–376.

Birbaumer N, Elbert R, Canavan A, Rockstroh B (1990): Slow potentials of the cerebral cortex and behaviour. *Physiol Rev* 70:1–41.

Birbaumer N, Roberts LE, Lutzenberger W, Rockstroh B, Elbert R: (1992) Area-specific self-regulation of slow cortical potentials on the sagittal midline and its effects on behavior. *Electroencephalogr Clin Neurophysiol* 84, 353–361.

Birbaumer N, Elbert T, Rockstroh B, Daum I, Wolf P, Canavan A: Clinical-psychological treatment of epileptic seizures: A controlled study. In: *Perspectives and Promises of Clinical Psychology*, Ehlers, A. et al., eds. New York: Plenum.

Braitenberg V (1978): Cell assemblies in the cerebral cortex. In: *Theoretical Approach to Complex Systems*, Heim R, Palm G, eds., pp. 171–188. Berlin: Springer-Verlag.

Braitenberg V (1984): *Vehicles—Experiments in Synthetic Psychology*. Cambridge: MIT Press.

Braitenberg V, Shüz A (1991): *Anatomy of the Cortex. A Statistical Approach to Its Structure*. Berlin: Springer-Verlag.

Donchin E, Coles MGH (1988): P300 and context updating—. Reply to Verleger. *Behav Brain Sci* 11:357–374.

Elbert T (1987): Regulation kortikaler Erregbarkit: Im EEG ein deterministisches Chaos? In: *Zugang zum Verständnis höherer Hirnfunktionen durch das EEG*, Weinmann H, ed., pp. 93–107. München: Zuckerschwerdt-Verlag.

Elbert T, Rockstroh B (1987): Threshold regulation—a key to the understanding of the combined dynamics of EEG and event-related potentials. *J Psychophysiol* 4:317–333.

Elbert T, Rockstroh B (1990): Psychopharmakologie. Ein einführendes Lehrbuch. Heidelberg: Springer-Verlag.

Elbert T, Lutzenberger W, Rockstroh B, Birbaumer N (1983): When regulation of slow brain potentials fails—A contribution to the psychophysiology of perceptual aberration and anhedonia. In: *Advances in Biological Psychiatry*, Vol. 13, Mendlewicz J, van Praag HM, eds., Basel: Karger, pp. 98–106.

Elbert T, Rockstroh B, Lutzenberger W, Birbaumer N. (1980): Biofeedback of slow cortical potentials. *J Electroencephalogr Clin Neurophysiol* 48:293–301.

Elbert T, Rockstroh B, Lutzenberger W, Birbaumer N (1982): Slow brain potentials after withdrawal of control. *Arch Psychiatr Nervenkr* 232:201–214.

Elbert T, Rockstroh B, Birbaumer N, Canavan A, Lutzenberger W, van Bülow I (1991): Self-regulation of slow cortical potentials and its role in epileptogenesis. In: *International Perspectives on Self-Regulation And Health*, Carson JG; Seifert R, eds., pp. 65–94. New York: Plenum.

Freeman W (1991): The physiology of perception. *Sci Am* 264:78–85.

Gustafson B, Wigstrøm H, Abraham WC, Huang YY (1987): Long term potentiation in the hippocampus using depolarizing current pulses as the conditioning stimulus to single volley synaptic potentials. *J Neurosci* 7(3):774–780.

Haag G, Gerber WD, Birbaumer N, Mayer K, Lutzenberger W, Schroth G (1982): Differentielle Indikation zur Psychotherapie der Migräne [Differential indication for behavior therapy of migraine]. In: *Migräne Fortschritte der Klinischen Psychologie*, Huber H, ed., pp. 205–232. München: Urban & Schwarzenberg.

Hebb DO (1949): *The Organization of Behavior*. New York: Wiley.

Hebb DO (1961): Distinctive features of learning in the higher animal. In: *Brain Mechanisms and Learning*, Delafresnaye JF, ed. New York: Oxford University Press.

Kutas M, Hillyard SA (1980): Reading senseless sentences: Brain potentials reflect semantic incongruity. *Science* 207:203–204.

Lorente de No R (1943): Cerebral cortex: Architecture, intracortical connections, motor projections. In: *Physiology of the Nervous System*, Fulton FJ, ed. New York, Oxford University Press.

Lutzenberger W, Elbert T, Ray WJ, Birbaumer N (1992) The scalp distribution of the fractal dimension of the EEG and its variation with mental tasks *Brain Topography* 5, 27–34.

Lutzenberger W, Elbert T, Rockstroh B, Birbaumer N (1979): The effects of self-regulation of slow cortical potentials in a signal detection task. *Int J Neurosci* 9:175–183.

Lutzenberger W, Haag G, Birbaumer N, Stegagno L (1980a): Biofeedback langsamer kortikaler Potentiale (LKP): Zusammenhang von LKP und Reaktionslatenz bei Patienten mit psychosomatischen Störungen. *Med Psychol* 6:140–151.

Lutzenberger W, Birbaumer N, Elbert T, Rockstroh B, Bippus W, Breidt R (1980b): Self-regulation of slow cortical potentials in normal subjects and patients with frontal lobe lesions. In: *Motivation, Motor and Sensory Processes of the Brain, Progress in Brain Research*, Vol. 54, Kornhuber HH, Deecke L, eds., pp. 427–430. Amsterdam: Elsevier.

Mayer-Kress G, Yates FE, Benton L, Keidel M, Tirsch W, Pöppl SJ, Geist K (1988): Dimensional analysis of nonlinear oscillations in brain, heart, and muscle. *Math Biosci* 90:155–182.

McCallum WC (1988): Potentials related to expectancy, preparation and motor activity. In: *Human Event-Related Potentials, EEG Handbook* (revised series, Vol. 3), Picton TW ed., pp. 427–534. Amsterdam: Elsevier.

Mitzdorf U (1985): Current source-density method and application in cat cerebral cortex: Investigation of evoked potentials and EEG phenomena. *Physiol Rev* 65:37–99.

Palm G (1982): *Neural Assemblies: An Alternative Approach to Artificial Intelligence.* Berlin: Springer-Verlag.

Rockstroh B (1990): Hyperventilation-induced EEG-changes in humans and their modulation by an anti-convulsant drug. *Epilepsy Res* 7:146–154.

Rockstroh B, Elbert T, Lutzenberger W, Birbaumer N (1979): Slow cortical potentials under conditions of uncontrollability. *Psychophysiology* 16:374–380.

Rockstroh B, Müller M, Elbert T, Cohen R (1991): P300 and cortical dysfacilitation. In: *The First European Psychophysiology Conference*, Boelhouwer AJW, Brunia CHM, eds. Tilburg: Tilburg University Press.

Rockstroh B, Elbert T, Canavan A, Lutzenberger W, Birbaumer N (1989): *Slow Brain Potentials and Behaviour*, 2d Ed. München: Urban & Schwarzenberg.

Rockstroh B, Dworkin BR, Lutzenberger W, Ernst M, Elbert T, Birbaumer N (1988): The influence of baroreceptor activation on pain perception. In: *Behavioral Medicine in Cardiovascular Disorders*, Elbert T, Langosch W, Steptoe A, Vaitl D, eds. Chichester: Wiley.

Rockstroh B, Elbert T, Lutzenberger W, Altenmüller E, Birbaumer N, Diener H-C, Dichgans J (1987): Effects of the anticonvulsant Carbamazepine on event-related brain potentials in humans. In: *Evoked Potentials III*, Barber R, Blum T, eds., pp. 361–369. Boston: Butterworths.

Rockstroh B, Elbert T, Birbaumer N, Düchting-Röth A, Wolf P, Daum I, Lutzenberger W, Dichgans J: 1993. Cortical self-regulation in patients with epilepsies *Epilepsy Research* 14, 63–72.

Rohracher H (1976): *Einführung in die Psychologie. 12. Aufl.* München: Urban & Schwarzenberg.

Schneider F, Rockstroh B, Heimann H, Lutzenberger W, Mattes R, Elbert T, Birbaumer N, Bartels M (1992). Self-regulation of slow cortical potentials in psychiatric patients: Schizophrenia. *Biofeedback and Self-Regul* 17, 277–291.

Schüz A, Palm G (1989): Density of neurons and synapses in the cerebral cortex of the mouse. *J Comp Neurol* 286:442–455.

Skarda A, Freeman W (1987): How brains make chaos in order to make sense of the world. *Behav Brain Sci* 10:161–195.

von Bülow I, Elbert T, Rockstroh B, Lutzenberger W, Canavan A (1989): Effects of hyperventilation on EEG frequency and slow cortical potentials in relation to an anticonvulsant and epilepsy. *J Psychophysiol* 3:147–154.

Chapter 12

The Influence of Hand Movements on Cortical Negative DC Potentials

J. NIEMANN, T. WINKER, A. HUFSCHMIDT,
AND C. H. LÜCKING

In contrast to the numerous studies investigating changes of slow negative DC potentials before movement onset, caused by either the so-called Bereitschaftspotential (BP) preceding only self-paced movements (Kornhuber and Deecke, 1965) or by the contingent negative variation (CNV) in a conditioned, forewarned reaction task (Walter, 1964), very few studies have examined the relationship between slow negative DC potentials and different motor task conditions. Only recently has research interest in slow negative DC potentials shifted from the registration of events *preceding the movement* to the registration of events *during the movement* (Cooper, McCallum, and Cornthwaite, 1989; Lang et al. 1988a, 1988b). This is because, as Lang et al. (1989) pointed out in a study comparing simple and complex sequential movement tasks, performance-related DC shifts are more useful for separating motor tasks than is the preceding BP.

Given the greater potential of performance-related DC shifts for functional understanding of movement control in man, we undertook a number of investigations. We had three objectives: to confirm and clarify the relationship between motor performance and negative DC potential shifts recorded from scalp electrodes during different motor tasks; to investigate the influence of different types of movements (e.g., active versus passive, simple versus complex, phasic versus tonic muscle activity) on the distribution and amplitude of surface electronegativity; and to determine whether intraindi-

Cognitive Electrophysiology
H-J. Heinze, T.F. Münte, and G.R. Mangun, editors
© 1994 Birkhäuser Boston

vidual electrophysiological correlates of increasing motor skill can be demonstrated in man during the acquisition of an unfamiliar, complex motor task.

In particular, the following motor tasks were investigated:

1. Uni- and bilateral finger movements: to look for the correlation between recorded potential curves and activation of cortical areas involved in the motor tasks.
2. A complex, difficult-to-perform motor task in right-handed subjects using either the untrained left or the skilled right hand to find differences in surface electronegativity caused by hand preference. Investigations on this topic using the BP have yielded ambiguous results, regardless of whether or not the use of the untrained hand is linked with a larger amplitude or an earlier onset of BP (e.g., Brunia and van den Bosch 1984; Brunia, Voorn, and Berger, 1985; Grünewald et al., 1979; Kristeva and Deecke, 1980; Kutas and Donchin, 1974; Papakostopoulos, 1980; Tarkka and Hallett, 1990).
3. A simple finger flexion/extension movement performed in an active and passive manner: to distinguish between motor and somatosensory components and to investigate their contribution to the recorded potential curve.
4. A simple and a complex finger movement task, which differed mainly in the degree of skill and in the amount of sensory feedback required to perform the movements. These were compared to an isometric contraction task, to demonstrate the influence of task complexity and of phasic or tonic muscle activity on cortical negative DC shifts.
5. A complex, sequential finger movement task: to study whether electrophysiological correlates of increasing motor skill can be demonstrated in man.

Methods

Subjects

Forty-one right-handed subjects (Ss) (whose right-handedness was determined by a questionnaire) aged 18–32 years participated in the experiments. Most were students, naive about the purpose of the experiments, and free of neurological diseases and pathological EEG patterns.

Tasks

The following motor tasks were performed and used in more than one investigation:

1. Simple task: Ss had to press four buttons by the fingers II–V starting with the index finger in an alternating order (II, IV, III, V). The

movement was repeated three times per single run (altogether 12 buttons had to be pressed in a single trial). Correct performance was verified by the appropriate sequential closing of electrical circuits attached to the four different buttons.

2. Complex task: The task consisted of moving a matchstick forward and backward between the index (II) and small (V) finger under somatosensory guidance while maintaining visual fixation (Fig. 1). In each run the Ss were asked to perform the movement twice. This task was selected for its complexity and unfamiliarity to study electrophysiological correlates of different hand skill (right versus left), task complexity *and* motor learning.

3. Finger flexion/extension: A flexion/extension movement of the fingers II–V was performed either by the Ss themselves (active condition) or by

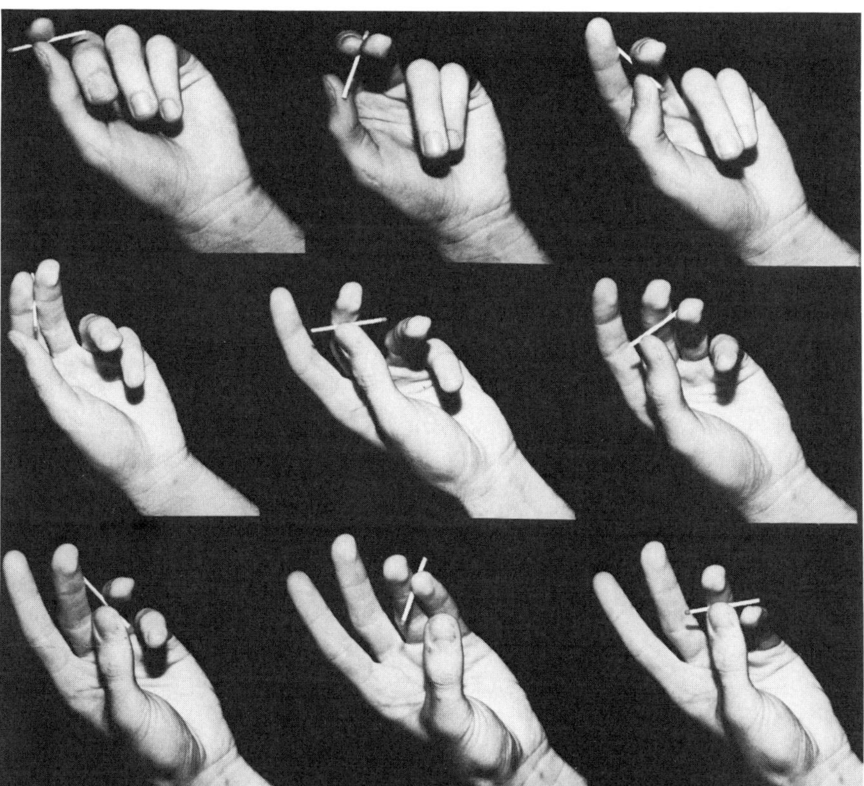

Figure 1.   Complex task consisted of moving matchstick from index finger to small finger of left hand while Ss maintained eye-fixation. In one single trial, Ss had to perform this forward and backward twice in succession. Sequence demonstrates execution of movement going from top left horizontally to bottom right.

the experimenter (passive condition). The task was chosen to be as simple as possible so that the movement was the same, whether it was performed actively or passively.

4.   Isometric contraction: The task consisted of pressing a 7-cm-diameter rigid dynamometer with a constant force level during the entire task period. To avoid any movements Ss were asked to grasp the dynamometer tightly during the whole task (also between trials). A fast contraction onset as well as termination was emphasized. The force level required amounted to about 20% of the maximum strength of the subject.

All movements were performed throughout the entire task period of 6 sec. Trials with a too early or too late termination of the movement were excluded from the averaging process. An EMG (left M interosseus dorsalis I) was recorded to monitor the exact onset and termination of the movement as well as to control for a correct passive performance. To avoid any extraneous movements, the arm was fixed in position during all tasks by an armrest and the head held steady by a headrest.

Experimental Setting

Subjects were seated comfortably and were asked to minimize eye movements by fixating a small black point. EEG potential shifts were recorded by nonpolarizable Ag/AgCl surface scalp electrodes using linked earlobes as a reference. Bipolar leads were recorded from symmetrical locations over the frontal (F1, F2 and F3, F4), central (C3, C4), and parietal (P3, P4) cortex. In addition, monopolar leads were recorded from midline (Fz, Cz, Pz) and central (C1, C2) positions according to the international 10/20 system (Jasper, 1957).

The electrical resistance of the skin was reduced to less than $2\ k\Omega$. The slow negative DC potentials were recorded with a time constant of 5 sec. Vertical and horizontal eye movements were recorded in separate channels to control artifacts caused by shifts of the retinocorneal dipole. The EEG signals were amplified, filtered (low pass filter, 2.5 Hz), and stored on magnetic tape (Ampex). Concurrently the potentials were registered on paper by an EEG mingograph (Siemens). This allowed a continuous control of artifacts caused mainly by eye, muscle, and tongue-movements and by the galvanic skin reflex. Artifact-free runs were identified offline by the experimenters and selected for the averaging process.

Experimental Procedure

One task consisted of a set of at least 20 artifact-free single runs. To study intraindividual changes during the process of motor learning, Ss were required to repeat the tasks 60–80 times. Each trial was divided into three periods of 6 sec each (pretask, task, posttask), indicated by different acoustic signals. Each run was started with an acoustic signal alerting the Ss to begin fixating the small black point.

The latency between the signal and the onset of the sampling period was varied (unknown to the Ss) to reduce the amplitude of a contingent negative variation during the pretask period (Kurtzberg and Vaughan, 1982; Kutas and Donchin, 1980). A stable baseline during the pretask period was essential for the subsequent computerized quantitative measurement procedure. Exactly 6 sec after the beginning of the sampling period another tone signaled the S to start performing the task. A third tone indicated the beginning of a final period, in which the S was required to maintain fixation but not to perform the motor task. After a fourth tone, the S was allowed to relax. Motor tasks, which were statistically compared, were performed at the same experimental session in separate series but in a counterbalanced order between the tasks. The averaged pretask period served as a baseline (mean level of activity) for measuring the increase in surface electronegativity during the motor task. A computer determined the integral between baseline and averaged potential curves as a parameter for potential size (measuring unit: microvoltseconds, μvs; also see Fig. 3).

On the basis of earlier results in our laboratory (Landwehrmeyer, 1990) a potential difference integral between homonymous leads exceeding 4 μvs for a task period of 6 sec (95% confidence interval) was considered as indicating an unequivocal preponderance of activity in one hemisphere ("lateralization").

In the figures, the polarity of the EEG in monopolar leads is plotted so that negativity is pointed upward and positivity downward; and the polarity in the bipolar leads is plotted so that an increase in left-hemispheric cortical activity is shown as an upward deflection and a right-hemispheric lateralization as a downward deflection. Data were digitized at a sampling rate of 1 point/25 msec per channel. The time interval between different single runs varied randomly.

Statistics

Statistical analyses employed either student t tests (using BMDP programs) for paired comparisons of motor tasks or an analysis of variance (ANOVA) for intraindividual changes in potential size during the process of motor learning.

Control for eye movements

Vertical and horizontal eye movements were averaged in a similar procedure as for the EEG signals. The group means for the averaged electrooculogram (= EOG) were less than 20 μvs for each of the various task conditions. As a control, student t tests (for the ANOVA, the influence of the EOG was tested as a covariable) were separately calculated for the vertical and horizontal components of the EOG. No statistically significant differences were found in the EOG in either the overall ANOVA or in the individual *t* tests.

Results and Discussion

Figures 2 and 3 demonstrate the close relationship between recorded potential curves and the movement performance. In a control condition, subjects listened to the acoustical signals, which indicated the different task periods, but did not perform any movement (Fig. 2). Neither an unequivocal lateralization nor a considerable surface electronegativity in monopolar leads was found. No CNVs were recorded during any of the experiments. We attribute this to the unimportance of the acoustic stimuli.

In contrast, performance of a motor task linked with a change from the right hand (task period I) to the left hand (task period II) revealed complementary alterations in bipolar leads. Similar changes can be demonstrated in monopolar leads (C1, C2), indicating that 'lateralization' means only a relative preponderance of contralateral as compared to ipsilateral cortical activity in motor tasks (Fig. 3).

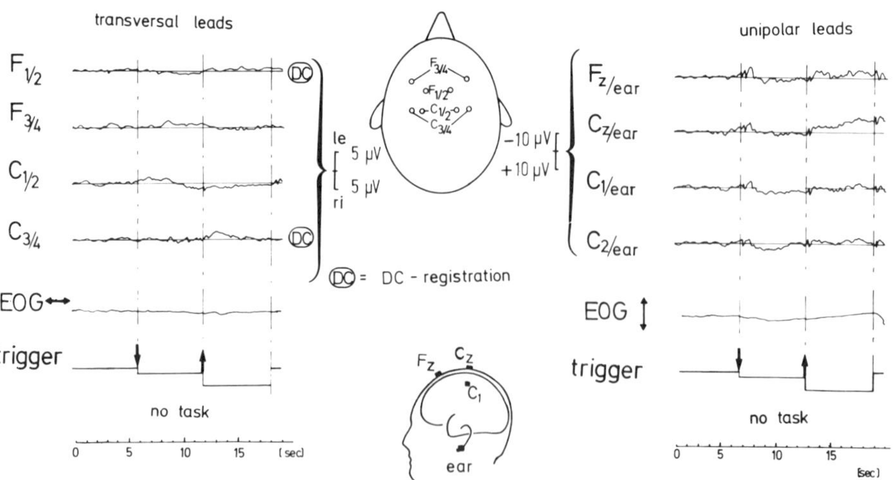

Figure 2.   In control condition, subject listened to acoustic stimuli indicating different task periods but did not perform any movement. Neither lateralization in bipolar nor considerable surface electronegativity in monopolar (= unipolar) leads (referenced to linked earlobes) was found.

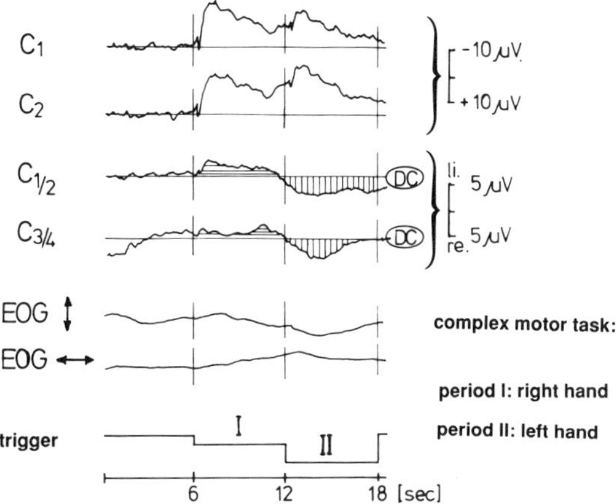

Figure 3.   Same complex finger movement (moving matchstick) was performed with right (period I) and subsequently with left (period II) hand. Bipolar leads reflect change of moving hand from period I to II with corresponding contralateral preponderance in lateralization. In monopolar leads, increase of surface electronegativity occurs bilaterally but is more pronounced over hemisphere contralateral to the moving hand.

Comparison of Unilateral Versus Bilateral Hand Use

The results for the simple motor task (pressing of 12 buttons in a specified order) for left hand only and simultaneous bilateral hand use are displayed in Fig. 4. For the left-hand-only condition, an unequivocal preponderance of the contralateral hemisphere (>4 µvs) in cortical activity occurs with maximum at C3/C4, overlying, according to Steinmetz et al. (1989), the lateral sensorimotor cortex. At the lateral parietal cortex (P3/P4) only a slight preponderance in surface electronegativity of the right hemisphere can be demonstrated.

The preponderance of one hemisphere disappeared (<4 µvs) when both hands performed the task simultaneously. Instead, a bilateral increase in surface electronegativity in monopolar leads with maximum at C1 ($p < .05$), contralateral to the additional new hand, occurred whereas C2 (contralateral to the already-used left hand) revealed the smallest increase in potential size. Whether the overall increased electronegativity of midline electrode locations (Cz, Fz, Pz) is an expression of a general increased cortical activation or just caused by volume conduction is debatable. However, it is worth noting that we did not find a significant increase in potential size at Cz overlying the mediofrontal cortex, including the supplementary motor area (SMA). This seems to be in contrast with the results of the Vienna group (Lang et al., 1988b), who reported a marked additional increase in surface electronegativity at the mediofrontal cortex during bilateral hand use. The different nature of the tasks in our experiments and those of Lang et al., however, may

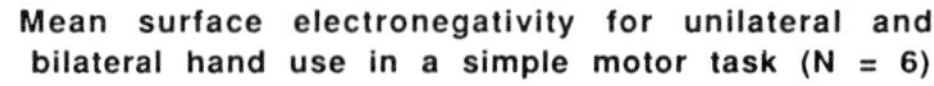

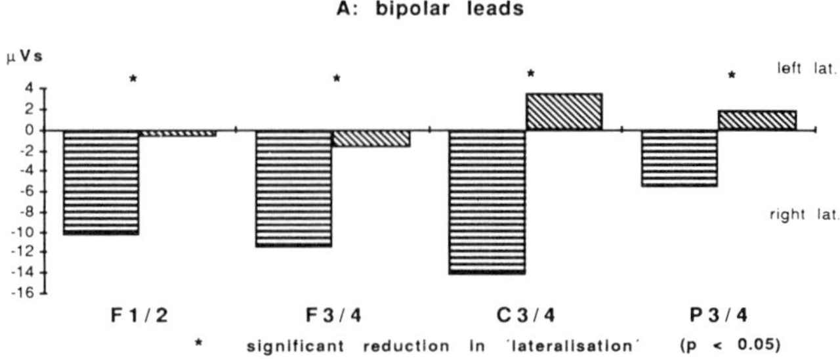

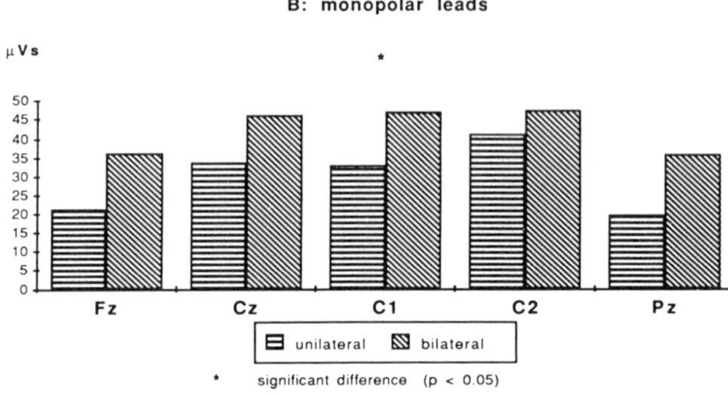

Figure 4. Simple motor task was performed with left hand only or both hands simultaneously. In bipolar leads (**A**), significant (*) decrease in preponderance of right-hemisphere activity during the bilateral condition as compared with left-hand-only condition occurs at all electrode positions. In monopolar leads, increase in potential size, significant at C1 (*) (contralateral to additional used hand), was found.

provide an explanation of the discrepancy. In our experiment, Ss performed the simultaneous task condition in a mirror-like manner, which a priori involves a high harmony and synchronization between both hands. In the Lang et al. experiments, on the other hand, the movement was linked with a higher degree of desynchronization and independence in the use of both hands. Therefore, it seems likely that they were specifically testing a function of the SMA.

The shift of DC potential maxima contralateral to the active hand in unilateral alternating motor performance, the lack of significant increase in surface electronegativity in the absence of motor acts, and the contralaterally pronounced bilateral increase if both hands are used simultaneously suggest that motor performance (and the underlying differential activity of cortical areas) is to a certain extent reflected by changes in surface electronegativity.

Thus the recording of surface electronegativity during performance may open up the possibility of investigating more closely the functional role of movement control in man.

Comparison of Right-Versus Left-Hand Use in Right-Handed Subjects

The next experiments were performed to determine whether differences in potential size can be associated with hand preference. During this task, right-handed Ss performed a complex motor task (moving a matchstick forward and backward between the index and small finger under somatosensory guidance) using the unskilled left and the trained right hand. The results are shown in Figure 5.

With the exception of P3/P4, all bipolar leads are lateralized ($>4\ \mu$vs) contralaterally to the performing right or left hand with potential maxima

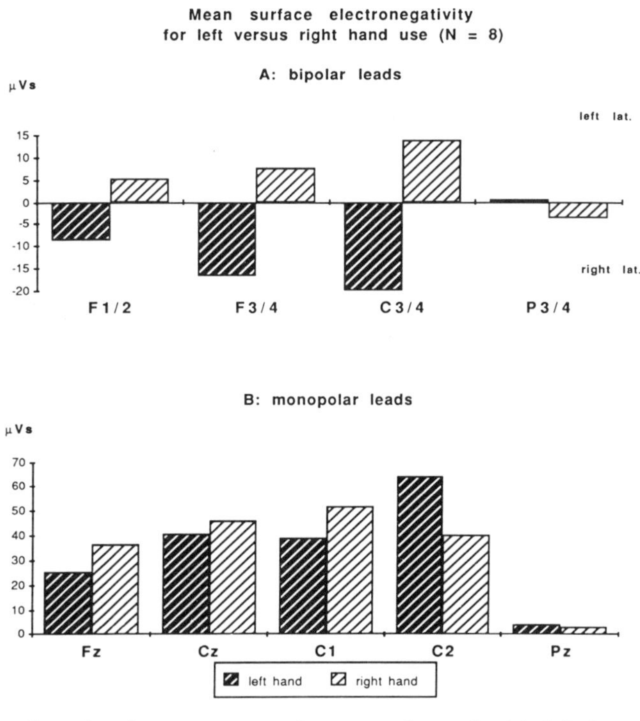

Figure 5.   Complex finger motor task was performed with left (untrained) and right (skilled) hand. Bipolar leads are lateralized contralaterally (according to used hand) with larger preponderance in cortical activity contralateral to unskilled left hand. Similarly, in monopolar leads increased surface electronegativity was found contralaterally for left-hand as compared with right hand condition.

at F3/F4 and C3/C4. Right-hand use (trained hand) reveals an overall smaller amount of lateralization as compared with left-hand use. However, these differences are not statistically significant, if one takes only the amount of lateralization into account and discounts the different direction.

In monopolar leads, the maximum in potential size accords with the cortical representation of the performing hand, either over the left (C1; for right-hand use) or right (C2; for left-hand use) lateral sensorimotor cortex.

Comparing the potential maxima for both task conditions, we find that the surface electronegativity recorded at C2 (contralateral to the untrained left hand) is about 10 μvs greater than at C1 although the ipsilateral cortical activation for each task condition seems to be the same. Although these differences were statistically not significant, they support a previous finding, indicating a different embossed increase in oxygen availability measured by chronically indwelling intracranial electrodes related to hand preference (Cooper and Crow, 1975). In addition, Halsey et al. (1979) have reported a significant increase in rCBF of the contralateral hemisphere for the unskilled left as compared with the trained right hand in a finger–thumb opposition task.

Whether an unspecific increased level of attention or the fact that the appropriate innervation has to be learned initially by the unskilled hand is the cause of this increased electronegativity cannot be determined from our data. Interestingly, no differences could be demonstrated between the skilled and untrained hand with a simple finger movement. Therefore, differences in potential size from the use of trained or untrained limbs may depend on the complexity and difficulty of the motor task studied.

Comparison of an Active Versus Passive Task Condition

In a third group of experiments, active versus passive movements were compared to study the contribution of motor and somatosensory components to the recorded potential curve. In a simple flexion/extension motor task of fingers II–V, we failed to find any significant differences, regardless of whether the task was performed actively or passively. This was true not only for the averaged results of the 16 Ss, but also for their individual potential curves (Figs. 6 and 7). Remarkably, the potential sizes are generally small as compared with those found with other motor tasks. Nevertheless a constant preponderance of contralateral cortical activity during the entire task period is found.

It is worth nothing, however, that the passive task reveals a larger electronegativity at the parietal cortex (Pz) as compared with the active motor task. Further, Fz reveals a larger potential size than Pz for active performance and vice versa for the passive task condition. This may correspond to the location of the electrodes relative to the pre- and postcentral cortices, which are, respectively, associated with functional dominance for motor and sensory processing. Support for this interpretation comes from

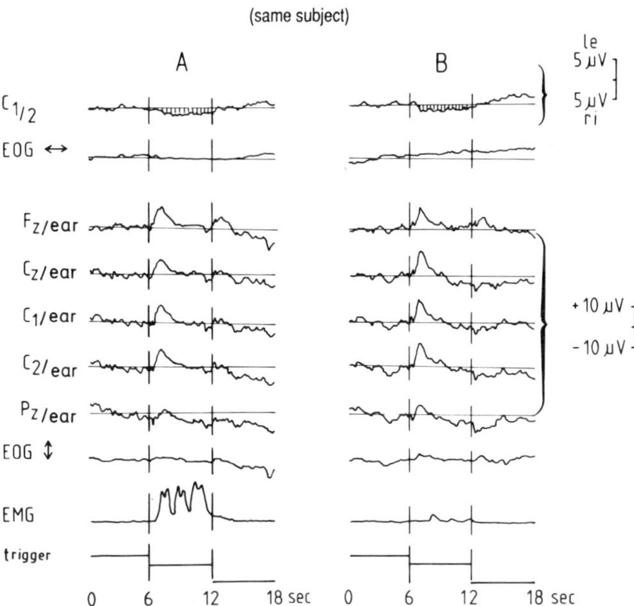

Figure 6.   Comparison of active and passive finger flexion/extension task the potential curve of single subject. Very similarly shaped curves were found. Note the unequivocal preponderance of right cortical surface electronegativity at C1/C2 in bipolar leads for both task conditions.

Roland and Larsen (1976), who also observed a more frontal or parietal increase in rCBF depending on the relative amount of somatosensory feedback necessary to perform different movements. Interestingly, the ipsilateral cortical activation (related to the used hand) at the lateral motor-sensory cortex (C1) appears to be the same in the active and passive task conditions. This may reflect a bilateral projection of somatosensory afferents, known to exist in the secondary somatosensory cortex (areas 40 and 43) and area 5 (Whitsel, Petrucello, and Werner, 1969).

Why did we find no differences in potential size between the active and passive task condition? Vaughan et al. (1970), who failed to find changes of cortical potentials recorded by epidural electrodes during the performance of a trained motor task (wrist extension) before and after upper-limb deafferentiation in monkeys, suggested that "kinesthetic feedback is not registered in motor cortex during the performance of skilled movements." Similarly, a reduced amplitude in somatosensory evoked potentials (SEP) has been observed during motor performance as compared with rest conditions (Coquery, Coulmance, and Leron, 1972; Papakostopoulos, Cooper, and Crow, 1975). At the level of single cells, quantitative comparisons of

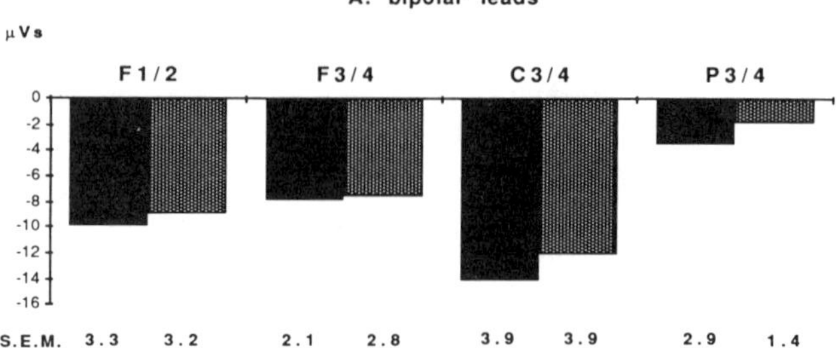

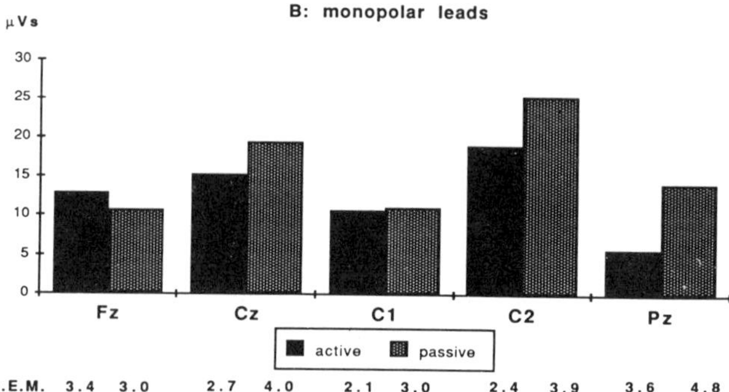

Figure 7.   Finger flexion/extension task was performed in active and passive manner. No significant difference between task conditions was found, in either bipolar or monopolar leads. Shown are the means (microvolts) and standard errors of means (SEM).

cortical neurons in the primary motor cortex of monkeys reveal that only approximately 5% of the 176 movement-related precentral cells studied were not influenced by passive manipulation of the forelimb (Lemon, Hanby, and Porter, 1976). Thus, it is difficult to distinguish between the motor and somatosensory system in electrical recordings, especially since they have numerous means of interaction along the neuraxis. Moreover, efferent pyramidal cells originate in deeper cortical layers (IV–V), and no strong correlation has been found between the discharge frequency of efferent fibers and the magnitude of surface electronegativity (Elger and Speckmann, 1980). The lack of difference in potential size between active and passive task conditions found here, therefore, accords with other reports in the literature and can be explained by several physiological factors.

An indication of the relative contribution of somatosensory afferents to the observed potential curve is given by passive simple flexion/extension

A passive finger movement task

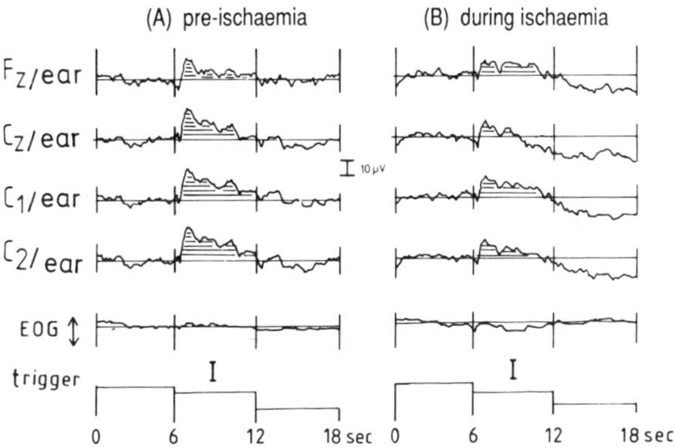

Figure 8. Potential curve of single subject before ischaemia (**A**) and during ischaemia (**B**) (onset after 25 min; RR, 250 mmHg) in a passive task condition. Note decrease in potential size during ischaemia.

finger movement before and during reversibly deafferentiation by ischaemia (onset after 25 min; RR, 250 mm Hg) (Fig. 8). A decrease in potential size is observable during ischaemia, probably because of elimination of somato-sensory afferents.

Comparison of Simple and Complex, Phasic Finger Movement with an Isometric Contraction Task

We investigated the influence of a change in muscle tonus (pressing of a 7-cm-diameter rigid dynamometer) on associated changes in cortical negative DC potentials in 17 subjects and compared these with motor tasks of phasic muscle activity that differed mainly in the degree of complexity and skill. We again used the movement of a matchstick between the index and small finger as the complex task and the pressing of buttons as the simple task.

The complex task revealed a constant high level of surface elec-tronegativity throughout the entire task period (see also Fig. 11), but the isometric contraction task revealed an increase in surface electronegativity only in close relationship with the onset and termination of the muscle contraction (Fig. 9). During the contraction with a constant level of force, the potential curve nearly approaches the basic level, but a small pre-ponderance of contralateral cortical activity is still detectable.

This typical potential course during isometric contraction with an in-creased electronegativity at the beginning as well as at the end of the contraction period probably represents an electrophysiological correlate of the change in muscle activity. In a similar finding, Jasper and Penfield (1949)

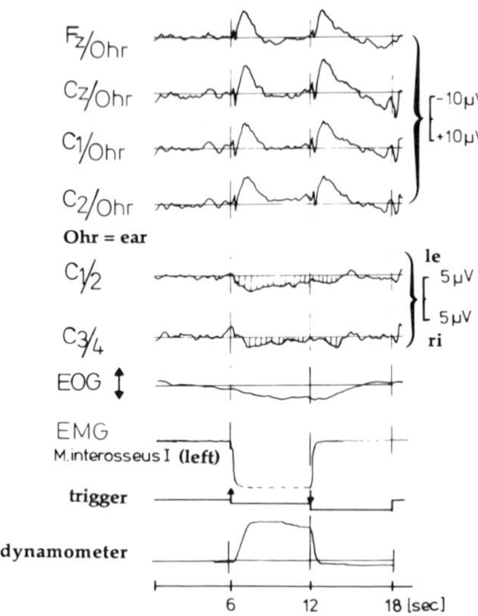

Figure 9.   Potential course during isometric contraction in single subject. In monopolar leads, increase in surface electronegativity occurs only in close relationship at beginning as well as at end of contraction period (EMG and dynamometer control). During the contraction with a constant level of force, the potential curve nearly approaches the basic level, but a preponderance of contralateral cortical activity is still detectable at bipolar leads (C1/C2; C3/C4).

observed during cortical EEG registration a blockage in β-rhythm at the beginning as well as at the end of a forceful finger grasping, but not during the time of contraction. Thus, more than 40 years ago they suggested that the motor cortex is activated mainly by phasic muscle activity and less so by tonic. This interpretation is also supported by our finding that the amplitude of DC potentials is more affected by the speed of contraction than by the strength (see also Grünewald and Grünewald-Zuberbier, 1983).

Roland et al. (1980), using the rCBF as a parameter for measuring cortical activity, compared a static compression of a spring with a force of 5.8 N by the thumb and index finger with finger flexions at a rate of 1 Hz using the same force and the same fingers. Although during the isometric contraction the rCBF increased in the contralateral hand area by 20% (compared to rest conditions), the rCBF was increased by 34% by the finger flexion task. Roland suggested that the rate of change in movement parameters was the factor that raised the rCBF in primary motor cortex most strongly. This is in accordance with our findings in a group of 17 Ss performing all three motor tasks at the same experimental session but in a

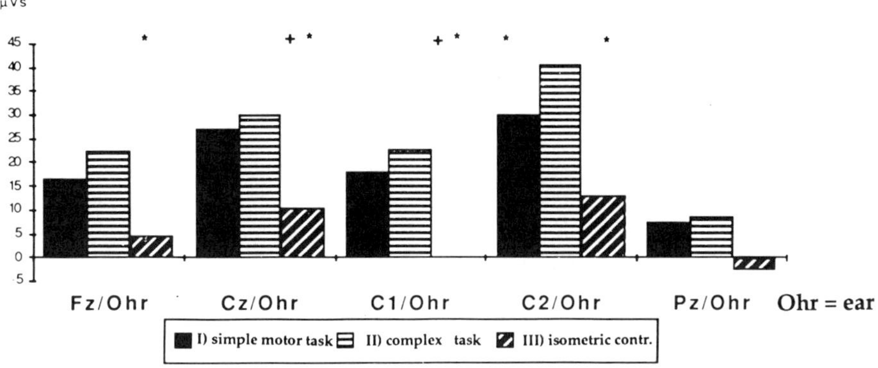

Figure 10.   Comparison of group of 17 subjects who performed all three motor tasks during same experimental session. Stars and crosses symbolize statistically significant differences between motor tasks. Note difference in potential size between tasks of involving predominantly phasic muscle activity and that involving isometric contraction.

counterbalanced order (Fig. 10). Looking at the monopolar leads, it is possible to distinguish statistically the simple and complex finger movement task from the isometric contraction by a larger potential size, probably because of the higher rate of movements. A significant difference in potential size between the simple and complex task consists only for the contralateral sensorimotor cortex (C2), although the complex task reveals an overall larger surface electronegativity.

Intraindividual Changes in Surface Electronegativity During Acquisition of a Complex Motor Task

So far, differences in surface electronegativity between different task conditions have been reported. To study whether electrophysiological correlates of increasing motor skill can be demonstrated in man, surface electronegativity was recorded during the acquisition of a complex finger movement in 21 subjects. Again, Ss had to move a matchstick under somatosensory guidance with the untrained left hand. Ss were asked to perform the motor tasks without increasing the performance speed along the training session. However, the movement had to be performed completely within the task period of 6 sec throughout the duration of the experiment.

In a control condition, Ss performed either the simple movement ($n = 7$)

280   J. Niemann et al.

or the passive finger flexion/extension motor task ($n = 3$) during the same experimental session but in a counterbalanced order. All tasks were repeated 60–80 times, and averages of the first and the last 15 artifact-free single runs were statistically compared. An ANOVA using within-subject factors time (two levels) and location (five levels) was calculated to test for time- and location-specific alterations in potential size for the complex task ($n = 21$). In addition, an ANOVA comparing the simple and complex task ($n = 7$) using within-subject factors complexity (two levels) and time to test for task specificity of time-dependent variations was calculated. (For more details, see Niemann et al., 1991).

An example from a single subject indicating the comparative reduction in potential size during repetition of the simple and complex motor task is displayed in Figure 11. For the simple motor task, no obvious changes during the training session are detectable. The complex task, however, reveals a marked decrease in potential size for the last as compared to the first 15 averaged runs during a comparable long-lasting training session. Initially

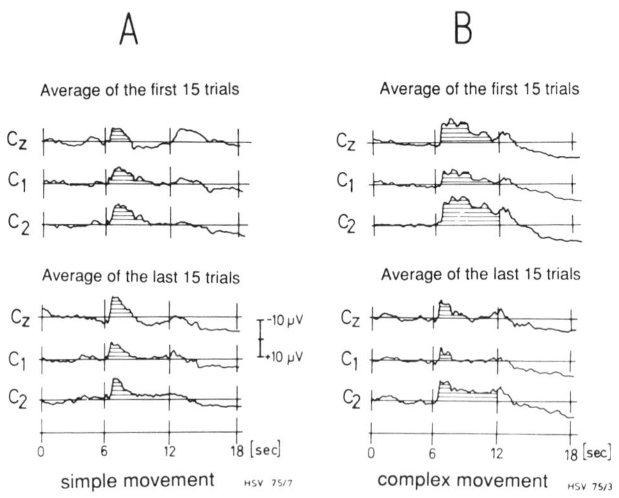

Figure 11.   Condition (**A**) revealed no signs of reduction in surface electronegativity during training period, whereas complex motor task (**B**) showed marked decrease in potential size over sensory-motor cortex. Number of averaged single trials: $n = 80$, (**A**); $n = 75$, (**B**). Note that both movements were performed during entire task period. Duration of slow negative DC shifts is longer in unfamiliar as compared with trained motor tasks, whereas amplitude in both task conditions has almost same maximum.

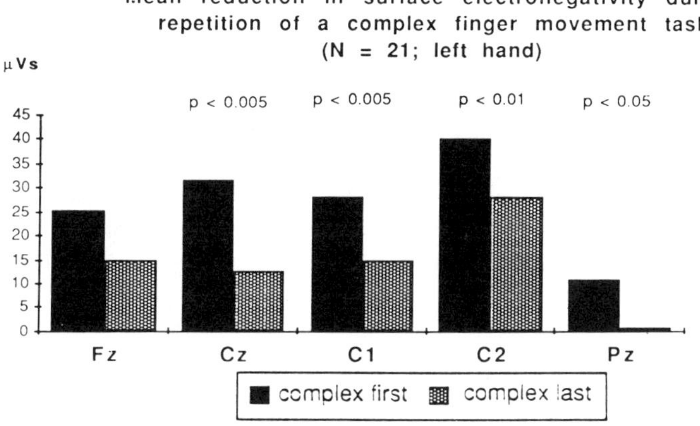

Figure 12.   Changes in mean surface electronegativity during acquisition (comparison between first and last trials) of complex motor task for various electrode positions. With exception of Fz, all electrodes reveal significant decrease in potential size. An ANOVA revealed a significant difference in amount of reduction among electrode positions.

the negativity extends throughout the whole task period, whereas later the negativity occurs mainly at the beginning of the interval, although the movement is being continuously performed. The later pattern resembles that of the simple motor task. That this reduction in potential size for the complex task is not only true for a single subject is demonstrated in Fig. 12.

With the exception of Fz all electrode positions reveal a significant decrease in potential size. The most pronounced decreases occur at Cz ($\sim 60\%$), C1 ($\sim 50\%$) and Pz. It could be assumed that the decrease in potential size is a consequence of an increase in performance velocity resulting in an earlier termination of the movement. However, this possibility can be excluded by our experimental design: Ss were asked to perform the motor tasks without increasing the performance speed throughout the entire task period of six seconds (EMG-control). Trials that did not fulfill this criterion were excluded from further analysis. Part of the observed reduction in potential size may be caused by dependence of the amplitude of surface electronegativity on the subject's attention necessary for performing the task, as described previously for the BP and CNV (McCallum and Papakostopoulos, 1973; McAdam and Seales, 1969). However, a reduction in potential size because a lower level of attention is necessary for still performing the complex finger movement at the end of the training session in a correct manner can already be regarded as motor learning. This accords with the observation that performance in our complex motor task did not deteriorate at the end of the training session.

In addition, all tested Ss confirmed by self-reports that they had experienced more difficulties performing the complex motor task at the beginning of the training period. Comparing the reduction in potential size for the complex and simple motor task ($n = 7$), we found a statistically significant difference in the amount of decrease for Cz, indicating a task-specific alteration. Further, the simple motor task revealed no statistical significant differences in potential size with time, whereas the complex task clearly showed a significant reduction in surface electronegativity for the electrode locations of Cz, C1, and C2 during the process of motor learning. No changes at all could be demonstrated in potential size during numerous repetitions of the passive flexion/extension movement (Fig. 13B), indicating that an altered pattern of somatosensory afferents is not responsible for the observed decrease in potential for the complex motor task.

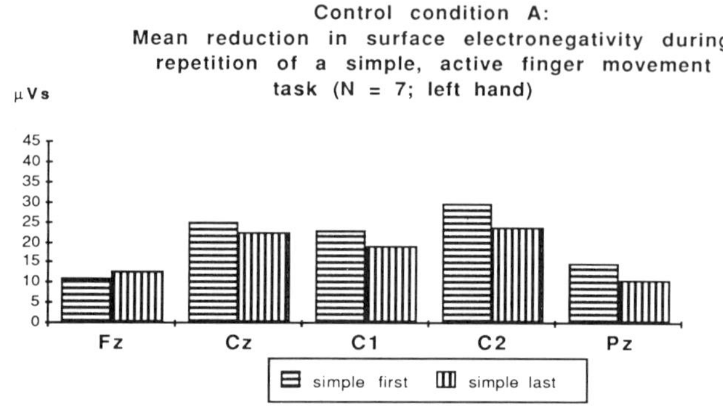

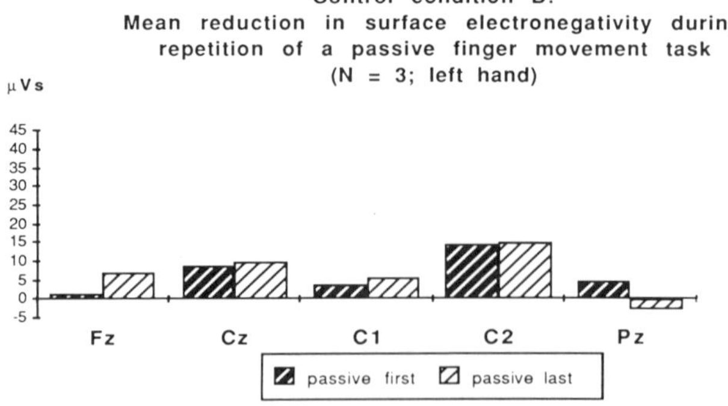

Figure 13. Changes in mean surface electronegativity during numerous repetitions of simple (**A**) and passive (**B**) task. No significant differences in potential size between first and last averaged single trials are detectable. For complex task (not shown), similar reduction (already significant for Cz, C1, and C2) as is shown in Figure 12 occurs.

Although the reduction occurs at all electrode positions for the complex task ($n = 21$), the ANOVA testing location specificity of the time-dependent decrease in surface electronegativity shows significant differences ($p = .02$) between the various electrode positions. On the basis of post hoc t tests, a significantly smaller decrease in surface electronegativity was found for Fz as compared with Ca and C1 ($p < .05$). We suggest that these statistically significant differences in the amount of reduction between various electrode positions during the process of motor learning cannot merely be explained by an unspecific decrease in attention or motivation; instead, they may reflect an altered cortical organization of movement control during the acquisition of a complex motor task.

In the following discussion, we interpret the different decrease in surface electronegativity with respect to the functions of cortical areas corresponding to the various electrode positions. The most marked decrease in electronegativity occurs at Cz overlying the mediofrontal cortex including the SMA. The reduction at Cz is larger than the simultaneous decrease observed at the surrounding electrodes, resulting in a smaller electronegativity for Cz at the end of the training period compared with those for Fz, C1, and C2.

Some studies, for example those by Roland et al. (1980), Roland (1984), Deecke et al. (1985), and Lang et al. (1988b, 1989, 1990), suggest that the SMA is specifically involved in the voluntary initiation and timing control of complex, sequential movements. In accordance with this suggested role it appears that the relative importance of the SMA is reduced as the movement is performed more and more automatically and the subject becomes more skillful. It is worth noting that although we did not set out to study self-paced movements we nevertheless observed a large decrease in surface electronegativity. This seems to imply a major function of the SMA for consciously controlled complex motor tasks independent of whether these movements are self-paced or triggered by external signals.

In contrast to Cz, the Fz electrode, overlying prefrontal and premotor cortical areas (Steinmetz, Fürst and Meyer, 1989), reveals only a non-significant reduction in surface electronegativity during the training period (-5.5 μvs). In addition, a significantly smaller decrease in potential size for Fz was found compared with those for Cz and C1. We suggest that this may indicate a relatively increased importance of premotor and prefrontal cortical areas for the performance of skilled complex motor tasks. This accords with other reports in which an important role of the frontal cortex for motor learning has been suggested (Lang et al., 1988a; Sasaki and Gemba, 1982, 1986).

The changes in potential size at C1 and C2 covering the lateral sensorimotor cortex deserve some attention. C1 (-13.7 μvs) located over the ipsilateral (in relation to the used hand) sensorimotor cortex shows a larger decrease in surface electronegativity than the corresponding electrode over the contralateral (C2) hemisphere (-11.2 μvs). Although the difference of reduction in surface electronegativity between both electrodes is insignificant, it is worth noting that C1 shows a smaller electronegativity overall. The difference between both electrodes is more obvious in the proportional reduction in potential size related to the first averaged trials: C2 decreases during the training period to 70% of its basic potential size, whereas C1 diminishes nearly to 50%. It seems that with progress in motor learning the ipsilateral activation diminishes, resulting in a larger preponderance of the right motor cortical activity. Similarly, Busk and Galbraith (1975) reported that during motor learning the correlation of coherence levels in the EEG-power spectrum decreased between various cortical areas including the ipsi- and contralateral sensorimotor cortex. This agrees with our finding that the proportional ipsilateral cortical activation at C1, compared with that measured contralaterally at C2, increases with greater task complexity. For the passive flexion/extension task condition, the quotient C1/C2 amounted to 0.44; for the same active task, the quotient increased to 0.57 and for the complex task of moving a matchstick, the quotient was 0.67. Interestingly, for the averaged result of the isometric contraction task no ipsilateral cortical activation at all could be detected (see Fig. 10). This could imply a sort of inhibition which occurs at the ipsilateral sensorimotor cortex during tonic muscle activity.

Finally, the small cortical activity at the parietal cortex found for all motor tasks studied here should be remarked. It is perhaps relevant that all our motor tasks were either of a ballistic character or relied on a high degree of somatosensory feedback processing. In any case, a visual control was excluded by the experimental design. When tracking tasks are studied, which rely on a high degree of visual-motor coordination, a more pronounced cortical activation at the parietal cortex is found (e.g., Hufschmidt et al., 1992). Performance of a tracking task with and without visual control ("blind" tracking) reveals a significant difference in potential size at the parietal cortex. This would seem to imply a major function of the parietal cortex for visual-guided motor tasks, a possibility that will be studied in more detail by our laboratory in the future.

Acknowledgment. This work was supported by grants SFB 70 and SFB 325 from the Deutsche Forschungsgemeinschaft.

References

Brunia CHM, van den Bosch WEJ (1984): Movement related slow potentials. (I) A contrast between finger and foot movements in right-handed subjects. *Electroencephalogr Clin Neurophysiol* 57:515–527.
Brunia CHM, Voorn FJ, Berger MPF (1985): Movement related slow potentials. (II) A contrast between finger and foot movements in left-handed subjects. *Electroencephalogr Clin Neurophysiol* 60:135–145.

Busk J, Galbraith CG (1975): EEG correlates of visual-motor practice in man. *Electroencephalogr Clin Neurophysiol* 38:415–422.

Cooper R, Crow HJ (1975): Changes of cerebral oxygenation during motor and mental tasks. In: *Brainwork*, Ingvar H, Lassen NA, eds., pp. 389–392. Copenhagen: Munksgaard.

Cooper R, McCallum WC, Cornthwaite SP (1989): Slow potential changes related to the velocity of target movement in a tracking task. *Electroencephalogr Clin Neurophysiol* 72:232–239.

Coquery JM, Coulmance M, Leron MC (1972): Modification of somaesthetic cortical evoked potentials during active and passive movements in man. *Electroencephalogr Clin Neurophysiol* 33:269–276.

Deecke L, Kornhuber HH, Lang W, Lang M, Schreiber H (1985): Timing function of the frontal cortex in sequential motor and learning tasks. *Hum Neurobiol* 4:143–154.

Elger CE, Speckmann EJ (1980): Focal interictal epileptiform discharges in the epicortical EEG and their relations to spinal field potentials in the rat. *Electroencephalogr Clin Neurophysiol* 48:447–460.

Grünewald G, Grünewald-Zuberbier E (1983): Cerebral potentials during voluntary ramp movements in aiming tasks. In: *Tutorials in ERP Research: Endogenous Components*, Gaillard AWK, Ritter W, eds., pp. 311–327. Amsterdam: Elsevier.

Grünewald G, Grünewald-Zuberbier E, Hömberg V, Netz J (1979) Cerebral potentials during smooth goal-directed hand movements in right-handed and left-handed subjects. *Pflügers Arch Physiol* 381:39–46.

Halsey JH, Blauenstein UW, Wilson EM, Wills EH (1979): Regional cerebral blood flow in comparison of right and left hand movements. *Neurology* 29:21–28.

Hufschmidt A, Lücking CH, Winker T, Niemann J (1992): Functional components of slow brain potentials during visuo-manual tracking. In: *Slow Brain Potentials and Magnetic Fields*, Haschke W, Speckmann EJ, eds. Schiller University Press, Jena.

Jasper HH (1957): The ten-twenty electrode system of the international federation. *Electroencephalogr Clin Neurophysiol* 10:371–375.

Jasper HH, Penfield W (1949): Electrocorticograms in man: Effect of voluntary movement upon electrical activity of the precentral gyrus. *Arch Psychiatr Neurol* 183:163–174.

Kornhuber HH, Deecke L (1965): Hirnpotentialänderungen bei Willkürbewegungen und passiven Bewegungen des Menschen: Bereitschaftspotential und reafferente Potentiale. *Pflügers Arch Ges Physiol* 284:1–17.

Kristeva R, Deecke L (1980): Cerebral potentials preceding right and left unilateral and bilateral finger movements in sinistrals. In: *Motivation, Motor and Sensory Processes of the Brain*, Kornhuber HH, Deecke L, eds., *Prog Brain Res* 54:748–754.

Kurtzberg D, Vaughan HG (1982): Topographic analysis of human cortical potentials preceding self-initiated and visually triggered saccades. *Brain Res* 243:1–9.

Kutas M, Donchin E (1974): Studies of squeezing: Handedness, responding hand, response force and asymmetry of readiness potential. *Science* 186:545–548.

Kutas M, Donchin E (1980): Preparation to respond as manifested by movement-related brain potentials. *Brain Res* 202:95–115.

Landwehrmeyer B (1990): *Hirnelektrische Korrelate einer Rechtsdominanz der Raumwahrnehmung*. Medical Dissertation, Albert-Ludwigs-Universität, Freiburg i. Brsg.

Lang W, Lang M, Podreka I, Steiner M, Uhl F, Suess E, Müller CH, Deecke L (1988a): DC-potential shifts and regional cerebral blood flow reveal frontal cortex involvement in human visuomotor learning. *Exp Brain Res* 71:353–364.

Lang W, Lang M, Uhl F, Koska CH, Kornhuber A, Deecke L (1988b): Negative cortical DC-shifts preceding and accompanying simultaneous and sequential finger movements. *Exp Brain Res* 71:579–587.

Lang W, Zilch O, Koska A, Lindinger G, Deecke L (1989): Negative cortical DC-shifts preceding and accompanying simple and complex sequential movements. *Exp Brain Res* 74:99–104.

Lang W, Obrig H, Lindinger G, Cheyne D, Deecke L (1990): Supplementary motor area activation while tapping bimanually different rhythms in musicians. *Exp Brain Res* 79:504–514.

Lemon RN, Hanby JA, Porter R (1976): Relationship between the activity of precentral neurons during active and passive movements in conscious monkeys. *Proc R Soc Lond B* 194:341–373.

McAdam DW, Seales DM (1969): Bereitschaftspotential enhancement with increased level of motivation. *Electroencephalogr Clin Neurophysiol* 27:73–75.

McCallum WC, Papakostopoulos D (1973): The CNV and reaction time in situations of increasing complexity. In: *Event-Related Slow Potentials of the Brain: Their Relations to Behaviour* McCallum WC, Knott JC, eds. *Electroencephalogr Clin Neurophysiol Suppl* 33:179–185.

McCallum WC, Cooper R, Pocock PV (1988): Brain slow potential and ERP changes associated with operator load in a visual tracking task. *Electroencephalogr Clin Neurophysiol* 69:453–468.

Niemann J, Winker T, Gerling J, Landwehrmeyer B, Jung R (1991): Changes of slow cortical negative DC-potentials during the acquisition of a complex finger motor task. *Exp Brain Res* 85:417–422.

Niemann J, Winker T, Jung R (1992): Changes in Cortical Negative DC Shifts Due to Different Motor Test Conditions. *Electroencephalogr Clin Neurophysiol* 83:297–305.

Papakostopoulos D (1980): The Bereitschaftspotential in left- and right-handed subjects. In: *Motivation, Motor and Sensory Processes of the Brain*, Kornhuber HH, Deecke L, eds., *Prog Brain Res* 54:742–747.

Papakostopoulos D, Cooper R, Crow HJ (1975): Inhibition of cortical evoked potentials and sensation by self-initiated movement in man. *Nature (London)* 258:321–324.

Roland PE (1984): Organization of motor control by the normal human brain. *Hum Neurobiol* 2:205–216.

Roland PE, Larsen B (1976): Focal increase of cerebral blood flow during stereognostic testing in man. *Arch Neurol* 33:551–558.

Roland PE, Larsen B, Lassen NA, Skinhoj E (1980): Supplementary motor area and other cortical areas in organization of voluntary movements in man. *J Neurophysiol* 43:118–136.

Sasaki K, Gemba H (1982): Development and change of cortical field potentials during learning processes of visually initiated hand movements in the monkey. *Exp Brain Res* 48:429–437.

Sasaki K, Gemba H (1986): Effects of premotor cortex cooling upon visually initiated hand movements in the monkey. *Brain Res* 374:278–286.

Steinmetz H, Fürst G, Meyer BU (1989): Craniocerebral topography within the international 10-20 system. *Electroencephalogr Clin Neurophysiol* 72:499–506.

Tarkka IM, Hallett M (1990): Cortical topography of premotor and motor potentials preceding self-paced, voluntary movement of dominant and non-dominant hands. *Electroencephalogr Clin Neurophysiol* 75:36–43.

Vaughn Jr. HG, Bossom J, Gross EG (1970): Cortical motor potentials in monkeys before and after upper limb deafferentiation. *Exp Neurol* 26:253–262.

Walter WG (1964): Slow potential waves in the human brain associated with expectancy, attention and decision. *Arch Psychiatr Nervenkr* 206:435–449.

Whitsel BL, Petrucello LM, Werner G (1969): Symmetry and connectivity in the map of the body surface in somato-sensory area II of primates. *J Neurophysiol* 32:170–183.

Chapter 13

Principles of Electrogenesis of Slow Field Potentials in the Brain

E.-J. SPECKMANN, U. ALTRUP, A. LÜCKE AND R. KÖHLING

Field potentials, generated in the extracellular space of the brain, consist of rapid potential fluctuations and slow baseline shifts. The rapid waves represent the conventional electroencephalogram (EEG). Both phenomena—the slow baseline shifts and the rapid waves—can be recorded as so-called direct current (DC) potential (Andersen and Andersson, 1968; Caspers, 1974; Caspers, Speckmann, and Lehmenkühler, 1980, 1984; Creutzfeldt and Houchin, 1974; Speckmann and Caspers, 1979a; Speckmann and Walden, 1991).

Extracellular field potentials are mainly caused by changes in membrane potential of neurons and glial cells. In this context, the following processes are of special importance. In a first step, transmembraneous ion fluxes lead to localized excitatory or inhibitory membrane potential changes. Between these localized potential changes and the unaffected membrane areas, a potential gradient is established. In a second step, these potential gradients give rise to intracellular and extracellular current flows along the cell membrane. The intracellular current flow is the basis for magnetic fields that can be recorded as magnetoencephalogram (MEG). In a third step, the extracellular current flow establishes, across the resistance of the extracellular space, potential gradients that are recorded as field potentials (EEG, DC). These principles of field potential generation apply to neurons as well as to glial cells. The latter are coupled by gap junctions building up a functional

Cognitive Electrophysiology
H-J. Heinze, T.F. Münte, and G.R. Mangun, editors
© 1994 Birkhäuser Boston

network (De Robertis and Carrea, 1965; Kuffler and Nicholls, 1966; Kuffler, Nicholls, and Orkand, 1964; Palay and Chan-Palay, 1977; Somjen and Trachtenberg, 1979).

The changes of membrane potential of glial cells and neurons are not independent from each other but are functionally linked by way of the extracellular potassium concentration gradient. In the course of grouped action potentials and of excitatory as well as of inhibitory postsynaptic potentials, potassium ions flow from the intracellular space into the extracellular space. This leads to an extracellular accumulation of potassium ions, which in turn depolarizes glial cells in the vicinity of active neurons. With slow extracellular potential shifts, further sources such as the blood–brain barrier and meninges must be taken into account (Kuffler and Nicholls, 1966; Kuffler, Nicholls, and Orkand, 1966; Lehmenkühler, 1988; Somjen and Trachtenberg, 1979).

This chapter concerns slow field potential shifts recorded epicranially and epicortically and their relationships to membrane potential changes of neurons and glial cells. As experimental models, focal and generalized epileptic activity, as well as changes of gas pressure in tissue during hypercapnia and hypoxia, have been selected.

Focal and Generalized Epileptic Activity

Focal epileptic activity can be elicited by local epicortical application of penicillin. A typical experiment is displayed in Figure 1. The focal epileptic

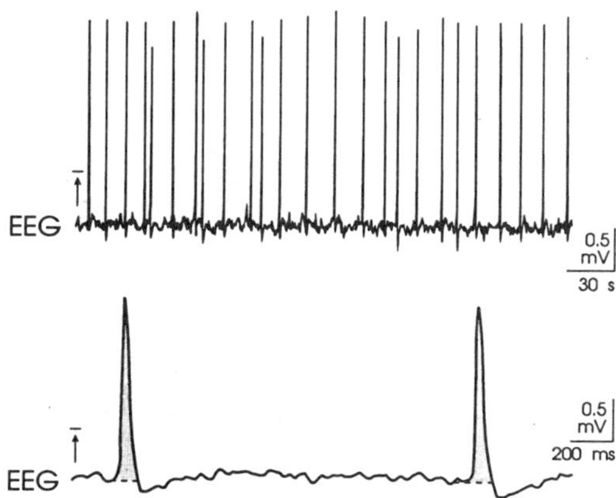

Figure 1.   Epicortical recordings of EEG during focal interictal epileptic activity, anesthetized and artificially ventilated rat. Focal interictal epileptic activity was elicited by local application of penicillin. Upper and lower tracings with different temporal resolution. (Modified from Speckmann, E.-J., *Experimentelle Epilepsieforschung*, Wissenschaftliche Buchgesellschaft, Darmstadt, 1986.)

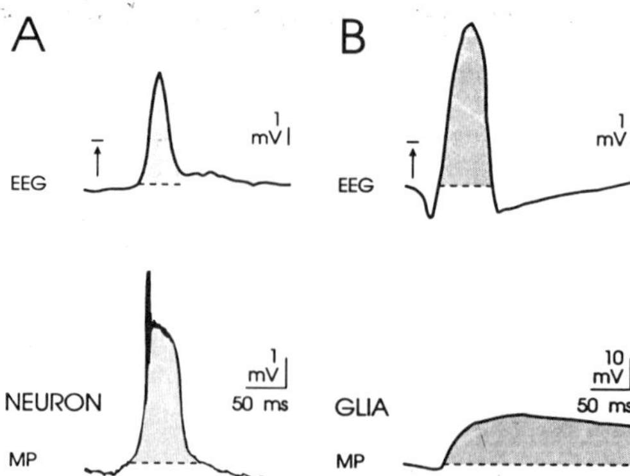

Figure 2.   Epicortical EEG recordings and simultaneous neuronal (**A**) and glial (**B**) membrane potential changes during focal interictal epileptic activity, anesthetized and artificially ventilated rat. Focal interictal epileptic activity was elicited by local application of penicillin. (Modified from Speckmann, E.-J., *Experimentelle Epilepsieforschung*, Wissenschaftliche Buchgesellschaft, Darmstadt, 1986.)

activity is represented by sharp negative fluctuations of the field potentials, which appear regularly and are stereotyped in shape (Elger and Speckmann, 1980, 1983a, 1983b; Elger et al., 1981, 1982; Jasper, Ward, and Pope, 1969; Klee, Lux, and Speckmann, 1982, 1991; Kuffler and Nicholls, 1966; Petsche et al., 1978, 1981; Speckmann, 1986; Speckmann and Elger, 1983; Wieser, 1983).

The focal epileptic field potentials are related to typical changes in membrane potential of neurons and glial cells (Fig. 2). The membrane potential changes of neurons in the vicinity of the generators of field potentials consist of a steep depolarization that triggers a burst of action potentials (Fig. 2A). This is followed by a plateau-like diminution of the membrane potential and a steep repolarization. The events change over to an afterhyperpolarization or an afterdepolarization. The described changes of membrane potential are labeled paroxysmal depolarization shifts (PDS). As can be derived from Figure 2, the PDS is in close temporal relationship to the epicortically recorded field potential. The membrane potential changes of glial cells also represent depolarizations (Fig. 2B). In comparison to those of neurons, they show a slow rise and outlast the EEG events for a long time. Thus, the focal interictal epileptic field potentials are thought to be predominantly caused by neuronal activity (Andersen and Andersson, 1968; Elger and Speckmann, 1980, 1983a, 1983b; Elger et al., 1981, 1982; Gumnit, Matsumoto, and Vasconetto, 1970; Speckmann, 1986; Speckmann and Elger, 1983, 1984).

Generalized tonic-clonic epileptic activity can be induced by repeated intraperitoneal injections of pentylenetetrazol. A typical experiment is dis-

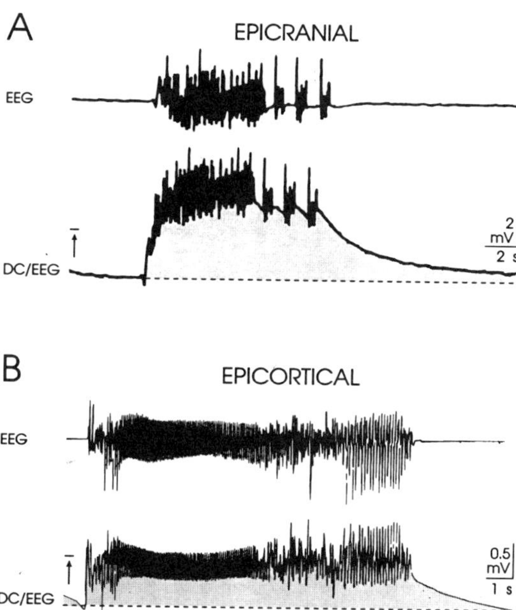

Figure 3.   Epicranial (**A**) and epicortical (**B**) recordings of EEG and DC potentials during tonic-clonic epileptic activity, anesthetized and artificially ventilated rat. Tonic-clonic epileptic activity was elicited by repeated injections of pentylenetetrazol. (**A**: Modified from: Lehmenkühler, A., Änderungen des Mikromilieus von Nervenzellen in der Hirnrinde bei epileptischen Anfällen. *Experimentelle Beobachtungen, EEG-Labor* 10:145–161, 1988. **B**: Modified from Speckmann, E.-J., *Experimentelle Epilepsieforschung*, Wissenschaftliche Buchgesellschaft, Darmstadt, 1986).

played in Figure 3. The EEG and DC potentials are recorded from the surface of the scalp (Fig. 3A) as well as the cortex (Fig. 3B). The clonic and tonic phases can clearly be distinguished from the EEG recording. Concomitant with the events in the conventional EEG, the epicranial and epicortical DC potentials show negative shifts that are characterized by a steep rise and a slow postictal decay. As a whole, epicranial and epicortical DC shifts during tonic-clonic activity are of same polarity and similar in time course (Caspers, 1963, 1974; Caspers and Speckmann, 1969; Caspers, Speckmann, and Lehmenkühler, 1987; Goldring, 1974; Gumnit, 1974; Speckmann, 1986; Wieser, 1983).

The negative epicranial and epicortical DC shifts during tonic-clonic activity are coupled with a series of paroxysmal depolarization shifts (Fig. 4A). Thus, there is a close relationship between the mean neuronal depolarization and the negative DC displacement. There are, however, differences with the commencement and termination of the seizure. At the neuronal level, the epileptic activity commences with a paroxysmal depolarization in

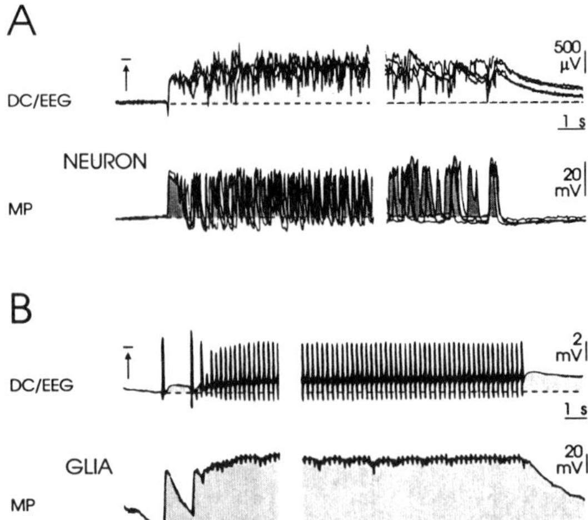

Figure 4.   Epicortical DC/EEG recordings and simultaneous neuronal (**A**) and glial (**B**) membrane potential changes during tonic-clonic (**A**) and high-frequency focal interictal epileptic activity (**B**); anesthetized and artificially ventilated cat (**A**) and rat (**B**). Tonic-clonic epileptic activity was elicited by repeated injections of pentylenetetrazol. Focal interictal epileptic activity was elicited by local application of penicillin and driven by high-frequency epicortical electrical stimulation (stimulation artifacts). (Modified from Speckmann, E.-J., *Experimentelle Epilepsieforschung*, Wissenschaftliche Buchgesellschaft, Darmstadt, 1986.)

any case. This first neuronal depolarization is associated with a monophasic negative and positive or a biphasic positive negative fluctuation in the field potential recordings. At the termination of the seizure activity, the neurons show an afterhyperpolarization whereas the field potentials are still negative. These discrepancies demonstrate that, as generators for field potentials, sources other than neuronal ones have to be taken into account; glial cells may be considered primarily among those. There is in fact a close correlation between the repolarization of glial cells and the decay of the negative DC shift in the postictal phase (Fig. 4B; Klee, Lux, and Speckmann, 1982, 1991; Speckmann, 1986; Speckmann and Caspers, 1979b; Speckmann, Caspers, and Elger, 1984; Speckmann, Caspers, and Janzen, 1972, 1978).

In summary, epileptic events are associated with negative DC shifts at the surface of the scalp and of the neocortex. These negative displacements go in parallel with neuronal depolarizations during the seizures.

Hypercapnia

Hypercapnia is achieved by elevation of the inspiratory CO_2 content or by a ventilatory arrest after breathing 100% O_2 for 15–30 min (apnea tech-

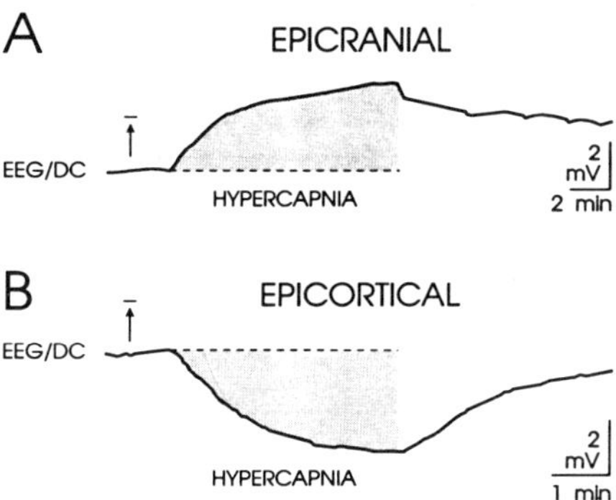

Figure 5.   Epicranial (**A**) and epicortical (**B**) recordings of DC potentials during hypercapnia, anesthetized and artificially ventilated rat (**A**) and cat (**B**). Hypercapnia was induced by elevation of the inspiratory CO_2 content to 10%. EEG is not seen on tracing because of reduced upper-frequency limit. (**A**: Modified from Pöppelmann, T., DC-Potentiale der Schädeloberfläche bei Hyperkapnie, Hypoxie und Asphyxie. Doctoral thesis, Münster, 1988. **B**: Modified from Caspers, H., Speckmann, E.-J., Cortical DC shifts associated with changes of gas tension in blood and tissue. In: *Handbook of Electroencephalography and Clinical Neurophysiology*, Vol. 10A, *Direct, Cortical and Depth Evaluation of the Brain*, Remond A., ed., pp. 42–65. Amsterdam: Elsevier, 1974).

nique). A typical experiment is displayed in Figure 5. The epicranial recording shows a marked negative DC shift during hypercapnia. In contrast to this, the epicortical tracing is characterized by a positive DC displacement. As a whole, epicranial and epicortical DC shifts during hypercapnia are similar in time course but opposite in polarity (Caspers, 1974; Caspers and Speckmann, 1974; Caspers, Speckmann, and Lehmenkühler, 1987).

The negative epicranial and the positive epicortical DC shift during hypercapnia go in parallel with a hyperpolarization of cortical neurons (Fig. 6A). This hyperpolarization is caused by an increase in potassium conductance leading to an outward flow of potassium ions. The resulting increase in the extracellular potassium concentration gives rise to a depolarization of glial cells during the hypercapnic period (Fig. 6B; Caspers and Speckmann, 1974; Caspers, Speckmann, and Lehmenkühler, 1979; Pöppelmann, 1988; Speckmann and Caspers, 1974; Speckmann, Caspers, and Elger, 1984).

The findings indicate that the positive epicortical DC shift during hypercapnia mainly results from a hyperpolarization of neurons. Participation of the membrane potential changes of glial cells can be neglected. As far as the epicranial negative shift is concerned, generators other than neurons, for example, the blood–brain barrier, predominate (Caspers, 1963).

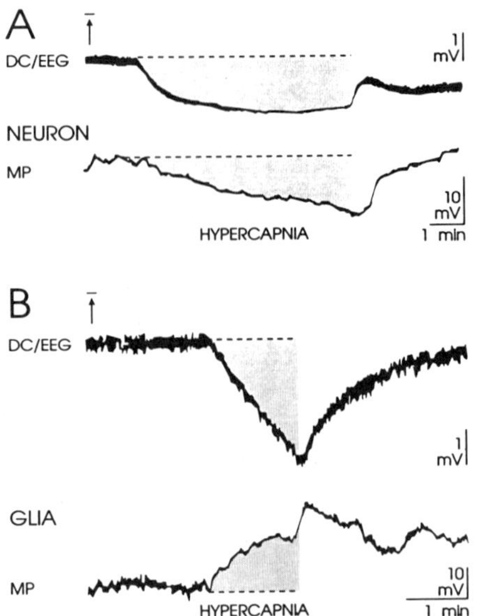

Figure 6.   Epicortical DC/EEG recordings and simultaneous neuronal (**A**) and glial (**B**) membrane potential changes during hypercapnia, anesthetized and artificially ventilated cat (**A**) and rat (**B**). Hypercapnia was induced by ventilatory arrest after breathing 100% O_2 for about 15 min (apnea technique). (Modified from Caspers, H., Speckmann, E.-J., Lehmenkühler, A., Electrogenesis of slow potentials of the brain. In: *Self-Regulation of the Brain and Behaviour*, Elbert, N., Rockstroh, B., Lutzenberger, W., Birbaumer, N., eds., pp. 26–41. Berlin: Springer-Verlag, 1984).

Hypoxia

Hypoxia is elicited by a decrease of the inspiratory O_2 content or by a short ventilatory arrest after breathing air. A typical experiment is displayed in Figure 7. The epicranial recording shows a marked positive DC shift during hypoxia. In contrast to this, the epicortical tracing is characterized by a negative DC displacement. As a whole, epicranial and epicortical DC shifts during hypoxia are similar in time course and opposite in polarity. The positive epicranial and the negative epicortical (Fig. 8A) DC shift during hypoxia go in parallel with a depolarization of cortical neurons (Fig. 8B-1). The depolarization is superimposed by frequent excitatory postsynaptic potentials of high amplitude. As with the hyperpolarization during hypercapnia the depolarization with hypoxia goes in parallel with an outflow of potassium ions; this in turn depolarizes glial cells (Fig. 8B-2; Caspers, 1974; Caspers and Speckmann, 1974; Caspers, Speckmann, and Lehmenkühler, 1987; Pöppelmann, 1988; Speckmann and Caspers, 1974; Speckmann, Caspers, and Elger, 1984).

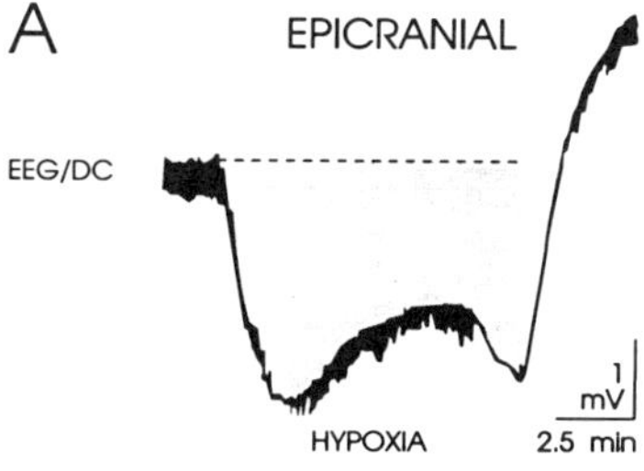

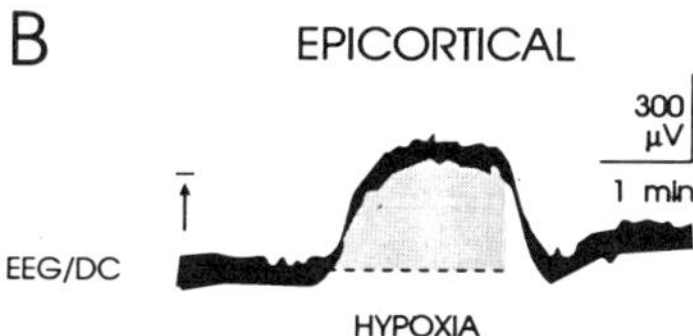

Figure 7.   Epicranial (**A**) and epicortical (**B**) recordings of DC potentials during hypoxia, anesthetized and artificially ventilated rat. Hypoxia was induced by decrease of inspiratory O_2 content to 10% (**A**) and 12% (**B**). **A**: Modified from Pöppelmann, T., DC-Potentiale der Schädeloberfläche bei Hyperkapnie, Hypoxie und Asphyxie, Doctoral thesis, Münster, 1988. **B**: Modified from Caspers, H., Speckmann, E.-J., Cortical DC shifts associated with changes of gas tension in blood and tissue. In: *Handbook of Electroencephalography and Clinical Neurophysiology, Vol. 10A, Direct, Cortical and Depth Evaluation of the Brain*, Remond A., ed., pp. 42–65. Amsterdam: Elsevier, 1974.)

The findings indicate that the negative epicortical DC shift during hypoxia is mainly caused by depolarization of neurons. This may be supported by glial depolarizations. As far as the positive epicranial DC shift is concerned, as in the case of hypercapnia other generator structures must be taken into account (Caspers and Speckmann, 1974; Caspers, Speckmann, and Lehmenkühler, 1980).

Conclusions

There is a close correspondence between the epicortical DC shift and membrane potential changes of neurons with epileptic activity and gas pressure changes. The glial membrane potential changes contribute to the sustained field potential shifts at the cortical surface only to a minor extent. There is no general parallelism with respect to polarity between the DC potentials recorded from the surface of the scalp and that of the cortex.

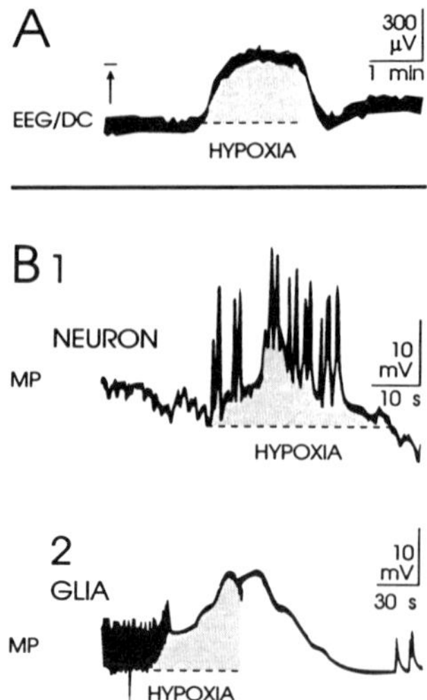

Figure 8.   Epicortical DC/EEG recording (**A**) and corresponding neuronal (**B-1**) and glial (**B-2**) membrane potential changes during hypoxia, anesthetized and artificially ventilated rat. Hypoxia was induced by decrease of the inspiratory O_2 content to 12% (**A**) and by short ventilatory arrest after breathing air (**B-1, B-2**). (**A**: Modified from Caspers, H., Speckmann, E.-J., Cortical DC shifts associated with changes of gas tension in blood and tissue. In: *Handbook of Electroencephalography and Clinical Neurophysiology, Vol. 10A, Direct, Cortical and Depth Evaluation of the Brain*, Remond A, ed., pp. 42–65. Amsterdam: Elsevier, 1974. **B-1**: Modified from: Speckmann, E.-J., Caspers, H., The effect of O_2- and CO_2-tensions in the nervous tissue on neuronal activity and DC potentials. In: *Handbook of Electroencephalography and Clinical Neurophysiology, Vol. 2C, Electrical Activity from the Neuron to the EEG and EMG.*, Remond A., ed., pp. 71–89. Amsterdam: Elsevier, 1974. **B-2**: Modified from: Caspers, H., Speckmann, E.-J., Lehmenkühler, A., Electrogenesis of slow potentials of the brain. In: *Self-Regulation of the Brain and Behaviour*, Elbert, N., Rockstroh, B., Lutzenberger, W., Birbaumer, N., eds., pp. 26–41. Berlin: Springer-Verlag, 1984.)

References

Andersen P, Andersson SA (1968): *Physiological Basis of the Alpha Rhythm*. New York: Meredith.

Caspers H (1963): Relations of steady potential shifts in the cortex to the wakefulness-sleep spectrum. In: *Brain Function*, Brazier, MAB, ed., pp. 177–200. Berkeley: University of California Press.

Caspers H, ed. (1974): DC potentials recorded directly from the cortex. In: *Handbook of Electroencephalography and Clinical Neurophysiology*, Vol. 10, Part A, Remond A, ed., p. 3. Amsterdam: Elsevier.

Caspers H, Speckmann E-J (1969): DC potential shifts in paroxysmal states. In: *Basic Mechanisms of the Epilepsies*, Jasper HH, Ward AA, Jr., Pope A, eds., pp. 375–395. Boston: Little, Brown.

Caspers H, Speckmann E-J (1974): Cortical DC shifts associated with changes of gas tensions in blood and tissue. In: *Handbook of Electroencephalography and Clinical Neurophysiology*, Vol. 10, Part A, Remond A, ed., pp. 41–65. Amsterdam: Elsevier.

Caspers H, Speckmann E-J, Lehmenkühler A (1979): Effects of CO_2 on cortical field potentials in relation to neuronal activity. In: *Origin of Cerebral Field Potentials*, Speckmann E-J, Caspers H, eds., pp. 151–163. Stuttgart: Thieme.

Caspers H, Speckmann E-J, Lehmenkühler A (1980): Electrogenesis of cortical DC potentials. In: *Motivation, Motor and Sensory Processes of the Brain: Electrical Potentials, Behaviour and Clinical Use, Progress in Brain Research*, Vol. 54, Kornhuber HH, Deecke L, eds., pp. 3–15. New York: Elsevier.

Caspers H, Speckmann E-J, Lehmenkühler A (1984): Electrogenesis of slow potentials of the brain. In: *Self-Regulation of the Brain and Behavior*, Elbert T, Rockstroh B, Lützenberger W, Birbaumer N, eds., pp. 26–41. New York: Springer.

Caspers H, Speckmann E-J, Lehmenkühler A (1987): DC potentials of the cerebral cortex. Seizure activity and changes in gas pressures. *Rev Physiol Biochem Pharmacol*, 106:127–178.

Creutzfeldt O, Houchin J (1974): Neuronal basis of EEG waves. In: *Handbook of Electroencephalography and Clinical Neurophysiology*, Vol. 2, Part C, Remond A, ed., pp. 5–55. Amsterdam: Elsevier.

De Robertis EDP, Carrea R, eds. (1965): *Biology of Neuroglia. Progress in Brain Research*, p. 15. New York: Elsevier.

Elger CE, Speckmann E-J (1980): Focal interictal epileptiform discharges (FIED) in the epicortical EEG and their relations to spinal field potentials in the rat. *Electroencephalogr Clin Neurophysiol* 48:447–460.

Elger CE, Speckmann E-J (1983a): Penicillin-induced epileptic foci in the motor cortex: Vertical inhibition. *Electroencephalogr Clin Neurophysiol* 56:604–622.

Elger CE, Speckmann E-J (1983b): Vertical inhibition in motor cortical epileptic foci and its consequences for descending neuronal activity to the spinal cord. In: *Epilepsy and Motor System*, Speckmann E-J, Elger CE, eds., pp. 152–160. Baltimore: Urban & Schwarzenberg.

Elger CE, Speckmann E-J, Caspers H, Prohaska O (1982): Focal interictal epileptiform discharges in the cortex of the rat: Laminar restriction and its consequences for activity descending to the spinal cord. In: *Physiology and Pharmacology of Epileptogenic Phenomena*, Klee MR, Lux HD, Speckmann E-J, pp. 13–20. New York: Raven Press.

Elger CE, Speckmann E-J, Prohaska O, Caspers H (1981): Pattern of intracortical potential distribution during focal interictal epileptiform discharges (FIED) and its relation to spinal field potentials in the rat. *Electroencephalogr Clin Neurophysiol* 51:393–402.

Goldring S (1974): DC shifts released by direct and afferent stimulation. In: *Handbook of Electroencephalography and Clinical Neurophysiology*, Vol. 10, Part A, Remond A, ed., pp. 12–24. Amsterdam: Elsevier.

Gumnit R (1974): DC shifts accompanying seizure activity. In: *Handbook of Electroencephalography and Clinical Neurophysiology*, Vol. 10, Part A, Remond A, ed., pp. 66–77. Amsterdam: Elsevier.

Gumnit RJ, Matsumoto H, Vasconetto C (1970): DC activity in the depth of an experimental epileptic focus. *Electroencephalogr Clin Neurophysiol* 28:333–339.

Jasper HH, Ward AA, Pope A, eds. (1969): *Basic Mechanisms of the Epilepsies*. Boston: Little, Brown.

Klee MR, Lux HD, Speckmann E-J, eds. (1982): *Physiology and Pharmacology of Epileptogenic Phenomena*. New York: Raven Press.

Klee MR, Lux HD, Speckmann E-J, eds. (1991): *Physiology, Pharmacology and Development of Epileptogenic Phenomena*. Experimental Brain Research Series 20. Berlin: Springer.

Kuffler SW, Nicholls JG (1966): The physiology of neuroglial cells. *Ergeb Physiol* 57:1–90.

Kuffler SW, Nicholls JG, Orkand RK (1966): Physiological properties of glial cells in the central nervous system of amphibia. *J Neurophysiol* 29:768–780.

Lehmenkühler A (1988): Änderungen des Mikromilieus von Nervenzellen in der Hirnrinde bei epileptischen Anfällen. *Exp Beobachtung. EEG* 10:145–161.

Palay SL, Chan-Palay V (1977): General morphology of neurons and neuroglia. In: *Handbook of Physiology. The Nervous System*, Vol. 1, Part 1, Kandel ER, ed., pp. 5–37. Bethesda: American Physiological Society.

Petsche H, Müller-Paschinger I, Pockberger H, Prohaska O, Rappelsberger P, Vollmer R (1978): Depth profiles of electrocortical activities and cortical architectonics. In: *Architectonics of the Cerebral Cortex*, Vol. 3, Brazier MAB, Petsche H, eds., pp. 257–280. IBRO Monograph Series. New York: Raven Press.

Petsche H, Pockberger H, Rappelsberger P (1981): Current source density studies of epileptic phenomena and the morphology of the rabbit's striate cortex. In: *Physiology and Pharmacology of Epileptogenic Phenomena*, Klee MR, Lux HD, Speckmann E-J, eds., pp. 53–63. New York: Raven Press.

Pöppelmann T (1988): *DC-Potentiale der Schädeloberfläche bei Hyperkapnie, Hypoxie und Asphyxie*. Doctoral Thesis, Münster, Germany.

Somjen GG, Trachtenberg M (1979): Neuroglia as generator of extracellular current. In: *Origin of Cerebral Field Potentials*, Speckmann E-J, Caspers H, eds., pp. 21–32. Stuttgart: Thieme.

Speckmann E-J (1986): Experimentelle Epilepsieforschung. Darmstadt: Wissenschaftliche Buchgesellschaft.

Speckmann E-J, Caspers H (1974): The effect of O_2 and CO_2 tensions in the nervous tissue on neuronal activity and DC potentials. In: *Handbook of Electroencephalography and Clinical Neurophysiology*, Vol. 2, Part C, Remond A, ed., pp. 71–89. Amsterdam: Elsevier.

Speckmann E-J, Caspers H, eds. (1979a): *Origin of Cerebral Field Potentials*. Stuttgart: Thieme.

Speckmann E-J, Caspers H (1979b): Cortical field potentials in relation to neuronal activities in seizure conditions. In: *Origin of Cerebral Field Potentials*, Speckmann E-J, Caspers H, eds., pp. 205–213. Stuttgart: Thieme.

Speckmann E-J, Elger CE, eds. (1983): *Epilepsy and Motor System*. Baltimore: Urban & Schwarzenberg.

Speckmann E-J, Elger CE (1984): The neurophysiological basis of epileptic activity: A condensed overview. In: *Epilepsy, Sleep and Sleep Deprivation*, Degen R, Niedermeyer E, eds., pp. 23–34. Amsterdam: Elsevier.

Speckmann E-J, Walden J (1991): Mechanisms underlying the generation of cortical field potentials. *Acta Otolaryngol* (Suppl.) 491:17–24.

Speckmann EJ, Caspers H, Elger CE (1984): Neuronal mechanisms underlying the generation of field potentials. In: *Self-Regulation of the Brain and Behavior*, Elbert T, Rockstroh B, Lützenberger W, Birbaumer N, eds., pp. 9–25. New York: Springer.

Speckmann E-J, Caspers H, Janzen RWC (1972): Relations between cortical DC shifts and membrane potential changes of cortical neurons associated with seizure activity. In: *Synchronization of EEG Activity in Epilepsies*, Petsche H, Brazier MAB, eds., pp. 93–111. New York: Springer.

Speckmann E-J, Caspers H, Janzen RWC (1978): Laminar distribution of cortical field potentials in relation to neuronal activities during seizure discharges. In: *Architectonics of the Cerebral Cortex*, Vol. 3, Brazier MAB, Petsche H, eds., pp. 191–209. IBRO Monograph Series. New York: Raven Press.

Wieser HG (1983): *Electroclinical Features of the Psychomotor Seizure. A Stereo-encephalographic Study of Ictal Symptoms and Chronotopographical Seizure Patterns Including Clinical Effects of Intracerebral Stimulation*. New York: Gustav Fischer.

Chapter 14

The Neural Substrates of Cognitive Event-Related Potentials: A Review of Animal Models of P3

KEN A. PALLER

One reason to study the electrical activity of the brain is that the knowledge gained might be useful for understanding the neural basis of cognitive functions. In other words, *endogenous* or *cognitive* event-related potentials (ERPs), which are averaged field potentials that are sensitive to manipulations of psychological variables, could provide evidence to constrain theories on the brain mechanisms that mediate cognition. The success of this approach, however, depends on the extent to which two types of relationships can be substantiated: relationships between ERPs and theories of cognition and relationships between ERPs and particular neural events, structures, or systems. Although cognitive ERPs have been thoroughly characterized only in humans, critical evidence for clarifying the neurophysiological substrates of cognitive ERPs may come from studies in animals (Galambos and Hillyard, 1981). This review focuses on recent studies that have begun to develop animal models of a particularly well-characterized cognitive ERP, the P3 or P300 potential. To develop criteria for assessing these animal models, the characteristics of P3 in humans are first outlined. Evidence pertaining to the adequacy of each of the putative animal models is then discussed by species, along with evidence about the neural basis of these ERPs. The concluding section will reflect on prospects for understanding the neural basis of P3 and on the usefulness of the animal approach.

Cognitive Electrophysiology
H-J. Heinze, T.F. Münte, and G.R. Mangun, editors
© 1994 Birkhäuser Boston

Studies of P3 and Its Neutral Substrates in Humans

P3 was first described by Sutton and colleagues (Sutton et al., 1965, 1967), who showed that the potential could be elicited by an unexpected stimulus as well as by the unexpected absence of a stimulus. P3 is a positive ERP that typically peaks at about 300 msec or more after an informative stimulus that is delivered unexpectedly (see reviews by Fabiani et al., 1987; Hillyard and Kutas, 1983; Pritchard, 1981). P3 is most commonly recorded during a randomly ordered sequence of stimuli from two classes, such as high- and low-frequency tones. Stimuli from a relatively infrequent class ("oddball" stimuli) reliably elicit P3 if the subject actively discriminates the stimuli, for example, by counting stimuli from one class. In some cases, P3 can be elicited by oddball stimuli not given explicit relevance (Ritter, Vaughan, and Costa, 1968; Roth et al., 1976; Squires et al., 1977), but often the scalp topography then has a frontal maximum and the peak latency is shorter (e.g., Squires, Squires, and Hillyard, 1975). P3 amplitude tends to decrease when attention is diverted (Wickens et al., 1983), or when the subject is uncertain about having correctly perceived the stimulus (Hillyard et al., 1971; Ruchkin and Sutton, 1978). The latency of P3 is thought to vary with the time required for stimulus evaluation and to be independent of motor processing (Kutas, McCarthy, and Donchin, 1977; McCarthy and Donchin, 1981). For example, P3 to a soft click interspersed in a sequence of loud clicks may peak at a latency of about 300 msec, but P3 to a female name in a list of male names may occur at a latency of 500 msec or greater. P3 can be elicited by stimuli in auditory, visual, or somatosensory modalities (Desmedt and Debecker, 1979; Simson, Vaughan, and Ritter, 1977; Snyder, Hillyard, and Galambos, 1980).

Hypotheses about specific information processing events that are indexed by P3 have been developed by coopting a variety of psychological constructs, although a consensus on this issue has not been reached (e.g., see Verleger, 1988, and associated commentaries). We can nevertheless adopt the working hypothesis that P3 waves reflect neural events called into play as a part of some physiologically meaningful process or processes. At present, it is impossible to specify the nature or function of these neural events in either physiological or psychological terms, nor is it clear whether different types of P3 waves produced in different task situations reflect identical or different processes. Future advances may bring about a resolution of these issues, but it is likely that advances of this sort will require more knowledge about the neural bases of the potentials.

Empirical support for hypotheses about the anatomical and physiological substrates of P3 can be derived from detailed mappings of potential fields on the scalp, intracranial recordings, magnetoencephalographic recordings, and studies in neurological patients with known brain lesions. Evidence from field distributions of P3 on the scalp has been interpreted to implicate certain configurations of intracranial current generators, despite the ambiguity of this "inverse method." The scalp distribution of P3 is broad and bilaterally symmetric, reaches maximal amplitudes at central and parietal sites on the midline, and is relatively independent of stimulus modality. It has been hypothesized that P3 arises from bilateral generators in parietal association cortex with an additional contribution from frontal cortex (Simson, Vaughan, and Ritter, 1976, 1977). Another suggestion is that diffuse projections from brain stem to cortex play a key role (Desmedt, 1981; Galambos and Hillyard, 1981). A third hypothesis emphasizes the importance of subcortical structures, especially the hippocampus (see below). Extracranial recordings have proven insufficient to differentiate between these three hypotheses.

Intracranial recordings have been obtained from epileptic patients who were candidates for neurosurgery to relieve medically intractable seizures and from a few patients with other neurological disorders (Halgren et al., 1980; McCarthy et al., 1989; Prim, Ojemann, and Lettich, 1983; Smith et al., 1990; Wood et al., 1984; Yingling and Hosobuchi, 1984). The most pervasive intracranial finding has been that ERP waveforms that resemble P3 can be recorded from medial temporal locations under the same experimental conditions that are used to elicit P3 from the scalp. These high-amplitude responses showed steep potential gradients across closely spaced electrodes. This evidence, along with evidence from single unit responses, indicated that electrical activity was locally produced within this part of the brain at the time P3 occurred. It was thus suggested that this activity was volume conducted through brain tissue and made a major contribution to the scalp-recorded P3.

Intracranial data such as these face a number of interpretative problems. One problem is that it is difficult to determine precisely where the electrodes are located with respect to anatomical boundaries. Another problem is that the patients' histories of disordered brain electrical activity and anticonvulsant pharmacotherapy may limit the generalizability of the results. Most importantly, intracranial electrode placements are dictated by clinical considerations such that probes are targeted for areas of suspected pathology. Normal tissue may also be sampled in many cases, but in any given patient only a limited number of areas are examined. Such limited samplings pose difficulties for intracranial field mapping and for conclusively determining the sources of scalp-recorded potentials.

Magnetoencephalographic recordings in normal subjects have been interpreted to confirm the idea that P3 is generated in the hippocampus (Okada, Kaufman, and Williamson, 1983). In this study, the magnetic field measured using a single sensor moved to various locations around the head showed an event-related response presumed to correspond to the electrical P3. This field approximated that which would be produced by a dipole source in the hippocampus. The assumption of an equivalent dipole source, however, may have been misleading in this case. Further, magnetoencephalographic mapping results from other laboratories have diverged considerably (e.g., Gordon et al., 1987; Richer, Johnson, and Beatty, 1983; Trahms et al., 1990). Indeed, there are many difficulties in using magnetic event-related activity to investigate ERP sources, although the situation may be improved by multichannel recording and by advances in modeling techniques, such as using neuroimaging evidence to constrain localization in a multiple dipole model (e.g., Dale and Sereno, 1993). In some circumstances, magnetic and electrical techniques can be used in a complementary manner by exploiting their differential sensitivities (e.g., magnetic recordings are relatively more sensitive to tangential versus radial sources). Moreover, some sources of an electrical signal may not yield an appreciable magnetic signal. Therefore, the sources of an ERP are likely to differ from the sources of the corresponding magnetoencephalographic signal, which underscores the need for biologically accurate source modeling.

ERP studies in patients with known brain lesions can provide evidence that is more direct, although interpretations must take into account confounding factors such as conductivity changes from skull defects and scar tissue. Still, lesion evidence has convincingly implicated a selective role for the frontal lobe in modulating P3 (Knight, 1984). Patients with unilateral prefrontal lesions exhibited normal P3 waves to target stimuli but not to novel, nontarget stimuli. Another lesion study suggested that neocortex in the superior temporal plane plays an important role in the system that generates P3 (Knight et al., 1989). This study showed that P3 was unaffected by lesions of the lateral parietal lobe, whereas lesions of the parietotemporal junction were associated with large decrements in P3. However, it is presently unclear why unilateral lesions were associated with a bilateral disruption, given that callosal damage alone does not cause such a disruption (Kutas et al., 1990). Finally, several studies have shown that unilateral temporal lobectomies in epileptic patients have minimal effects on P3 (Johnson, 1988; Stapleton, Halgren, and Moreno, 1987; Wood et al., 1982). In addition to these findings from epileptic patients with unilateral damage, normal P3 potentials have also been recorded from a patient with unilateral medial temporal damage due to a tumor (Rugg et al., 1991) and from patients with bilateral medial temporal damage due to encephalitis (Onofrj et al., 1991, 1992; Potter et al., 1993). In sum, lesion studies have contributed to our understanding of the neural generation of P3, although the data are still consistent with a large number of alternative hypotheses.

Animal Models of P3

A clarification of the neurophysiological substrates of P3 may require some synergism between studies in humans and studies in nonhuman animals. Many types of evidence could be used to argue that homologous brain circuitry in different species produces the same sort of electrical fields. Such evidence could include intracranial field analyses, current source density analyses, correlations between ERPs and unit activity, stimulation experiments, pharmacological experiments, and tests with brain lesions. In animals, these techniques can often be applied in a more systematic manner than in humans, yet the decision to conduct invasive experimentation should not be taken lightly. Can adequate justification be achieved by weighing possible benefits against costs? Some benefits are readily anticipated. For example, an increase in the clinical utility of cognitive ERPs may develop from learning about their neural substrates. In contrast, long-term benefits can be difficult to foresee because they may depend on discoveries yet to be made. (A corollary of this idea is that future technical advances could obviate the need to rely on some of today's invasive techniques.) Given that the benefits of any particular research direction can seldom be determined with certainty a priori, limiting our goals to short-term, practical applications would stifle many scientific advances. Although the costs of using animals in research can similarly be difficult to determine, steps can be taken to minimize discomfort and other disadvantages of living in captivity. Indeed, the key ingredient for dealing with the costs may be compassion. Further, advances in comparative psychology are giving us a deeper understanding of the complexity of the cognitive capabilities of animals (e.g., Cheney and Seyfarth, 1992), which underscores the need to adequately consider their psychological well-being when used as subjects. It is conceivable that ERP measures themselves may eventually be useful tools for furthering our understanding of cognition in animals (Hillyard and Bloom, 1982).

Rigorously validating a P3-like potential in a nonhuman species would require demonstrating each of the important characteristics of the human P3. Although many characteristics can be listed, there is little consensus about exactly which ones are necessary or sufficient for unambiguously identifying the P3 component. Still, it may be helpful to subdivide the features of P3 into three categories. (1) The *functional characteristics* strive toward integrating the measure into models of information processing and generally constitute task manipulations known to influence P3. (2) The *waveform characteristics* include latency, polarity, amplitude, and wave shape. (3) The *neural characteristics* refer more directly to the particular neural elements responsible for the ERP. There is room for a great deal of improvement in our understanding of the neural characteristics of P3, and this knowledge will ultimately be necessary for confirming the validity of the animal models.

Of course, assessing the waveform characteristics of P3-like ERPs is the most straightforward way to begin. Whenever feasible, however, neural and functional characteristics should be weighed more heavily in evaluating the animal models, because waveforms may vary with anatomical or morphological differences between species that may be irrelevant to brain function. A tentative list of functional characteristics that may be useful for this purpose is given in Table 1. Because these characteristics have been derived from

Table 1. Functional criteria for P3-like ERPs in animals

1. Elicited by relatively unpredictable stimuli
    a. amplitude proportional to degree of unpredictability
    b. amplitude responsive to sequential patterns
2. Requires that attention is allocated to stimulus processing
    a. elicited by imperative stimuli in categorization tasks
    b. amplitude can be controlled by task requirements
3. Stimulus modality not critical
    a. elicited by auditory stimuli
    b. elicited by visual stimuli
    c. elicited by somatosensory stimuli
    d. elicited by semantically defined categories
    e. elicited by stimulus omissions
4. Latency increases with complexity of discrimination
    a. latency can vary with modality
    b. latency relatively independent of motor processing

manipulations of experimental parameters in P3 experiments, parallel tasks are important for assessing the animal models. The oddball task, in particular, has been central to the study of P3. In this task, a discriminative response is made to one type of stimulus. An animal can be trained to perform such a task through the use of operant conditioning. The behavioral response can then be used to infer that the animal is devoting attention to processing the stimuli. This sort of task, here termed a *categorization task*, can be distinguished from two other types of tasks that have also been used in animal studies. In a *conditioning task*, animals undergo procedures that may lead them to pay attention to stimuli, but, unlike the categorization paradigm, classical conditioning is used and the behavioral response initially occurs automatically. After training, the conditioned response is monitored to infer that the animal is processing the stimulus to some extent, but it may not be safe to assume that stimulus processing is equivalent to that in a categorization task. Because conditioning paradigms have rarely been used in studies of P3 in humans, they provide relatively weaker evidence for validating an animal model. Nevertheless, further evidence (such as waveform similarities across tasks) may be used to substantiate the P3-like responses that are recorded in conditioning tasks. A parallel argument applies to the last type of task, a *passive task*, in which no behavioral responses are made and the stimuli presented have no conditioned significance. Despite these uncontrolled circumstances, some investigators have found that animals may nonetheless exhibit P3-like responses.

When referring to the putative animal models, the phrase "P3-like ERPs" is used to avoid prejudging the relationship between the ERPs and P3 in humans. Incidentally, some early experiments in animals revealed P3-like ERPs that were not explicitly recognized as such (e.g., Donchin et al., 1971; Fuster et al., 1982; Galambos and Sheatz, 1962; John, 1963). The

Table 2.

| | Paradigm used to elicit P3-like ERPs | | |
Animal	Categorization	Conditioning	Passive
Macaque monkey	Arthur and Starr, 1984 Paller et al., 1988, 1992 Glover et al., 1991	Glover et al., 1986, 1988 Onofrj et al., 1987	Paller et al., 1982, 1988
Squirrel monkey	Pineda et al., 1988		Neville and Foote, 1984 Pineda et al., 1987, 1989 Swick et al., 1991
Cat		Wilder et al., 1981 Buckwald and Squires, 1982 O'Conner and Starr, 1985 Katayama et al., 1985a, b Harrison et al., 1985, etc. Csépe et al., 1987	Başar-Eroglu et al., 1987, 1991a, 1991b
Rabbit	Aleksandrov and Maksimova, 1985	Weisz et al., 1983 Stolar et al., 1989	
Rat	Brenner et al., 1987 Voorn et al., 1987	Hurlbut et al., 1987	O'Brien, 1982 Hurlbut et al., 1987 Yamaguchi et al., 1993
Dolphin		Woods et al., 1986	Woods et al., 1986
Dog		Koslov and Pirogov, 1988	

focus here is on studies explicitly designed to examine animal models of P3. Recordings were made from chronically implanted epidural electrodes unless otherwise specified. Table 2 shows which paradigms have been used to elicit P3-like ERPs in animals. Methodological differences between studies were such that any ERP differences across species are difficult to interpret, so rather than highlight these differences, emphasis is given to the high degree of similarity between ERP results from different species.

Macaque Monkey

The steps toward establishing an animal model of P3 have been approximated most closely in studies with several species of Old World monkey, all from the genus *Macaca*. In general, P3-like responses could be elicited in each of the three types of paradigms.

Categorization tasks were first used in the auditory modality. A pitch discrimination was required in these tasks, as monkeys were trained to respond to a pure tone of a particular frequency by releasing a lever (Fig. 1). These target tones occurred relatively infrequently ($p = .1$) in a sequence of nontarget tones ($p = .9$). Late positive ERPs were elicited by target tones

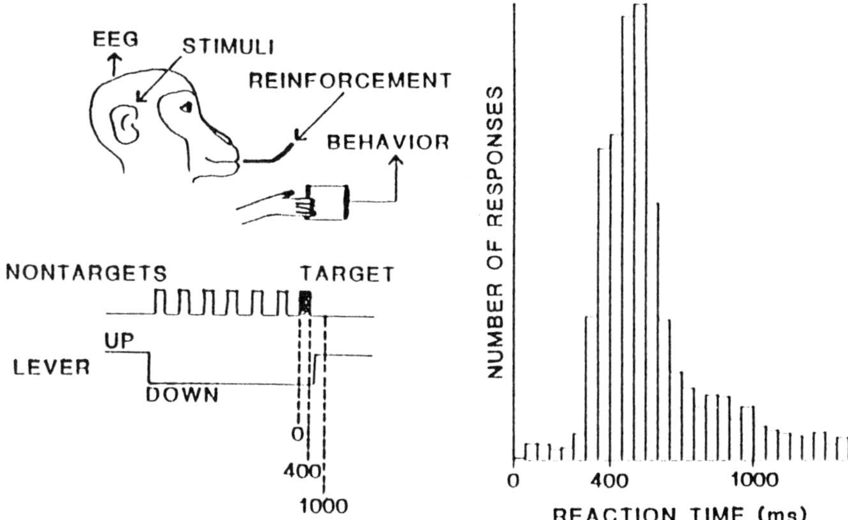

Figure 1.   Schematic diagram of auditory categorization task. Monkey pulled lever to initiate sequence of nontarget tones. After pseudorandom number of tones, target tone of different frequency occurred. Monkey released lever in 400- to 1000-msec interval to receive juice reinforcement. Bar graph shows that most responses occurred in this interval. (From Paller, 1986.)

in *Macaca nemestrina* (Arthur and Starr, 1984) as well as in *Macaca fasicularis* (Paller et al., 1988, Experiment 2). As shown in Figure 2, the amplitude of this ERP decreased when the probability of the rare tone was increased. The amplitude also decreased when the stimuli were presented while the response could not be executed. Other manipulations showed that the P3-like ERPs were contingent on the infrequent nature of stimuli rather than on particular physical stimulus parameters. Comparisons between ERPs from intact monkeys and monkeys with brain lesions showed that the hippocampus, amygdala, and overlying neocortical areas were not critical for the elicitation of normal P3-like responses (Paller et al., 1988).

Categorization paradigms in the visual modality have also been used. Surface and depth recordings were made in monkeys (*Macaca mulatta*) trained to touch stimuli presented on a video monitor with a touch-sensitive screen (Paller et al., 1992). Three types of stimuli were presented: a black-and-white checkerboard target ($p = .1$) and two nontarget stimuli, a gray square and a red-and-white checkerboard ($p = .8$ and .1, respectively). These stimuli were selected such that the target differed from one nontarget primarily in form and from the other nontarget primarily in color, which may promote the allocation of attention to making the discrimination (Treisman and Gelade, 1980). Targets elicited P3-like potentials from

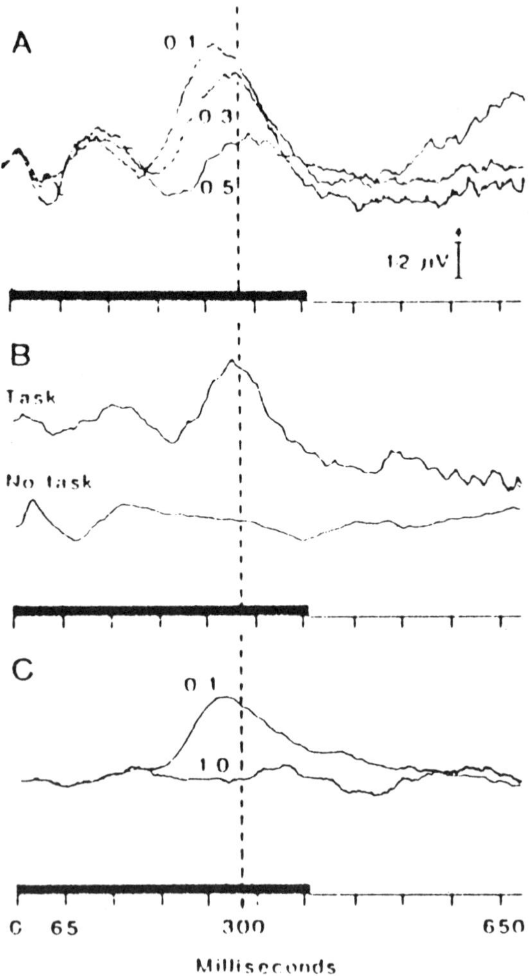

Figure 2.   ERPs elicited by target in an auditory categorization paradigm. **A.** ERPs averaged from two monkeys show a late positive component that increased in amplitude inversely with probability of target. **B.** ERPs averaged across three midline electrodes in one monkey show that P3-like ERP was elicited during performance of task but not when task could not be performed. **C.** ERPs averaged across three midline electrodes in one monkey show that P3-like ERP was not elicited when 100% of tones were targets. Reprinted with permission of the AAAS from Arthur DL, Starr A (1984): Task-relevant late positive component of the auditory event-related potential in monkeys resembles P300 in humans. *Science* 223: 186–188. Copyright 1984 by the AAAS.

epidural contacts and corresponding negative potentials from depth contacts in the medial temporal lobe (Fig. 3). The epidural ERPs reached positive peaks between 260 and 300 msec, while negative peaks in the medial temporal lobe tended to occur slightly later. This correspondence between surface and depth responses has also been reported in P3 studies conducted in epileptic patients with electrodes implanted for seizure monitoring (McCarthy et al., 1989). However, the resolution with which recording electrodes could be localized was much greater in monkeys, suggesting that questions about the physiological substrates of the potentials may be resolved by further field mapping in monkeys.

In another visual paradigm, monkeys (*Macaca fasicularis*) were trained to respond to sinusoidal gratings of a particular orientation (Glover et al., 1991). P3-like ERPs were elicited by target stimuli that were infrequent

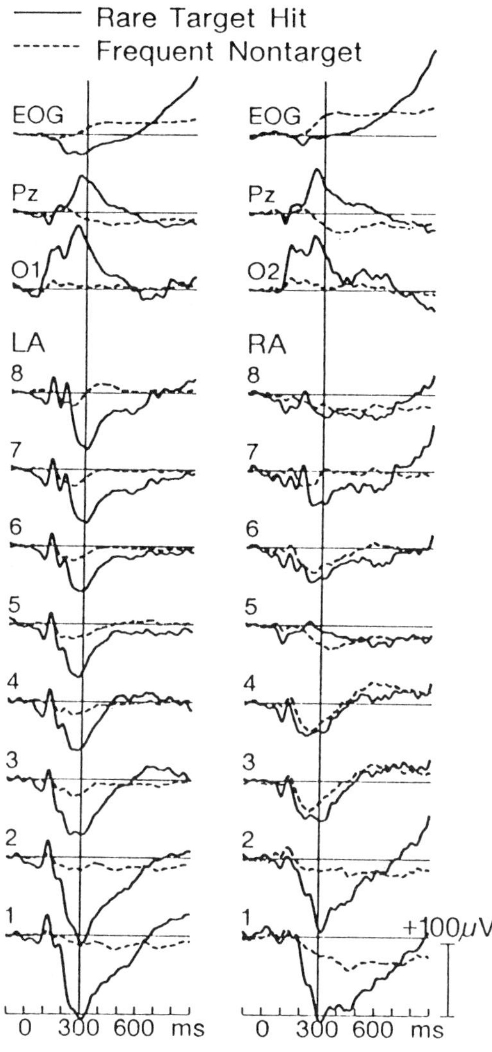

Figure 3. ERPs elicited from a monkey during a visual categorization paradigm. From epidural electrodes, a late positive ERP was elicited by rare targets that were detected (i.e., touched) but not by frequent nontargets. Left column: recordings from first half of session; right column: recordings from second half of session. ERPs from Pz are quite similar between half-sessions. Recordings from many contacts in the left anterior (LA) medial temporal lobe and the right anterior (RA) medial temporal lobe show negative deflections during latency of P3-like ERPs. Reprinted with permission from Elsevier Science Publishers, Inc. from Paller et al. (1992).

($p = .3$) or equiprobable ($p = .5$) but not when they were continual ($p = 1.0$). Results from 1 monkey trained to make a delayed response showed that P3-like ERPs were similar whether the mean reaction time was near 600 msec or 900 msec. Similar P3-like ERPs in visual categorization paradigms have also been recorded by other investigators (Antal et al., 1993; Arthur, 1985; S. Grant, personal communication).

Monkey ERPs have also been studied using a conditioning paradigm (Glover et al., 1986). Tones of two frequencies were presented, and the rare tone ($p = .2$) was followed 700 msec later by a mild shock (the intensity of which was adjusted to a level that produced a hind-paw twitch "without obvious autonomic effects"). The late positive ERP showed a maximum amplitude at 314 msec at a central midline electrode. This ERP appeared after 3–5 training sessions, and its amplitude was proportional to the probability of the rare tone (across manipulations from $p = .5$ to $p = .1$). The ERP decreased in amplitude during extinction sessions when conditioning was discontinued. The authors also noted that after 12 conditioning sessions the amplitude of P3-like ERPs tended to decrease during a single session even though hind-paw movements in response to the shock were stable. Further studies with this paradigm have shown that the P3-like ERPs were abolished in animals suffering from a temporary Parkinsonian syndrome caused by the administration of the neurotoxin MPTP (Glover et al., 1988). Levodopa treatment was used to alleviate the motor symptoms, but it did not bring back the P3-like ERPs. In clinical studies with humans, alterations of P3 have been associated with idiopathic Parkinson's disease, although these alterations may only occur when the disease includes a dementia component (Goodin and Aminoff, 1987). Other evidence also supports the hypothesis that P3 alterations can index Parkinsonian dementia by reflecting a cholinergic rather than a dopaminergic deficiency (see Bodis-Wollner, 1990). In a study with the drug L-acetylcarnitine, which may lead to improvements in patients with dementia via a cholinergic mechanism, the amplitude of P3-like ERPs in the conditioning paradigm was found to increase, along with a less consistent decrease in latency (Onofrj et al., 1987, but see Antal et al., 1993). Overall, the conditioning paradigm is useful because P3-like ERPs can be recorded after very few training sessions, but longitudinal studies may be problematic because responses tend to habituate.

Auditory stimuli in passive paradigms have also been used to elicit P3-like ERPs (Paller et al., 1982, 1988). In one paradigm, three types of tone were presented: a frequent pure tone ($p = .8$), a rare pure tone ($p = .1$), and a rare complex tone ($p = .1$). Tones were randomly ordered in a sequence presented at a 1-sec interstimulus interval (ISI). Figure 4 juxtaposes late

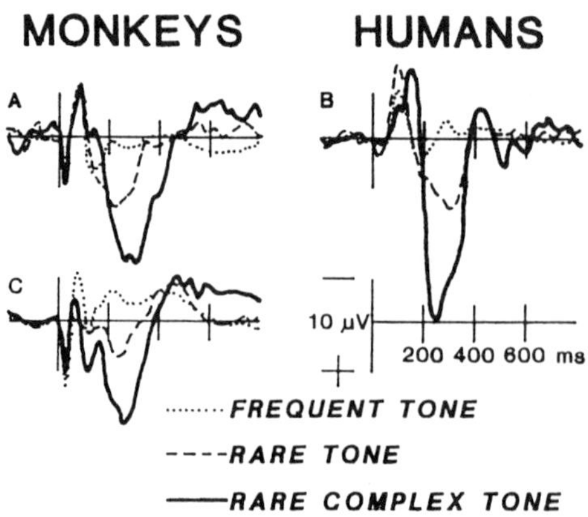

Figure 4. ERPs elicited during a passive paradigm from groups of five intact monkeys (**A**), five human subjects (**B**), and five monkeys with medial temporal lobectomies (**C**). P3-like ERPs were elicited by rare tones in all groups. Medial temporal lobe structures appear to be unnecessary for generation of normal P3-like ERPs. (See Paller et al., 1988.)

positive ERPs elicited from monkeys (*Macaca fasicularis*) with and without medial temporal lobectomies, and ERPs from human subjects. In the intact monkeys, the P3-like response to the rare complex tone reached a peak at 278 msec at Cz. Waveform characteristics were highly similar between monkeys and humans. The topography of the P3-like ERPs across electrodes (e.g., maximal amplitude at the central midline) also resembled the distribution of P3 recorded from scalp electrodes in human subjects under analogous circumstances. The authors hypothesized that monkeys tended to pay attention to the stimuli, although there were no task requirements to do so.

In summary, P3-like ERPs have been found in several macaque species. The peak amplitude of these positive responses generally occurred at a latency between 260 and 350 msec. Similar ERPs were elicited by auditory and visual stimuli, although responses to visual stimuli tended to be somewhat later than those to auditory stimuli, as is the case for P3 in humans. In direct comparisons, P3-like ERPs in monkeys bore a strong resemblance to P3 waves recorded from human subjects under comparable circumstances.

P3-like ERPs were found in each of the three types of experimental paradigm. However, some results in the passive paradigm showed that P3-like ERPs tended to habituate across sessions (Paller, 1986), whereas in other cases P3-like ERPs were never found (Glover et al., 1986). Similarly, passive conditions have led to variable results in human subjects (Donchin and Cohen, 1967; Ford, Roth, and Kopell, 1976; Polich, 1989b; Ritter, Vaughan, and Costa, 1968; Roth et al., 1976; Smith et al., 1970; Squires et al., 1977). The reliability of P3 may depend on the extent to which attentional resources are devoted to processing the stimuli. Such a dependence on attention is strongly supported by results from experiments in which human subjects performed two tasks concurrently (Wickens et al., 1983). In monkeys, P3-like ERPs may be elicited in passive conditions only if circumstances are such that monkeys tend to pay attention. Factors such as prior experience in testing situations and the monkeys' level of stress may play a role in determining whether they are attentive to the stimuli. Similarities between ERPs in the different paradigms nevertheless support the hypothesis that the underlying processes have much in common.

Response habituation can occur in passive and conditioning paradigms because of the lack of sufficient attentional requirements, whereas categorization paradigms minimize this problem and also duplicate the standard conditions for eliciting P3 in humans. Although P3 amplitude in humans has been shown to decline slightly during the course of a prolonged categorization paradigm (Polich, 1989a), the available evidence suggests that robust P3-like ERPs can still be elicited after multiple training sessions in monkeys. On the negative side, the use of reinforcement can exacerbate problems with response artifacts. Elaborate techniques to control eye movements are available, but without these, animals can produce electroocular and other artifacts that are time-locked to the stimulus of interest. It is therefore crucial either to eliminate these artifacts through training or to record them in a manner that allows contaminated trials to be eliminated. In general, the evidence from the studies cited above suggests that P3-like responses did not arise artifactually.

Squirrel Monkey

In the squirrel monkey (*Saimiri sciureus*), a New World monkey, P3-like ERPs have been recorded in passive and categorization paradigms. In passive auditory paradigms, positive ERPs in the 300- to 800-msec latency range were elicited by infrequent auditory stimuli with no conditioned significance (Neville and Foote, 1984; Pineda, Foote, and Neville, 1987; see also Ehlers, 1988, 1989). In one paradigm, rare tones ($p = .08$) in a sequence of frequent tones ($p = .92$) elicited a broad positivity beginning at about 300 msec and continuing for several hundred milliseconds. In another paradigm, a second type of rare stimulus (an edited dog bark that resembled the complex tone referred to above, $p = .08$) elicited a larger positivity with a sharper peak at 470 msec (Fig. 5). When two equiprobable tones were presented, the amplitude of P3-like ERPs was shown to depend on the sequential ordering of the stimuli as has been reported for P3 in humans

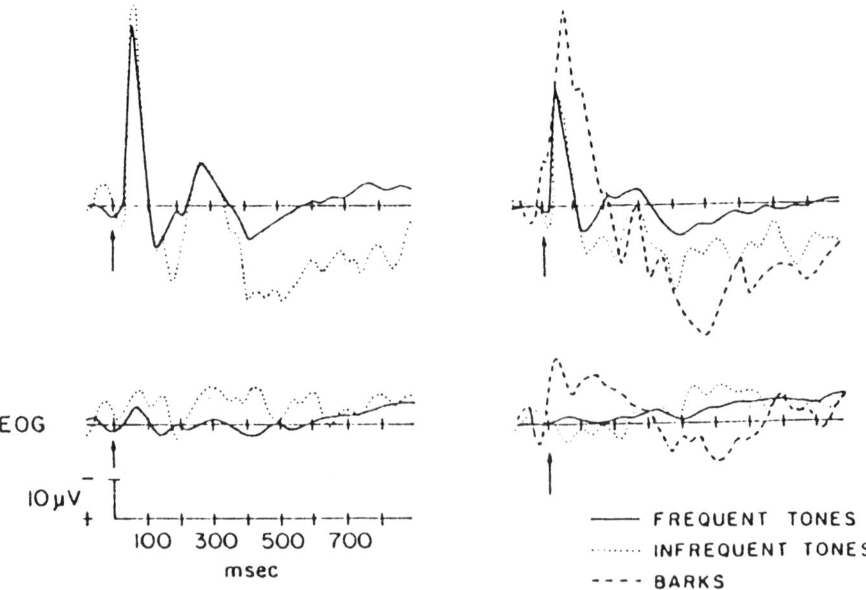

Figure 5.   ERPs elicited during two passive paradigms from squirrel monkeys. Left: in P3-like ERPs were elicited by an infrequent tone; right: in P3-like ERPs were elicited by an infrequent tone as well as by an infrequent stimulus resembling a bark. Reprinted with permission of Elsevier Science Publishers from Neville HJ, Foote SL (1984).

(Squires et al., 1976; similar sequential effects have been found in macaques, see Arthur, 1985; Paller, 1986). The positivity resembled P3 waves elicited by the same stimuli in human subjects, except that the latency was greater and the distribution was somewhat different. Unlike the distribution of P3 in humans, P3-like ERPs in squirrel monkeys were very small at midline electrodes and maximal at lateral parietal electrodes, with a tendency to be larger over the right hemisphere. These distributional differences could arise either because of different generators in the two species or because homologous generators are arranged differently in the squirrel monkey. A relevant observation may be that the squirrel monkey cortex is less gyrencephalic than macaque or human cortex.

ERPs in the squirrel monkey have also been studied in an auditory discrimination task similar to the one used in macaques (Pineda et al., 1988). In this paradigm, monkeys were recorded for pulling a lever following rare target tones ($p = .1$) but not frequent tones ($p = .9$). The target elicited a broad P3-like ERP in the 200- to 450-msec latency range. The distribution along the cortical surface was similar to that in the passive paradigm, with larger amplitudes at lateral locations than at midline locations, except that in the categorization paradigm, large P3-like potentials were also found at frontal midline locations. Interestingly, ERPs were extremely different in a variation of the task in which frequent tones were omitted. Target tones occurred at approximately the same ISIs (i.e., 4–8 sec), and responses were only rewarded during periods in which a small light was illuminated. Targets elicited a broad negative deflection that began at about 200 msec and continued for 250–550 msec; this is the same latency range in which positive, P3-like responses were elicited in the other paradigms. In human subjects, P3 potentials elicited by targets presented in the absence of nontargets were highly similar to P3 potentials in a conventional categorization task (Becker and Shapiro, 1980; Scott et al., 1989). It is presently unclear whether this difference between results in squirrel monkeys and results in humans reflects a species difference or critical differences in the paradigms used.

P3-like ERPs in squirrel monkeys have also been studied to determine whether they are dependent on the noradrenergic nucleus locus coeruleus. In a lesion study using the passive paradigm, conjoint damage to cell bodies in the nucleus and ascending axons from the nucleus ($n = 3$) resulted in a large decrease in the amplitude of the P3-like response, whereas damage to the axons alone ($n = 2$) had no consistent effect (Pineda, Foote, and Neville, 1989). Convergent results were found in one study using a pharmacological manipulation in which clonidine, which suppresses locus coeruleus firing, produced a decrement in P3-like ERPs elicited by auditory stimuli (Swick, Pineda, and Foote, in manuscript). In a subsequent study, however, the same dose of clonidine had no effect on P3-like ERPs elicited by visual stimuli (Pineda and Swick, 1992). Evidence from unit recordings in *Macaca fasicularis* also suggests a dissociation between locus caruleus activity and P3-like potentials (S. Grant, personal communication; Swick, Pineda and

Foote, 1991). Although the notion of a direct relationship between P3 and locus coeruleus firing is intriguing, strong empirical support for the hypothesis is lacking.

Cat

Conditioning paradigms have been relied on in several studies with cats. In cats paralyzed with gallamine, a late positive ERP was elicited by auditory or visual stimuli that had been paired with tail-shock (O'Conner and Starr, 1985; Wilder, Farley, and Starr, 1981). Tones of two frequencies were presented at a 2-sec ISI, and the low-frequency tone ($p = .2$) was designated the signal. Conditioning began after an habituation session (a passive paradigm), during which P3-like ERPs were not apparent. During conditioning sessions, a 300-msec shock was delivered 700 msec after the signal. After 200–300 signal presentations, the signal was associated with a conditioned pupillary response that began at a latency of about 300 msec (Fig. 6). P3-like ERPs peaked at a latency somewhat less than 300 msec with the tone signal and at 330 msec in subsequent experiments with the light signal. The amplitude of these ERPs decreased with increased signal probability. Intracranial recordings were also made from various locations using two single-contact probes concurrently. One consistent observation was that negative ERPs in the latency of P3 were observed at locations in the marginal gyrus and suprasylvian gyrus (the two gyri that run in the anterior-posterior direction near the midline). Late positive and negative ERPs were also found at other intracranial locations, including some in the hippocampus. The distribution of these potentials, however, was difficult to determine precisely because of the lack of concurrent recordings from multiple recording contacts or of evidence to confirm the stability of the intracranial responses over time.

Experiments in a very similar conditioning paradigm were also conducted using multicontact electrode probes that passed through the feline thalamus, as well as in two human patients in whom electrodes were implanted for stimulation therapy to relieve intractable pain (Katayama, Tsukiyama, and Tsubokawa, 1985a). The same conditioning paradigm was used in the two species. Following an habituation session, sequences of high- and low-frequency tones were presented at a 2-sec ISI. The rare low-frequency tone ($p = .2$) was paired with a shock that occurred 700 msec after tone onset (delivered to a finger in humans and to the tail in cats). Pupillary responses to the rare tone were used to infer that the two tones were being discriminated at some level by the cats. A P3-like ERP to the rare tone was recorded from the cortical surface. At roughly the same latency, negative ERPs were recorded from electrodes in the thalamus. Negative ERPs with smaller amplitudes were recorded from dorsal contacts in neocortical white matter, while no clear deflections were apparent in recordings from nearby electrodes located ventral to the thalamus. In the patients, the P3 potential elicited by the rare tone reached a peak at about 350 msec at

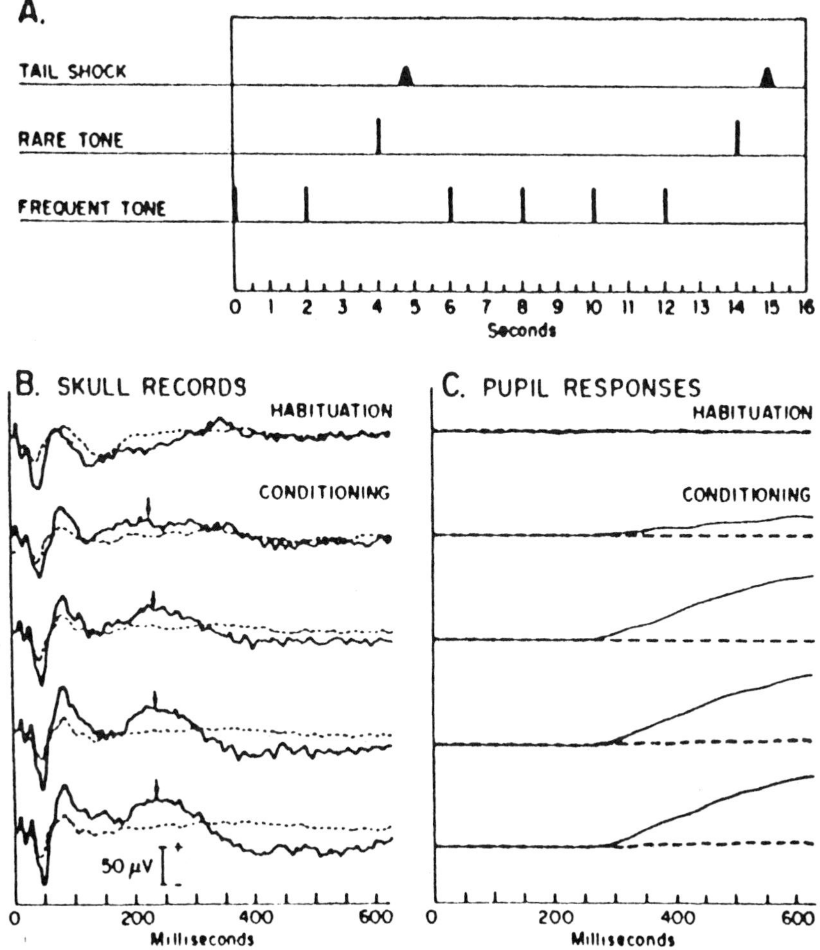

Figure 6.   ERPs elicited during a conditioning paradigm in cats. **A.** Paradigm involved pairing the rare tone with shock. **B.** P3-like ERPs developed over course of conditioning, as seen in sequential blocks of 400 trials each. **C.** Conditioning also led to pupillary dilation that anticipated shock. Reprinted with permission of Elsevier Publishers from O'Conner T, Starr A (1985).

Cz. Over roughly the same latency range, recordings from contacts in and near the thalamus showed large negative deflections (see Yingling and Hosobuchi, 1984, for similar results in another patient). Only six intracranial recording sites were used in the two patients, so the results in cats provided more thorough distributional information. It is also notable that this study included the best evidence that P3 potentials can be recorded in humans using a conditioning paradigm of the sort used in animal studies.

In another study with the same paradigm, the animal model was applied to study the effects of carbachol microinjection into the pontine reticular formation (Katayama et al., 1985b). This procedure is known to produce a state resembling coma, but it is unclear whether the behavioral unresponsiveness arises simply from a suppression of motor capabilities or whether information processing is also impaired. The results supported the latter hypothesis, in that P3-like ERPs disappeared concurrently with the carbachol-induced unresponsiveness. Given that carbachol is a cholinergic agonist, the authors concluded that the cholinoceptive pontine inhibitory area "may normally function to suppress excessive attendance to external stimuli." Furthermore, pilot data (cited by DeSalles et al., 1987, who used the conditioning paradigm in a study of cerebral concussion) suggested that P3-like ERPs were enhanced by the cholinergic antagonist atropine.

P3-like ERPs in the awake cat have been studied using a slightly different classical conditioning paradigm (Buchwald and Squires, 1982). Stimulus sequences presented at a 1.5-sec ISI included a tone ($p = .05$) as well as a soft click and a loud click ($p = .8$ and $.15$, or $.15$ and $.8$, in alternating blocks). During conditioning, the tone was paired with a shock to the supraorbital margin that began 1 sec after tone onset. Conditioned responses to the tone in the form of eye blinks were noted. Before conditioning (i.e., in a passive condition), no ERPs with latencies greater than 75 msec were consistently recorded. After conditioning, positive ERPs to the rare click were generally observed in the 200- to 500-msec latency range. P3-like ERPs were also found when stimulus omissions were substituted for the rare click. Subsequent work showed that whereas these P3-like ERPs could be elicited in young cats between 1 and 3 years old, little or no differential activity was exhibited in a group of aged cats between 11 and 23 years old (Harrison and Buchwald, 1985). Similarly, P3 in human subjects is known to decrease in amplitude as a function of age (e.g., Pfefferbaum et al., 1984; Picton et al., 1984). It should also be noted that abnormal ERP responses in aged cats were often associated with sustained conditioned responses, suggesting that these measures were somewhat independent. Further studies in young cats showed that the P3-like responses to rare clicks in this paradigm were unaffected by bilateral ablations of primary auditory cortex (Harrison, Buchwald, and Kaga, 1986) or of several areas of polysensory association cortex (Harrison et al., 1990). When lesions were made in the caudal hippocampus, P3-like ERPs were enhanced postoperatively in 5 cats, but abolished in 2 other cats (Kaga et al., 1992). In another study, however, highly consistent effects were reported after septal lesions (Harrison et al., 1987). In six cats, damaged areas included the medial septum and the vertical limb of the diagonal band of Broca and, as a result, acetylcholinesterase in other brain areas was depleted. P3-like ERPs were enhanced and delayed during the first week of recording, but in the second week the responses were absent. The authors related the time course of this effect to the time course of cholinergic terminal degeneration. Results from two other cats showed that P3-like ERPs did not change after areas rostral to the septum were damaged. Intracranial recording studies showed that P3-like ERPs with multiple polarity inversions were appearent in the medial temporal region (Kaga et al., 1992), paralleling results from humans and monkeys, but with more detailed mapping. In another study, P3-like ERPs were found in recordings from electrodes in the dorsal part of the medial septal area (Harrison and Buchwald, 1987), lending additional support to the hypothesis that cholinergic systems are critical for P3.

P3-like ERPs have also been recorded by another group using classical conditioning (Csépe, Karmos, and Molnár, 1987). Frequent and rare clicks were presented and the latter were followed 750 msec later by a shock to the neck. Conditioning was monitored by heart rate and blink responses occurring in the interval between the rare click ($p = .1$ or .03) and the shock. As the conditioned responses developed, the rare click also elicited P3-like responses in the 200- to 300-msec latency range. This ERP was not elicited during an initial habituation phase (passive condition) and tended to decline in amplitude during a posttraining extinction phase. However, the authors noted in two of the four cats tested that amplitudes declined before the extinction phase while shock reinforcement was still being delivered, perhaps because of the decreased novelty of overtraining.

Instead of using conditioning, one group has recorded ERPs from freely moving cats using a passive paradigm (Başar-Eroglu and Başar, 1987). Tones were presented at a 2.6-msec ISI and every fifth tone was omitted. ERPs were averaged by selecting responses to omitted tones on the basis of wakefulness and lack of movement. Peaks with latencies near 300 msec were noted in recordings from the dorsal hippocampus in four cats. The authors used further analyses in both the time and frequency domains to relate the hippocampal responses to theta rhythms and to EEG synchronization occurring just prior to stimuli (see Başar and Stampfer, 1985). Further studies in this type of passive paradigm have shown that P3-like ERPs can be recorded in auditory cortex and reticular formation, but most reliably in the hippocampus, especially in the CA3 region (Başar-Eroglu, Başar, and Schmielau, 1991; Başar-Eroglu et al., 1991b).

P3-like ERPs in the cat, in sum, appeared sensitive to stimulus probability and significance as manipulated by classical conditioning. These cross-species comparisons, however, are weakened by the restricted nature of the behavioral measures and by the divergence from conventional paradigms used to elicit P3 in humans. Pupillary or blink responses recorded from cats monitored the general effects of conditioning but may not have been highly correlated with the allocation of attention to the experimental stimuli. A manipulation could conceivably disrupt P3-like ERPs in a conditioning or passive paradigm, but not alter P3-like ERPs in a categorization paradigm. This weakness thus limits the conclusions that can be drawn from some of the experimental results, especially those found using repetitive testing sessions.

Rabbit

ERPs have been studied in rabbits using a discriminative conditioning paradigm that has traditionally been used to study the neural mechanisms of conditioning (Stolar et al., 1989). (Discriminative nictitating membrane conditioning has also been used; see Weisz et al., 1983). In this avoidance task, rabbits learned to avoid shocks by locomoting in an activity wheel (the conditioned response). Shocks were delivered 5 sec after a tone unless the conditioned response was made. Tones of two frequencies were used, only one of which was paired with shock. Recordings were made after training during conditions in which the probabilities of the two tones were manipulated. In one condition, the two tones ($p = .2$ and $.8$) were presented at a 1-sec ISI in the absence of shocks and with no wheel movement possible to parallel conventional oddball paradigms (Fig. 7). Epidural ERPs differed as a function of probability in that a positive peak between about 100 to 200 msec was more positive for the rare tone than for the frequent tone. These differences were evident when the rare tone had been paired with shock but not when the frequent tone had been paired with shock. This finding parallels evidence from humans showing target effects on P3 (Duncan-Johnson and Donchin, 1977), although such target effects are generally much smaller; unfortunately, data are not available to determine whether P3 in humans would show this strong of a target effect under exactly the same conditions. ERP differences were also found in recordings from the dentate gyrus and the medial dorsal nucleus of the thalamus. Enhanced unit responses to the rare tone that had been paired with shock were found at several brain locations but at a somewhat longer latency. The authors interpreted their results to implicate an involvement of hippocampal and thalamic brain areas, but also noted that ERPs differed substantially from human ERPs in that the enhanced positive potentials occurred at a much earlier latency. Indeed, it is curious that the latencies of positive ERPs recorded in this study differed from those found in the following study.

Another group recorded ERPs from rabbits during a signal-detection task (Aleksandrov and Maksimova, 1985, 1987). Freely moving rabbits were trained to press a lever after a light flash to receive a food reinforcement. Each trial was initiated when the rabbit was positioned appropriately on a platform; position was determined by means of an infrared light attached to the rabbit's head. Within 0.5–10 sec later, a series of four to seven light flashes was presented at a 1-sec ISI. Signal intensity began at a subthreshold level and was gradually increased. The rabbit was able to obtain food by running from the platform to press a level within 700 msec after a flash. P3-like ERPs were elicited by detected flashes but not by undetected flashes. Somewhat smaller potentials were associated with false alarms. In ERPs averaged from 10 rabbits, the onset latency was 113 msec and the peak latency was 414 msec. Concurrent recordings of unit activity during this paradigm showed that units in visual cortex and motor cortex tended to be

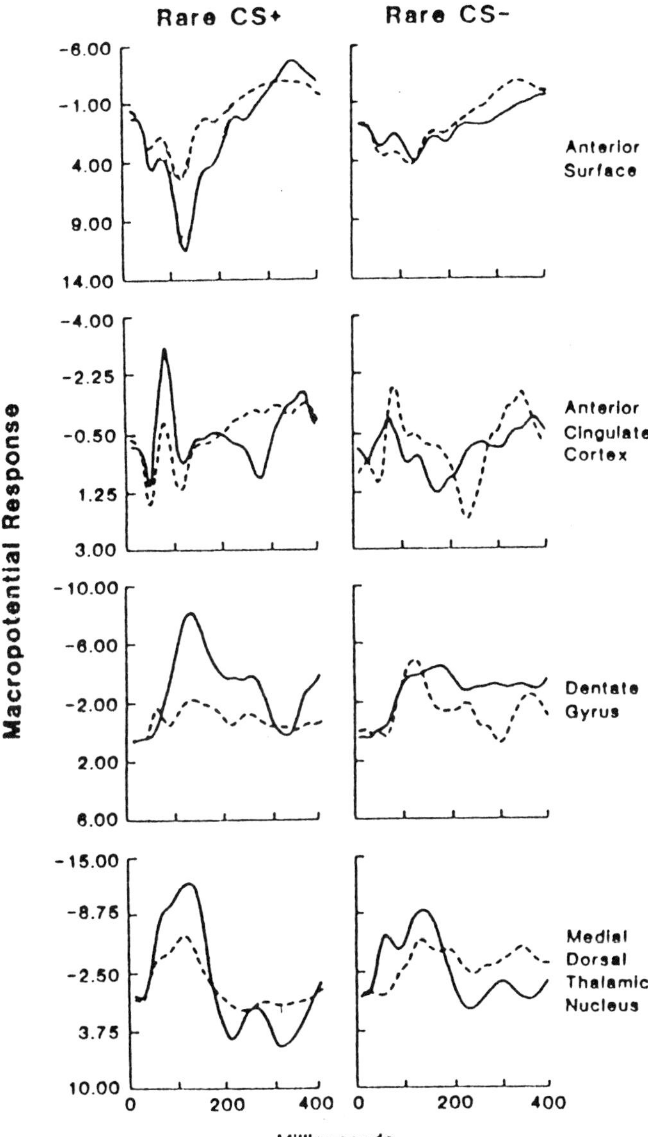

Figure 7.   ERPs in form of z scores derived from recordings in rabbits. P3-like ERPs were elicited after conditioning in which a rare tone had been paired with shock ($CS+$, $p = .2$). (See Stolar et al., 1989.) (*Figure continued on next page.*)

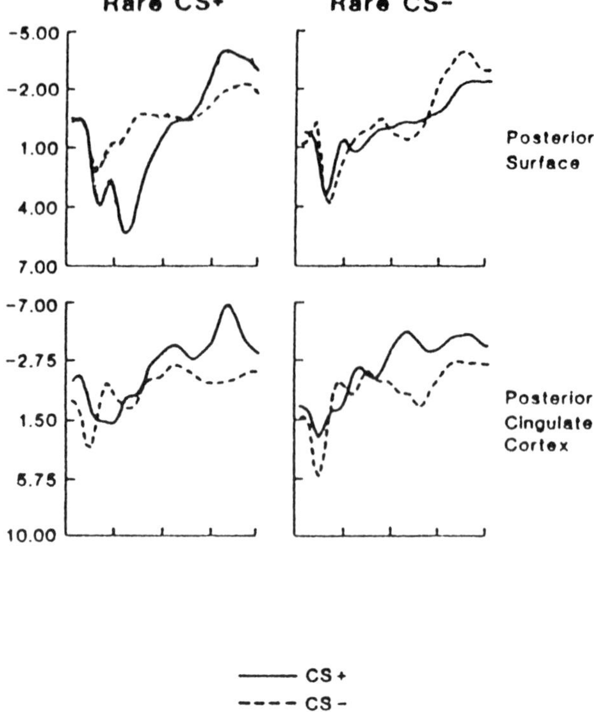

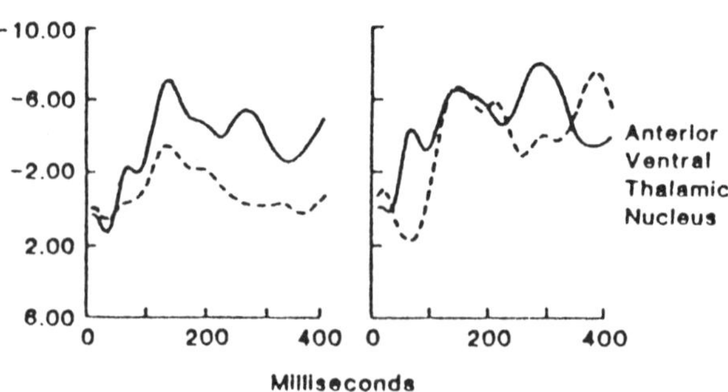

Figure 7.   (*continued*)

activated simultaneously during the time that the amplitude of the P3-like response was increasing (Aleksandrov and Maksimova, 1987). Further, human subjects tested in a parallel paradigm exhibited P3 waves in response to detected flashes. Indeed, signal-detection paradigms comparable to this paradigm have been used in many studies of P3 in humans (e.g., Cooper et al., 1977; Squires, Squires, and Hillyard, 1975). Whereas this paradigm differs somewhat from the usual categorization paradigm, it appears to be well-suited for studying P3-like ERPs in animals and thus deserves further study.

Rat

Promising evidence for the existence of P3-like ERPs in rats has been obtained using several different paradigms. In one report, P3-like ERPs with peaks between 300 and 400 msec were elicited by a rare tone ($p = .1$) in a passive paradigm as well as in a conditioning paradigm (Hurlbut, Lubar, and Satterfield, 1987). In the conditioning paradigm, the rare tone was paired with a foot shock that the rat could avoid by moving to the opposite side of the experimental chamber. After criterion performance in avoidance conditioning was reached, ERPs were recorded in the absence of reinforcement. However, the frequent tone ($p = .9$) was delivered at a frequency that was not as audible as that of the rare tone, thus confounding the comparison. This underscores the importance of taking into account the sensory and behavioral capabilities of the subjects when selecting stimuli and tasks.

In another passive paradigm, the frequency and intensity of three stimuli were adjusted such that the early ERP components were similar (Yamaguchi, Globus, and Knight, 1993). P3-like ERPs were elicited by two types of rare tones (Fig. 8). This paradigm thus appears to be a suitable one for further studies of the neural substrates of these potentials.

In another experiment, occasional auditory stimuli were interspersed in sequences of tactile stimuli, and vice versa (O'Brien, 1982; see also Wirtz-Brugger et al., 1986). However, these results are difficult to interpret, because recordings were made under anesthesia, which can eliminate P3 in human subjects (Fowler et al., 1988), and because this sort of bimodal paradigm has seldom been used to elicit P3 in human subjects.

ERPs have also been elicited using a categorization paradigm (Brenner et al., 1987). Rats were trained to respond to brief (10-msec) light flashes by pressing one of two levers to receive food. Lights over one lever or the other lever were flashed on a random basis whenever 4 sec had elapsed since the last lever-press. Positive peaks near 140 msec were elicited by detected flashes in sessions in which performance was near perfect but not in sessions in which performance was poor. Positive ERPs were also elicited by bright flashes from another source that were delivered either predictably (rate 1 per sec) or unpredictably (average rate 1 per min). Results obtained after noradrenergic denervation achieved by an injection of the neurotoxin DSP4 showed no changes in behavioral performance or ERPs, except that the responses to unpredictable flashes included an additional positive peak at about 280 msec. The authors concluded that the drug effect reflected a role of noradrenergic systems in inhibiting responses to unexpected stimuli.

In another categorization paradigm, rats were trained to obtain food by pressing a lever after one of two lights was illuminated (Voorn et al.,

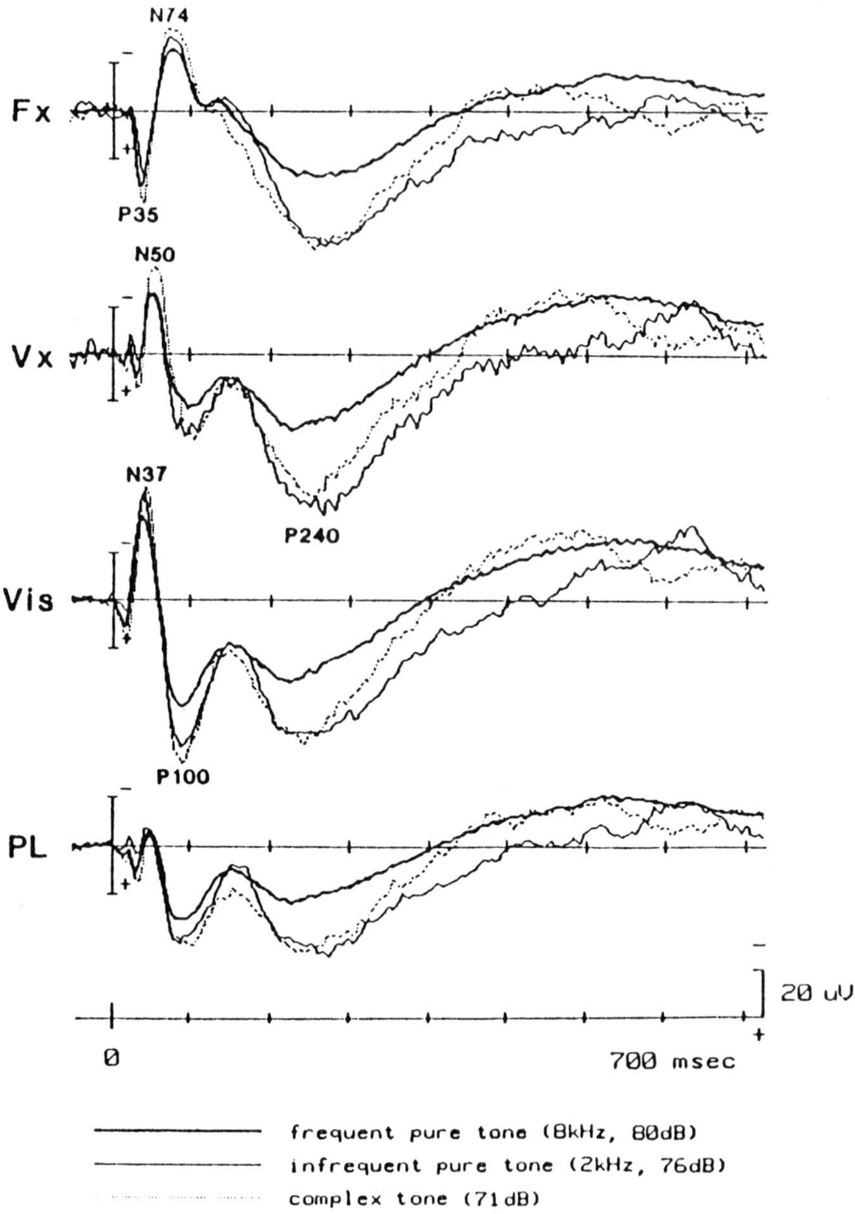

Figure 8.   ERPs elicited in a passive condition in rats. P3-like ERPs were elicited by infrequent tones at epidural electrodes near frontal cortex (Fx), vertex (Vx), visual cortex (Vis), and posterior-lateral skull (PL). Reprinted with permission from Elsevier Science Publishers, Inc. from Yamaguchi et al. (1993).

1987). Presses while the other light was illuminated or while neither light was illuminated gave rise to a delay. ERPs were averaged for trials in which a correct response occurred within 600 msec of light onset. Positive potentials with a peak between 300 and 500 msec were exhibited in most subjects, preceded by a negative peak at 120 msec. Recordings within and near the hippocampus also revealed late potentials of both polarities and sporadic distribution. However, there was a great deal of unexplained variability across different recordings. Also, it is difficult to explain the differences between the potentials elicited in these categorization tasks and those elicited in passive paradigms (e.g., Fig. 8), given that the modality of stimulation was not the same. It should be noted that many other studies in rats have examined potentials elicited in the hippocampus. In particular, analyses of potentials elicited in the dentate gyrus during classical discriminative conditioning with two tones showed features somewhat analogous to those of P3 in humans, in that early responses (i.e., latencies <150 msec) were responsive to the acquired significance of tones as well as to the pattern of immediately preceding tones (West et al., 1982; see also Hampson and Deadwyler, 1988).

Other Animals

Experiments with a bottlenose dolphin have revealed P3-like ERPs to auditory stimuli in two paradigms (Woods et al., 1986). Tones and digitized dolphin vocalizations were presented and stimulus types were balanced across conditions. In a passive paradigm, an ERP peak at a latency of 550 msec was elicited by deviant stimuli and was enhanced as a function of probability (Fig. 9). In a conditioning paradigm in which one of several tones were differentially reinforced with food, the P550 deflection was elicited by the reinforced tone but also by some of the nonreinforced tones. The authors suggested that the prolonged latency of the P3-like ERP may have been associated with the advanced age of the dolphin.

ERPs were recorded from the frontal cortex in dogs in experiments using an appetitive conditioning paradigm (Koslov and Pirogov, 1988). Dogs were trained in classical conditioning using two tones, one of which was followed by a meat powder reinforcement. The reinforced tone elicited a series of deflections that appeared to be locally generated in an area in the medial part of the frontal lobe. The amplitude of these potentials decreased during extinction and varied as a function of the stimulus preceding the evoking stimulus, and thus may very well represent a P3-like phonomenon.

ERP recordings from chimpanzees and a gorilla have also been made in a passive paradigm (Boysen and Berntson, 1985). However, late positivity in scalp recordings was not readily apparent. This finding could reflect the young age of most of the animals (less than 11 weeks of age), the use of ketamine sedation in the older animals (2 juvenile chimpanzees aged 3–3.5 years), or shortcomings of the paradigm. The fact that flashes were presented

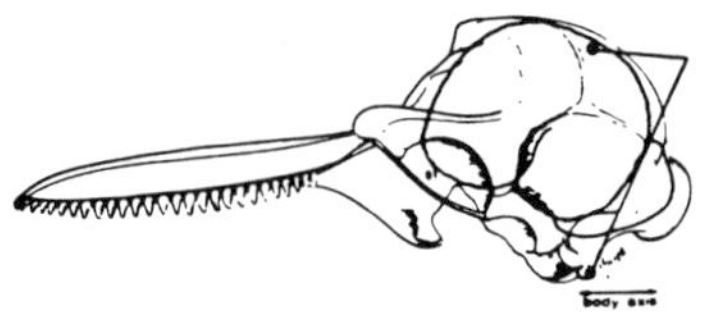

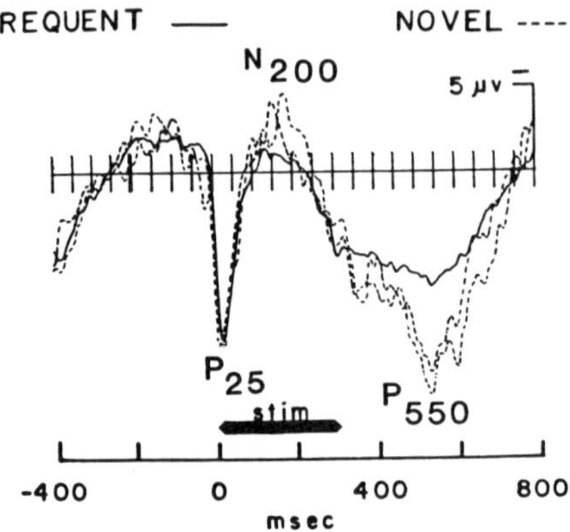

Figure 9.   ERPs elicited in a passive condition in a dolphin. Rare auditory stimuli elicited a P3-like P550 deflection. Reprinted with permission of Lawrence Erlbaum Associates, Inc., from Woods DL, Ridgway SH, Carder DA, Bullock TH (1986): Middle- and long-latency auditory event-related potentials in dolphins. In: *Dolphin Cognition and Behavior: A Comparative Approach*, Schusterman RJ, Thomas JA, Wood FG, eds. Hillsdale, New Jersey: Lawrence Erlbaum Associates, Inc.

at a 1.4-sec ISI meant that they were highly predictable and unlikely to elicit P3 waves.

Diffuse flash stimuli, however, were used in more appropriate and inventive ways in experiments with elasmobranch and teleost fish (Bullock et al., 1990) and pond turtles (Prechtl and Bullock, 1990, 1992). ERPs recorded in the retina, optic tectum, and cortex were studied as a function of changes in flash intensity and duration. In particular, ERPs were elicited by stimulus omissions that occurred after a long train of flashes. These ERPs began 50–100 msec after an omitted flash, were apparent in single trials, and lasted for as long as 1 sec. The ERPs were responsive to small changes in ISI, *as if* animals were developing expectations in some sense. The authors concluded that the ERPs reflected a rebound disinhibition and were not homologous to P3 in mammals, thus providing a cautionary note that paradigms designed to elicit cognitive ERPs may also elicit reflexlike responses because of slowly decaying neuronal activity that is unrelated to cognition (See Bullock, Karamürsal, and Achimowicz, 1993).

Evaluating the Approach

The variety of animal species in which P3-like responses have been recorded suggests that the underlying processes may be a universal feature of mammalian brains. The available evidence is consistent with the notion that many animals, not just humans, exhibit P3 responses, although the criteria by which these responses could be evaluated are far from exhausted. Further work will be needed to show that the neural characteristics of these potentials are indeed the same across species.

Given the goals of using ERPs to study the neural basis of cognition, it follows that waveform criteria are less important than functional criteria for validating P3-like ERPs. The most convincing support for an animal model of P3 would be evidence that homologous neural elements are activated during the very same circumstances. However, at present there is a high level of ignorance about the functional and neural properties of P3, compared to what may ultimately be learned. Studies of the functional and neural properties of P3 should thus progress in parallel with the development of animal models of P3.

Methodological concerns play a prominent role in this work, among them the following five issues.

1. The choice of species for these studies was in many cases pragmatic rather than reasoned. Initially, arguments for preferring some species over others can be made. Neuroanatomical similarities to humans and the ease with which behavioral tasks can be performed, for example, are important considerations. Also, close waveform similarities between species can give additional weight to the argument that parallel processes are being indexed.

2. A shortcoming of many studies was the lack of parallel results from humans. Studies should include human ERPs from comparable paradigms, when possible, to guarantee that the parameters used are appropriate for eliciting P3. This was particularly problematic for the conditioning paradigms because they have seldom been used with human subjects.

3. Much effort has been spent to assure that the ERPs recorded were not contaminated by artifacts, particularly those produced by eye movements. Such steps are crucial because artifacts can masquerade as P3-like responses under certain circumstances.

4. In order to study the functional properties of P3-like ERPs, it is often useful to compare ERPs from different stimulus conditions. In particular, it can be helpful to show that P3-like ERPs are not elicited by frequent stimuli in an oddball paradigm, which requires an appropriately short ISI (e.g., 1 sec) and stimuli that are suitable for the animal under study.

5. Caution is required in interpreting results from conditioning paradigms because of the possibility that the behavioral measures used are insufficient to verify that subjects are attending to the stimuli. In addition, evidence was cited suggesting that P3-like ERPs in conditioning paradigms as well as in passive paradigms tend to habituate, which poses problems for the use of longitudinal experimental manipulations such as in lesion and drug studies. Further, task manipulations in categoriza-

tion paradigms, such as comparisons as a function of whether the subjects are attending to the stimuli, are critical for demonstrating the relationship between P3-like ERPs and cognition.

The primary usefulness of the animal models will arise from the application of techniques for exploring the neural basis of P3, and from building connections between our conceptions of the neural events underlying P3 and neural events that can be studied using the full repertoire of available neurophysiological techniques. This approach may still be in its infancy, but much relevant evidence has already been obtained. Intracranial recordings have shown that many brain areas are active at the time that P3 is produced. The hippocampus has held a prominent role in these studies for several reasons. Extremely high-amplitude potentials are produced there, perhaps because of synchronous activity in large numbers of hippocampal cells that are spatially aligned. The hippocampus has also been related to P3 because it receives input from all modalities, it plays an important role in memory functions (as has been suggested for P3), and it can be affected by diseases known to be associated with abnormal P3 potentials. Nevertheless, ample evidence implies that the integrity of the hippocampus is not required for normal P3 waves to be produced at the scalp (e.g., Onofrj et al., 1991, 1992; Paller et al., 1988; Potter et al., 1993) even though P3-like potentials can be elicited in the hippocampus (Halgren et al., 1980; McCarthy et al., 1989; Paller et al., 1992). This state of affairs leads naturally to the hypothesis that P3 arises from diffuse systems that innervate both the hippocampus and widespread neocortical areas. The norepinephrine system of the locus coeruleus and the acetylcholine system of the basal forebrain, in particular, have been investigated using animal models of P3. In squirrel monkeys, lesion and drug studies have accrued evidence that the locus coeruleus may be important for P3 (Pineda, Foote, and Neville, 1989; Swick, Pineda, and Foote, 1991). Several lines of evidence from studies in cats (e.g., Harrison and Buchwald, 1987) have provided very convincing support for the notion that acetylcholine systems of the basal forebrain are instrumental in the generation of P3. Converging evidence implicating particular neurochemical systems in human subjects can be obtained via drug studies. For example, some results suggest that P3 is disrupted by clonidine, which suppresses locus coeruleus firing (Duncan and Kaye, 1987), as well as by scopolamine, which is a cholinergic antagonist (Meador et al., 1987, 1989; Potter et al., 1992; Rugg et al., 1989). The finding that some drug effects occurred for stimuli presented in the auditory modality but not the visual modality is not well understood, but it underscores the need to study P3-like ERPs in both modalities. More work is needed to clarify how the different mechanisms that are activated in the two modalities relate to P3, and studies in both humans and nonhuman animals may prove useful in this regard.

In conclusion, the clinical and experimental use of P3 is limited by the fact that alterations in P3 cannot be rigorously associated with specific neural processes. An enhanced understanding of the neural bases of P3 may thus have widespread clinical application, as well as important implications for basic research on P3. One particularly vexing problem in P3 research has been the difficulty of identifying P3 across different experimental tasks. This component identification problem may be solved only after neural characteristics of P3 can supplement the waveform characteristics that have proven so ambiguous. Although the past three decades of study have accumulated a large literature on P3, basic questions about its neural and functional underpinnings are still enigmatic. The question "what is P3?" cannot be answered in any profound way without a better understanding of its neural basis. Further study of the comparative psychophysiology of P3, however, holds much promise for solving this mystery by clarifying the significance of the neural mechanisms that are put into play whenever P3 is produced.

Acknowledgments. Work on this review was supported by NIMH grant MH-05286, the Dept. of Veterans Affairs, the Institute of Cognitive Studies at the University of California, Berkeley, and the Dept. of Cognitive Science at the University of California, San Diego. I thank Truett Allison, Ted Bullock, and Marta Kutas for their comments on the manuscript, Maria Stone for her help with translations from Russian, and Bob Galambos and Steve Hillyard for inspiring me to look at ERPs in animals in the first place.

References

Aleksandrov IO, Maksimova NE (1985): P300 and psychophysiological analysis of the structure of behavior. *Electroencephalogr Clin Neurophysiol* 61:548–558.

Aleksandrov IO, Maksimova NE (1987): Slow brain potentials and their relation to the structure of behavior: Data on cortical unit activity. *Electroencephalogr Clin Neurophysiol* (Suppl.) 40:3–7.

Antal A, Bodis–Wollner I, Ghilardi MF, Glover A, Mylin L, Toldi J (1993): The effects of levo–acetyl–carnitine in visual cognitive evoked potentials. *Electroencephalogr Clin Neurophysiol* 86:268–274.

Arthur DL (1985): Long-latency auditory event-related potentials in behaving monkeys (doctoral dissertation, University of California, Irvine). *Diss Abstr Int* 46B:3349.

Arthur DL, Starr A (1980): Task-relevant late positive component of the auditory event-related potential in monkeys resembles P300 in humans. *Science* 223:186–188.

Başar E, Stampfer HG (1985): Important associations among EEG-dynamics, event-related potentials, short-term memory and learning. *Int J Neurosci* 26:161–180.

Başar-Eroglu C, Basar E (1981): Endogenous components of event-related potentials in hippocampus: An analysis with freely moving cats. *Electroencephalogr Clin Neurophysiol* (Suppl.) 40:440–444.

Başar-Eroglu C, Başar E, Schmielau F (1991a): P300 in freely moving cats with intracranial electrodes. *Int J Neurosci* 60:215–226.

Başar-Eroglu C, Schmielau F, Schramm U, Schult J (1991b): P300 response of hippocampus analyzed by means of multielectrodes in cats. *Int J Neurosci* 60:239–248.

Becker DE, Shapiro D (1980): Directing attention toward stimuli affects the P300 but not the orienting response. *Psychophysiology* 17:385–389.

Bodis-Wollner I (1990): Physiological effects of acetyl-levo-carnitine in the central nervous system. *Int J Clin* Pharmacol Res 10:109–114.

Boysen ST, Berntson GG (1985): Visual evoked potentials in the great apes. *Electroencephalogr Clin Neurophysiol* 62:150–153.

Brenner E, Mirmiran M, Overdijk J, Timmerman M, Feenstra MGP (1987): Effect of noradrenergic denervation on task-related visual evoked potentials in rats. *Brain Res Bull* 18:297–302.

Buchwald JS, Squires NS (1982): Endogenous auditory potentials in the cat: A P300 model. In Woody CD, ed., *Conditioning: Representation of the Involved Neural Function*, pp. 503–515. New York: Plenum.

Bullock TH, Hofmann MH, Nahm FK, New JG, Prechtl JC (1990): Event-related potentials in the retina and optic tectum of fish. *J Physiol* 64:903–914.

Bullock TH, Karamürsel S, Achimowicz JZ (1993): Two types of event related potentials to omission of stimuli in humans. *Soc Neurosci Abstr* 19:1605.

Cheney D, Seyforth R (1992): How monkeys see the world. *Behav Brain Sci* 15:135–182.

Cooper R, McCallum WC, Newton P, Papakostopoulos D, Pocock PC, Warren WJ (1977): Cortical potentials associated with the detection of visual events. *Science* 196:74–77.

Csépe V, Karmos G, Molnár M (1987): Effects of signal probability on sensory evoked potentials in cats. *Int J Neurosci* 33:61–71.

Dale AM, Sereno MI (1993): Improved localization of cortical activity by combining EEG and MEG with MRI with cortical surface reconstruction: A Linear approach. *J Cognit Neuro* 5:162–176.

DeSalles AAF, Newlon PG, Katayama Y, Dixon E, Becker DP, Stonnington HH, Hayes RL (1987): Transient suppression of event-related evoked potentials produced by mild head injury in the cat. *J Neurosurg* 66:102–108.

Desmedt JE (1981): Scalp-recorded cerebral event-related potentials in man as point of entry into the analysis of cognitive processing. In: *The Organization of the Cerebral Cortex*, Schmitt FO, Worden FG, Adelman G, Dennis SD, eds., pp. 441–473. Cambridge: MIT Press.

Desmedt JE, Debecker J (1979): Waveform and neural mechanisms of the decision P350 elicited without prestimulus CNV or readiness potential in random sequences of near-threshold auditory clicks and finger stimuli. *Electroencephalogr Clin Neurophysiol* 47:648–670.

Donchin E, Cohen L (1967): Averaged evoked potentials and intramodality selective attention. *Electroencephalogr Clin Neurophysiol* 22:537–546.

Donchin E, Otto D, Gerbrandt LK, Pribram KH (1971): While a monkey waits: Electrocortical events recorded during the foreperiod of a reaction time study. *Electroencephalogr Clin Neurophysiol* 31:115–127.

Duncan CC, Kaye WH (1987): Effects of clonidine on event-related potential measures of information processing. *Electroencephalogr Clin Neurophysiol* (Suppl.) 40:527–531.

Duncan-Johnson CG, Donchin E (1971): On quantifying surprise: The variation of event-related potentials with subjective probability. *Psychophysiology* 14:456–467.

Ehlers CL (1988): ERP responses to ethanol and diazepam administration in squirrel monkeys. *Alcohol* 5:315–320.

Ehlers CL (1989): EEG and ERP responses to naloxone and ethanol in monkeys. *Psychopharmacol Biol Psychiatry* 12:217–228.

Fabiani M, Gratton G, Karis D, Donchin E (1987): The definition, identification and reliability of measurement of the P300 component of the event-related brain potential. In: *Advances in Psychophysiology*, Vol. 2, Acklis PK, Jennings JR, Coles MGH, eds., pp. 1–78. Greenwich: JAI Press.

Ford JM, Roth WT, Kopell BS (1976): Auditory evoked potentials to unpredictable shifts in pitch. *Psychophysiology* 13:32–39.

Fowler B, Kelso B, Landolt J, Porlier G (1988): The effects of nitrous oxide on P300 and reaction time. *Electroencephalogr Clin Neurophysiol* 69:171–178.

Fuster JM, Willey TJ, Riley DM, Ashford JW (1982): Effects of ethanol on visual evoked responses in monkeys performing a memory task. *Electroencephalogr Clin Neurophysiol* 53:621–633.

Galambos R, Hillyard SA (1981): Electrophysiological approaches to human cognitive processing. *Neurosci Res Program Bull* 20:141–265.

Galambos R, Sheatz GC (1962): An electroencephalographic study of classical conditioning. *Am J Physiol* 203:173–184.

Glover AA, Onofrj MC, Ghilardi MF, Bodis-Wollner I (1986): P300-like potentials in the normal monkey using classical conditioning and an auditory 'oddball' paradigm. *Electroencephalogr Clin Neurophysiol* 65:231–235.

Glover A, Ghilardi MF, Bodis-Wollner I, Onofrj M (1988): Alterations in event-related potentials (ERPs) of MPTP-treated monkeys. *Electroencephalogr Clin Neurophysiol* 71:461–468.

Glover A, Ghilardi MF, Bodis-Wollner I, Onofrj M, Mylin LH (1991): Visual 'cognitive' evoked potentials in the behaving monkey. *Electroencephalogr Clin Neurophysiol* 90:65–72.

Goodin DS, Aminoff MD (1987): Electrophysiological differences between demented and nondemented patients with Parkinson's disease. *Ann Neurol* 21:90–94.

Gordon E, Sloggett G, Harvey I, Kraiuhin C, Rennie C, Yiannikas C, Meares R (1987): Magnetoencephalography: Locating the source of P300 via magnetic field recording. *Clin Exp Neurol* 23:101–110.

Halgren E, Squires NK, Wilson CL, Rohrbaugh JW, Babb TL, Crandall PH (1980): Endogenous potentials generated in the human hippocampal formation and amygdala by infrequent events. *Science* 210:803–805.

Hampson RE, Deadwyler SA (1988): Reflections on closure and context, with a note on the hippocampus. *Behav Brain Sci* 11:385–386.

Harrison J, Buchwald H (1985): Aging changes in the cat P300 mimic the human. *Electroencephalogr Clin Neurophysiol* 62:227–234.

Harrison JB, Buchwald JS (1987): A cat model of the P300: Searching for generator substrates in the auditory cortex and medial septal area. *Electroencephalogr Clin Neurophysiol* (Suppl.) 40:473–480.

Harrison J, Buchwald J, Kaga K (1986): Cat P300 present after primary auditory cortex ablation. *Electroencephalogr Clin Neurophysiol* 63:180–187.

Harrison JB, Dickerson LW, Song S, Buchwald JS (1990): Cat-P300 present after association cortex ablation. *Brain Res Bull* 24:551–560.

Harrison JB, Buchwald JS, Kaga K, Woolf NJ, Butcher LL (1987): 'Cat P300' disappears after septal lesions. *Electroencephalogr Clin Neurophysiol* 69:55–64.

Hillyard SA, Bloom FE (1982): Brain functions and mental processes. In: *Animal Mind—Human Mind*, Griffin DR, ed., pp. 13–32. Berlin: Springer-Verlag.

Hillyard SA, Kutas M (1983): Electrophysiology of cognitive processing. *Annu Rev Psychol* 34:33–61.

Hillyard SA, Squires KC, Bauer JW, Lindsay PH (1971): Evoked potential correlates of auditory signal detection. *Science* 172:1357–1360.

Hurlbut BJ, Lubar JF, Satterfield SM (1987): Auditory elicitation of the P300 event-related potential in the rat. *Physiol Behav* 39:483–487.

John ER (1963): Neural mechanisms of decision making. In: *Information Storage and Neural Control*, Fields WS, Abbott W, eds., pp. 243–282. Springfield: Thomas.

Johnson R, Jr. (1988): Scalp-recorded P300 activity in patients following unilateral temporal lobectomy. *Brain* 111:1517–1529.

Kaga K, Harrison JB, Butcher LL, Woolf NJ, Buchwald JS (1992): Cat 'P300' and cholinergic septohippocampul neurons: Depth recordings, lesions, and choline acetyltransferase immuno-histochemistry. *Neurosci Res* 13:53–71.

Katayama Y, Tsukiyama T, Tsubokawa T (1985a): Thalamic negativity associated with the endogenous late positive component of cerebral evoked potentials (P300): Recordings using discriminative aversive conditioning in humans and cats. *Brain Res Bull* 14:223–226.

Katayama Y, Reuther S, Dixon CE, Becker DP, Hayes RL (1985b): Dissociation of endogenous components of auditory evoked potentials following carbachol micro-injection into the cholinoceptive pontine inhibitory area. *Brain Res* 334:366–371.

Knight RT (1984): Decreased responses to novel stimuli after prefrontal lesions in man. *Electroencephalogr Clin Neurophysiol* 59:9–20.

Knight RT, Scabini D, Woods DL, Clayworth CC (1989): Contributions of temporal-parietal junction to the human auditory P3. *Brain Res* 502:109–116.

Koslov AP, Pirogov AA, (1988): Medlennie potenciale prefrontalnoi kori sobak i klassicheskii secretornii yslovnii reflex. [Slow potentials of dogs prefrontal cortex and classical secretery conditioned reflex.] *Zh Vyssh Nervn Deyat Im IP Pavlova* 38:434–442.

Kutas M, McCarthy G, Donchin E (1971): Augmenting mental chronometry: The P300 as a measure of stimulus evaluation time. *Science* 197:792–795.

Kutas M, Hillyard SA, Volpe BT, Gazzaniga MS (1990): Late positive event-related potentials after commissural section in humans. *J Cognit Neurosci* 2:258–271.

McCarthy G, Donchin E (1981): A metric for thought: A comparison of P300 latency and reaction time. *Science* 211:77–80.

McCarthy G, Wood CC, Williamson PD, Spencer DS (1989): Task-dependent field potentials in human hippocampal formation. *J Neurosci* 9:4253–4268.

Meador KJ, Loring DW, Adams RJ, Patel BR, Davis HC (1987): Central cholinergic systems and the P3 evoked potential. *Int J Neurosci* 33:199:205.

Meador KJ, Loring DW, Davis HC, Sethi KD, Patel BR, Adams RJ, Hammond EJ (1989): Cholinergic and Serotonergic effects on P3 potential and recent memory. *J Clin Exp Neuropsych* 11:252–260.

Neville HJ, Foote SL (1984): Auditory event-related potentials in the squirrel monkey: Parallels to human late wave responses. *Brain Res* 298:107–116.

O'Brien JH (1982): P300 in the rat. *Physiol Behav* 28:318–321.

O'Conner T, Starr A (1985): Intracranial potentials correlated with an event-related potential, P300, in the cat. *Brain Res* 339:27–38.

Okada YC, Kaufman L, Williamson SJ (1983): The hippocampal formation as a source of the slow endogenous potentials. *Electroencephalogr Clin Neurophysiol* 55:417–426.

Onofrj M, Ghilardi MF, Faricelli A, Bodis-Wollner I, Calvani M (1987): Effect of levo-acetylcarnitine on P300-like potentials of the normal monkey. *Drugs Exp Clin Res* 13:407–415.

Onofrj M, Fulgente T, Nibilio D, Malatestz G, Bazzano S, Colamartino P, Gambi D (1992): P3 recordings in patients with bilateral temporal lobe lesions. *Neurol* 42:1762–1767.

Onofrj M, Gambi D, Fulgente T, Bazzano S, Colamartino P (1991): Persistence of P3 component in severe amnestic syndrome. *Electroencephalogr Clin Neurophysiol* 78:480–484.

Paller KA (1986): Effects of medial temporal lobectomy in monkeys on brain potentials related to memory (Doctoral dissertation, University of California, San Diego). *Diss Abstr Int* 47B:4428.

Paller KA, Zola-Morgan S, Squire LR, Hillyard SA (1982): Late positive event-related potentials in cynomolgus monkeys (*Macaca fasicularis*). *Soc Neurosci Abstr* 8:975.

Paller KA, Zola-Morgan S, Squire LR, Hillyard SA (1988): P3-like brain waves in normal monkeys and in monkeys with medial temporal lesions. *Behav Neurosci* 102:714–725.

Paller KA, McCarthy G, Roessler E, Allison T, Wood CC (1992): Potentials evoked in human and monkey medial temporal lobe during auditory and visual oddball paradigms. *Electroenceph Clin Neurophysiol* 84:269–279.

Pfefferbaum A, Ford JM, Wenegrat BG, Roth WT, Kopell BS (1984): Clinical application of the P3 component of event-related potentials. *Electroencephalogr Clin Neurophysiol* 59:85–103.

Picton TW, Stuss DT, Champagne SC, Nelson RF (1984): The effects of age on human event-related potentials. *Psychophysiology* 21:312–325.

Pineda J, Foote SL, Neville HJ (1987): Long-latency event-related potentials in squirrel monkeys: Further characterization of wave form morphology, topography, and functional properties. *Electroencephalogr Clin Neurophysiol* 67:77–90.

Pineda J, Foote SL, Neville HJ (1989): Effects of locus coeruleus lesions on auditory, long-latency event-related potentials in monkey. *J Neurosci* 9:81–93.

Pineda J, Foote SL, Neville HJ, Holmes TC (1988): Endogenous event-related potentials in monkey: The role of task relevance, stimulus probability and behavioral response. *Electroencephalogr Clin Neurophysiol* 70:155–171.

Pineda JA, Swick D (1992): Visual P3-like potentials in squirrel monkey: Effects of a nonadrenergic agonist. *Brain Res Bull* 28:485–491.

Polich J (1989a): Habituation of P300 from auditory stimuli. *Psychobiology* 17:19–28.

Polich J (1989b): P300 from a passive auditory paradigm. *Electroencephalogr Clin Neurophysiol* 74:312–320.

Potter DD, Pickles CD, Roberts RC, Rugg MD (1992): The effects of scopolamine on event-related potentials in a continuous recognition memory task. *Psychophysiology* 29:29–37.

Potter DD, Pickles CD, Roberts RC, Rugg MD, Paller KA, Mayes AR (1993): Visual and auditory P300 in a case of bilateral destruction of the medial temporal lobes following viral encephalitis. In: *New Developments in Event-Related Potentials*, Heinze HJ, Münte TF, Mangun GR, eds., pp. 319–327. Cambridge, MA: Birkhauser.

Prechtl JC, Bullock TH (1990): Event-related potentials in tectum and cortex of freely moving turtles to visual stimuli. *Soc Neurosci Abstr* 16:920.

Prechtl JC, Bullock TH (1992): Barbiturate sensitive components of visual ERPs in a reptile. *Neuro Report* 3:801–804.

Prim M, Ojemann G, Lettich E (1983): Human cortical patterns of "P300" potentials to novel visual items. *Soc Neurosci Abstr* 9:655.

Pritchard WS (1981): Psychophysiology of P300: A review. *Psychol Bull* 89:506–540.

Richer F, Johnson RA, Beatty J (1983): Sources of late components of the brain magnetic response. *Soc Neurosci Abstr* 9:656.

Ritter W, Vaughan HG, Jr., Costa LD (1968): Orienting and habituation to auditory stimuli: A study of short term changes in averaged evoked responses. *Electroencephalogr Clin Neurophysiol* 25:550–556.

Roth WT, Ford JM, Lewis SJ, Kopell BS (1976): Effects of stimulus probability and task-relevance on event-related potentials. *Psychophysiology* 13:311–317.

Ruchkin DE, Sutton S (1978): Emitted P300 potentials and temporal uncertainty. *Electroencephalogr Clin Neurophysiol* 45:268–277.

Rugg MD, Pickles CD, Potter DD, Roberts RC (1991): Normal P300 following extensive unilateral medial temporal damage. *J Neurol Neurosurg Psychiat* 54:217–222.

Rugg MD, Potter DD, Pickles CD, Roberts RC (1989): Effects of scopolamine on the modulation of event-related brain potentials by word repetition. *Soc Neurosci Abstr* 15:245.

Scott T, McCarthy G, Paller KA, Wood CC (1989): Event-related potentials recorded from scalp and hippocampal formation in humans performing detection tasks. *Soc Neurosci Abstr* 15:478.

Simson R, Vaughan HG, Jr., Ritter W (1976): The scalp topography of potentials associated with missing visual or auditory stimuli. *Electroencephalogr Clin Neurophysiol* 40:33–42.

Simson R, Vaughan HG, Jr., Ritter W (1977): The scalp topography of potentials in auditory and visual discrimination tasks. *Electroencephalogr Clin Neurophysiol* 42:528–535.

Smith DBD, Donchin E, Cohen L, Starr A (1970): Auditory averaged evoked potentials in man during selective binaural listening. *Electroencephalogr Clin Neurophysiol* 28:146–152.

Smith ME, Halgren E, Sokolik M, Baudena P, Musolino A, Liegeois-Chauvel C, Chauvel P (1990): The intracranial topography of the P3 event-related potential during auditory oddball. *Electroencephalogr Clin Neurophysiol* 76:235–248.

Snyder E, Hillyard SA, Galambos R (1980): Similarities and differences among the P3 waves to detected signals in three modalities. *Psychophysiology* 17:112–122.

Squires NK, Squires KC, Hillyard SA (1975): Two varieties of long-latency positive waves evoked by unpredictable auditory stimuli in man. *Electroencephalogr Clin Neurophysiol* 38:387–401.

Squires NK, Donchin E, Squires KC, Grossberg S (1977): Bisensory stimulation: Inferring decision-related processes from the P300 component. *J Exp Psychol Hum Percep Perform* 3:299–315.

Squires KC, Wickens C, Squires NK, Donchin E (1976): The effect of stimulus sequence on the waveform of the cortical event-related potential. *Science* 193:1142–1146.

Stapleton JM, Halgren E, Moreno KA (1987): Endogenous potentials after anterior temporal lobectomy. *Neuropsychologia* 25:549–557.

Stolar N, Sparenborg S, Donchin D, Gabriel M (1989): Conditional stimulus probability and activity of hippocampal, cingulate cortical and limbic thalamic neurons during avoidance conditioning in rabbits. *Behav Neurosci* 103:919–934.

Sutton S, Braren M, Zubin J, John ER (1965): Evoked potential correlates of stimulus uncertainty. *Science* 150:1187–1188.

Sutton S, Tueting P, Zubin J, John ER (1967): Information delivery and the sensory evoked potential. *Science* 155:1436–1439.

Swick D, Pineda JA, Foote SL (1991): Unit activity in the nucleus locus coeruleus related to P300-like potentials? *Soc Neurosci Abstr* 17:657.

Swick D, Pineda JA, Foote SL (in press): Effects of systemic clonidine on auditory event-related potentials in squirrel monkeys. *Brain Res Bull* (in press).

Trahms L, Stehr R, Erné SE, Seibertz E, Friederici AD (1990): Biomagnetic registration of P300 activity. *J Clin Exp Neuropsychol* 12:401.

Treisman A, Gelade G (1980): A feature-integration theory of attention. *Cognit Psychol* 12:97–136.

Verleger R (1988): Event-related potentials and cognition: A critique of the context-updating hypothesis and an alternative interpretation of P3. *Behav Brain Sci* 11:343–356.

Voorn FJ, Adamse H, Kop PFM, Brunia CHM (1987): Hippocampal potentials related to signal stimuli in unrestrained rats. *Electroencephalogr Clin Neurophysiol* (Suppl.) 40:493–498.

Weisz DJ, McCarthy G, Wood CC, Thompson DT (1983): Event-related potentials reflect stimulus significance during discriminative NM conditioning in the rabbit. *Soc Neurosci Abstr* 9:642.

West MO, Christian E, Robinson JH, Deadwyler SA (1982): Evoked potentials in the denate gyrus reflect the retention of past sensory events. *Neurosci Let* 28: 319–324.

Wickens C, Kramer A, Vanasse L, Donchin E (1983): Performance of concurrent tasks: A psychological analysis of the reciprocity of information-processing resources. *Science* 221:1080–1082.

Wilder MB, Farley GR, Starr A (1981): Endogenous late positive component of the evoked potential in cats corresponding to P300 in humans. *Science* 211:605–607.

Wirtz-Brugger F, McCormack K, Szemczak M, Fielding S, Cornfeldt M (1986): P300 in anesthetized rat: Possible model for detecting memory-enhancing drugs. *Soc Neurosci Abstr* 12:713.

Wood CC, McCarthy G, Allison T, Goff WR, Williamson PD, Spencer DD (1982): Endogenous event-related potentials following temporal lobe excisions in humans. *Soc Neurosci Abstr* 8:976.

Wood CC, McCarthy G, Squires NK, Vaughan HG, Woods DL, McCallum WC (1984): Anatomical and physiological substrates of event-related potentials: Two case studies. In: *Brain and Information: Event-Related Potentials*, Karra R, Cohen J, Tueting P, eds., pp. 681–721. New York: New York Academy of Sciences.

Woods DL, Ridgway SH, Carder DA, Bullock TH (1986): Middle- and long-latency auditory event-related potentials in dolphins. In: *Dolphin Cognition and Behavior: A comparative approach*, Schusterman RJ, Thomas JA, Wood FG, eds., pp. 61–78. New Jersey: Erlbaum.

Yingling CD, Hosobuchi Y (1984): A subcortical correlate of P300 in man. *Electroencephalogr Clin Neurophysiol* 59:72–76.

Yamaguchi S, Globus H, Knight RT (1993): P3-like potential in rats. *Electroencephalogr Clin Neurophysiol* 88:151–154.

Chapter 15

Theta and Delta Responses in Cognitive Event-Related Potential Paradigms and Their Possible Psychophysiological Correlates

EROL BAŞAR, MARTIN SCHÜRMANN,
CANAN BAŞAR-EROGLU, AND TAMER DEMIRALP

This chapter combines a review of event-related potentials (ERPs) with empirical data concerning the question: What are the differences between auditory evoked potentials (EPs) and two types of ERPs with respect to their frequency components? In this study, auditory EPs were elicited by 1500-Hz tones. The first type of ERPs was responses to third attended tones in an omitted stimulus paradigm where every fourth stimulus was omitted. The second type of ERPs was responses to rare 1600-Hz tones in an oddball paradigm. The amplitudes of delta and theta components of EPs and ERPs showed significant differences: In responses to third attended tones, there was a significant increase in the theta frequency band (frontal and parietal locations; 0–250 msec). In delta frequency band, there was no significant change. In contrast a diffuse delta increase occurred in oddball responses, and an additional prolongation of theta oscillations was observèd (late theta response, 250–500 msec). These results are discussed in the scope of ERPs as induced rhythmicities. The intracranial sources of ERPs, their psychological correlates, and the role of theta rhythms in the corticohippocampal interaction are reviewed. From these results and from the literature a working hypothesis is derived assuming that delta responses are mainly involved in signal matching, decision making, and surprise, whereas theta responses are more related to focused attention and signal detection.

In the past two decades the investigations showed that ERPs consist of

Cognitive Electrophysiology
H-J. Heinze, T.F. Münte, and G.R. Mangun, editors
© 1994 Birkhäuser Boston

"exogenous" and "endogenous" components (Başar and Stampfer, 1985; Hillyard and Picton, 1979; Näätänen, 1988; Picton and Hillyard, 1974; Picton and Stuss, 1980). The term "exogenous component" stands for a feature or component of the ERP that correlates with changing physical parameters of the stimuli, and "endogenous component" means a component which varies only in relation to the given tasks to the subject, which probably modulate the intrinsic brain mechanisms in perceptual processes. During the development of this type of research, the endogenous components have been found to correlate with higher hierarchial levels of information processing like expectation, short- and long-term memories as well as attention (see, e.g., Heinze et al., 1990, for an ERP study of focused visual attention).

A late ERP component following the early exogenous potentials with a latency shift of 250–400 msec, called the P300 wave, is an example of endogenous ERP components. The P300 response obtained by means of the oddball paradigm has been used successfully for psychological research and clinical diagnostics. Since the earlier applications there have been several excellent reviews concerning methodological, psychological, and clinical aspects of this paradigm (Birbaumer et al., 1990; Hillyard and Picton, 1979; Johnson, 1988; Näätänen, 1988, 1990; Regan, 1989; Rösler, 1982; Woods, 1990).

An important goal in the analysis of physiological correlates of the P300 response consists in searching sources of generators giving rise to the P300 component of the event-related potential. Accordingly, several investigators used intracranial electrodes to be able to localize the sources in human recordings and animal models (Başar-Eroglu et al., 1991a; Halgren et al., 1986; Harrison et al., 1990; Paller et al., 1988).

Several of these authors do indicate the existence of multiple generators, including sources in hippocampus, parietal, frontal, and several other areas of the association cortex (for review, see Knight et al., 1981; Paller et al., 1988; Smith et al., 1990).

In our earlier publications on P300 we used two types of approaches:

1. The analysis by means of frequency characteristics of the P300 response to emphasize the role of EEG synchronization and enhancement (Başar, 1988; Başar and Stampfer, 1985; Başar et al., 1984; Başar-Eroglu et al., 1992; Stampfer and Başar, 1985);
2. Experiments by means of a passive P300 paradigm with intracranial electrodes in the cat brain (Başar-Eroglu and Başar, 1987, 1991; Başar-Eroglu et al., 1991a, 1991b).

We concluded that the most prominent component of P300 is the so-called theta response in several structures of the cat brain, and the delta response in human brain.

The current study is on one hand an extension of the work by Stampfer and Başar (1985) with the oddball paradigm including results on various locations of the human scalp; on the other hand, it is a physiological approach to P300 by using the natural EEG frequencies. As Mountcastle (1992) has stated, the role of EEG as a most important physiological signal of the central nervous system is in reappraisal. It was a frequency domain approach that enabled Demiralp and Başar (1992) to analyze ERP components faster than the P300 wave: As these components overlap with early exogenous components, they cannot be detected in most paradigms. Demiralp and Başar used frequency domain analysis of ERPs to detect endogenous components of ERPs, which may not necessarily differ from the exogenous components in time and space, by means of a method that is sensitive to changes in frequency components of the ERPs. Their paradigm was based on earlier studies: Başar et al. (1989) carried out a series of ERP studies on human subjects by applying a modified form of the omitted stimulus paradigm of Sutton et al. (1967). The paradigm consisted of auditory or visual stimulations with regular interstimulus intervals in which some stimuli were omitted in a random or regular order with various degrees of probability. The subject's task was to mark mentally the time of the omitted stimulus. With this type of paradigm especially when the stimulus omission occurred in a regular manner (for example, every fourth stimulus was omitted) quasideterministic, reproducible patterns of EEG signal occurred anticipating the omitted stimulus. The subjects reported that they had paid attention to the rhythm of preceding stimuli to be able to fulfill the task. Demiralp and Başar (1992) applied this paradigm to test whether event-related changes occur in responses of different brain areas to the stimuli that precede or follow the omitted stimulus.

They showed that the frequency analysis approach may differentiate the responses to the stimuli that are coupled with a cognitive task from the standard EPs, detecting some specific changes in frequency components, whereas the time domain analysis of the same responses show no prominent differences.

In this study, frequency domain analysis is used as a tool to compare responses to auditory stimuli of three different types:

- stimuli in a standard EP paradigm,
- stimuli preceding the omitted one in the omitted stimulus paradigm described ('third attended stimuli')
- oddball stimuli in an oddball paradigm.

In this study we tentatively concluded, also in the light of the new extensive analysis and survey by Miller (1991), that theta and delta responses in the hippocampocortical system of the brain dominate or control the ERPs and might be interpreted as correlates of some functional states as selective attention and decision making.

A general review of animal P300 experiments and human P300 is also given in this study to bridge physiological states and psychological correlates.

Methods

Subjects

Each of the three parts of the experiments (auditory evoked potential, omitted stimulus paradigm, oddball paradigm) was carried out on 10 voluntary, right-handed, healthy subjects, 19–21 years of age. The subjects did not have any known neurological deficit and did not take any medicaments or drugs that are known to affect the EEG.

Environment

The subjects sat in a soundproof and echo-free room that was dimly illuminated. The room was also shielded to attenuate the environmental electromagnetic noise effects. After the electrode placement, a few minutes of rest time was given to subjects to get them familiar with the environment.

Data Acquisition and Equipment

ELECTRODE PLACEMENT The data were derived with Ag–AgCl disc electrodes placed on frontal, vertex, parietal, and occipital (F3, Cz, P3, O1) recording sites of the international 10–20 system against the reference of earlobes. All electrode impedances were maintained at less than 5 Kohms. The EOG was also registered to mark eye movement artifacts.

EEG AMPLIFICATION All data were amplified by means of a Schwarzer EEG apparatus with a time constant of 0.5 sec. A low-pass filter with cut-off frequency at 70 Hz (24 dB/octave) was applied to data to avoid aliasing in the following digitization step. An additional 50-Hz notch filter (36 dB/octave) is also applied to data to remove mains interference.

DIGITIZATION OF THE EEG After the application of the antialiasing filters to the analog signal, 1-sec prestimulus and 1-sec poststimulus EEGs were digitized with a sampling rate of 500 points/sec and stored on the hard disk of the computer. The recording of data and stimulation were controlled by a HP 1000 F computer that was also used for the offline analysis of the data.

ARTIFACT REJECTION For the elimination of artifactious trials, two online artifact rejection procedures are applied in addition to the manual offline selective averaging procedure:

- An automatic online artifact rejection procedure is used for the elimination of global artifactious EEG epochs. It is based on the rejection of trials with extremely high amplitudes.

338   E. Başar et al.

- The EEG is monitored and recorded continuously on paper during the experiments, and the subjects can be observed via closed circuit TV, so that the technician can mark the trials with artifacts during the recording. It is also possible to pause the recording procedure by a button press, if long-lasting artifacts occur in the EEG.

STIMULATIONS As auditory stimuli, 80 dB, 1500- or 1600-Hz tones with a 0.5-msec rise time and 800-msec duration were presented binaurally.

EXPERIMENTAL PARADIGMS In each recording session, first, the spontaneous EEG was registered for a few minutes to determine global characteristics of subject spontaneous EEG activity and arousal state at the beginning of the experiments. This period also helped subjects to become familiar with experimental conditions. Thereafter, the auditory EP, omitted stimulus, and oddball paradigms were applied with short resting periods between to the first group of subjects.

The auditory EP experiments consisted of the presentation of 1500-Hz tones with interstimulus intervals (ISI) randomly varying between 2.5 and 4 sec with a mean value of 3 sec.

The oddball and omitted stimulus paradigms are illustrated in Figures 1A and 1B. In the oddball paradigm (Fig. 1A), the tones were presented in a pseudorandom sequence with 1600-Hz tones occurring 20% of the time and

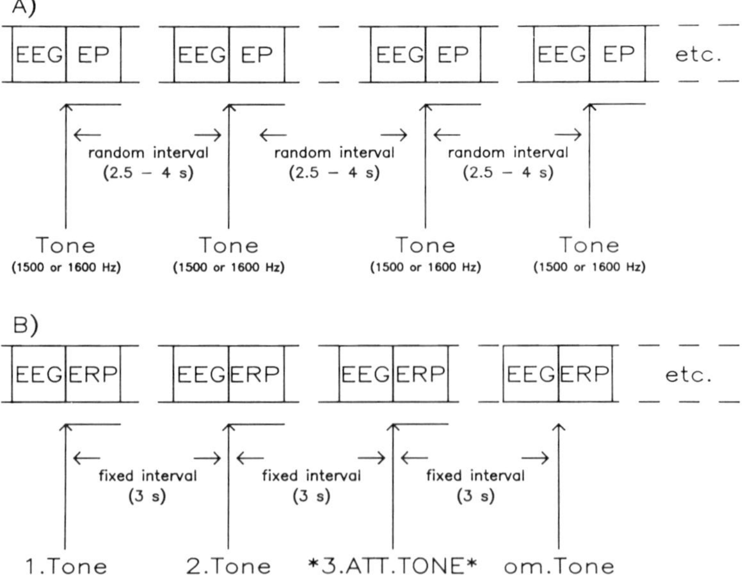

Figure 1.   Schematic illustration of (**A**) oddball and (**B**) omitted stimulus paradigms.

1500-Hz tones occurring 80% of the time. The interval between tones varied randomly from 2.5 to 4 sec with a mean value of 3 sec as in the auditory EP experiments. The subjects were instructed to keep a mental count of the number of 1600-Hz tones (nonfrequent target tones).

The omitted stimulus paradigm consisted of a series of 1500-Hz tones with a constant interstimulus interval of 3 sec, but every fourth stimulus was omitted, and this time the subject's task was to mark mentally the virtual onset time of the omitted stimuli (Fig. 1B). At the end of the experiment subjects also were to report about their performance in predicting the onset times of omitted stimuli. These subjective reports were taken into consideration in selecting the valid experiments. If they told that they were not successful, the experiment was either repeated or the subject's data were excluded from further analysis steps (for a more detailed description of the paradigm, see Demiralp and Başar, 1992).

Data Analysis

Before describing the method used, we want to explain the theoretical basis of the analyses carried out on the data. Resonance is the response that may be expected of underdamped systems when a periodic signal of a characteristic frequency is applied to the system. The response is characterized by a "surprisingly" large output amplitude for relatively small input amplitudes; that is, the gain is large.

Resonance phenomena or responses to forced oscillations can be analyzed in the direct empirical way as follows: A sinusoidal signal of a frequency f is applied to the system. After a certain period sufficient for the damping of the transient, only forced oscillations will remain that have the frequency of the signal. The amplitude of the applied signal (input), the amplitude of the forced oscillations (output), and the phase difference between input and output are then measured. Gradually increasing the frequency from $f = 0$ to $f = f_0$, the output amplitude relative to the input amplitude and the phase differences is measured as a function of frequency (amplitude characteristics and phase characteristics, respectively; Solodnikov, 1960).

Although this approach reveals the natural frequencies of a system, only a few workers have investigated the behavior of the EEG response using sinusoidally modulated light and sound signals (for details on pioneering experiments, see Van der Tweel, 1961). Difficulties result from the requirement for evoked responses to sinus signals of over at least three decades of stimulation frequencies, evoked responses in each stimulation frequency being averaged using at least 200 stimuli. Another difficulty comes from the frequent changes in brain activity stages: they may change within a few minutes and have a limited duration, which is not sufficient for the application of sinusoidal stimuli of different frequencies.

There is, however, another way of obtaining the frequency characteristics of a system, called the transient response frequency characteristics

(TRFC) method: according to general systems theory, all information concerning the frequency characteristics of a linear system is contained in the transient response of the system and vice versa. In other words, knowledge of the transient response of the system allows one to predict how this system would react to different stimulation frequencies, if the stimulating signal was sinusoidally modulated. If the step response $c(t)$ of the system—in our case, the sensory evoked potential—is known, the frequency characteristics, $G(j\omega)$, of this system can be obtained with a Laplace transform, that is, a one-sided Fourier transform:

$$G(j\omega) = \int_0^\infty \frac{d\{c(t)\}}{dt} \exp\{-j\omega t)dt$$

($\omega = 2\pi f$, where f is the frequency of the input signal).

The frequency characteristics $G(j\omega)$, including the information of amplitude changes of forced oscillations and the phase angle, is also called the frequency response function. It is a special case of the transfer function and is, in practice, identical with the transfer function (Bendat and Piersol, 1968). The amplitude frequency characteristics $|G(j\omega)|$ and the phase angle $\phi(\omega)$ can be obtained by numerical evaluation, using a fast Fourier transform, with the help of a digital computer.

Although this transform is valid only for linear systems, it can be applied to nonlinear systems as a first approach (Başar, 1980): the errors from system nonlinearities are smaller than errors resulting from the length of measurements in sinusoidal stimulation experiments given the rapid transitions of the brain's activity from one stage to another.

Finally, a limitation of this approach has to be mentioned: By application of sensory stimuli, the brain is not directly stimulated with the proper input signal—there are physiological transducers (cochlea, retina, skin) between the input signal and the measured electrical output. Therefore, a direct comparison of the input and output signal is impossible; instead, the relative output amplitudes, or the magnitude of the maxima in the amplitude characteristics, are to be compared.

The methodology to evaluate EPs, AFCs, and digitally filtered data was previously described (e.g., Başar, 1980, 1983). The essential steps are as follows:

- Recording of EEG-EP epochs: With every stimulus presented, a segment of EEG activity preceding and the EP or ERP following the stimulus were digitized and stored on computer disc memory. This operation was repeated about 100–200 times.
- Selective averaging of EPs: The stored raw single EEG-EP or EEG-ERP epochs were selected with specified criteria after the recording session: EEG segments showing movement artifacts, sleep spindles, or slow waves were eliminated.

- Amplitude frequency characteristics (AFC) were computed according to the formula given previously.
- Digital filtering: EP frequency components were computed using digital filters without phase shift (Başar and Ungan, 1973). The limits of the passband filters used were not arbitrarily chosen. Filters are applied only for selectivity channels or for tuning frequencies indicated by clear peakings in the amplitude frequency characteristics.

The essential mathematical procedures applied are schematically illustrated in Figure 2.

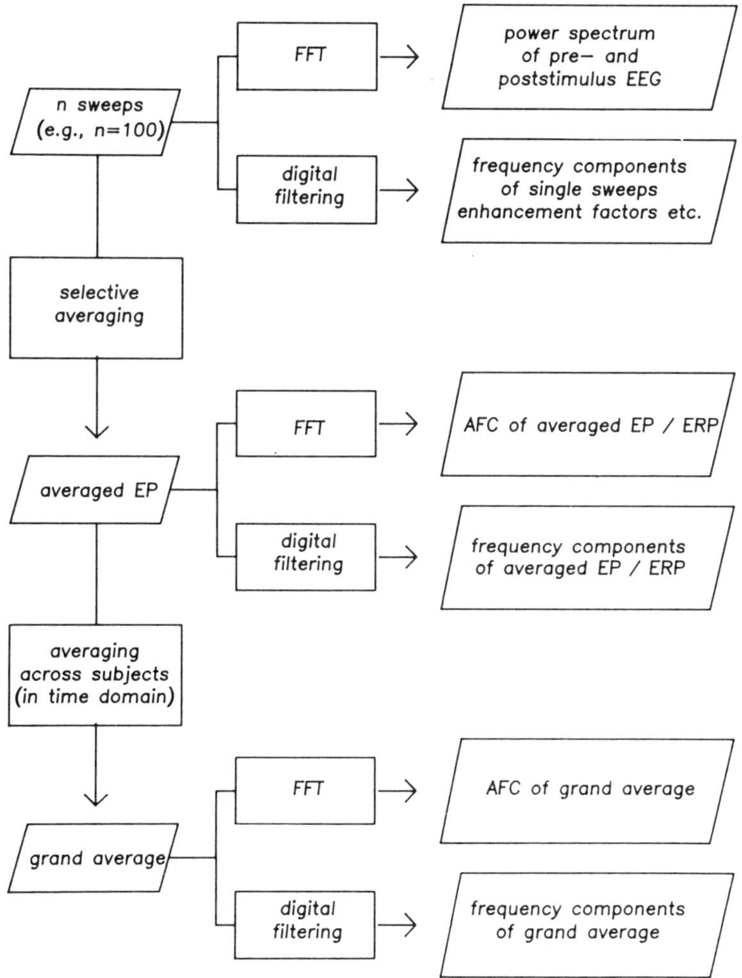

Figure 2.   Overview of combined time and frequency analysis of EEG-EP or EEG-ERP epochs (AFC, amplitude-frequency characteristics). For details see text.

Data Reduction and Statistical Evaluation

QUANTIFICATION OF FREQUENCY COMPONENTS OF TRANSIENT EP AND ERP. The averaged EPs or ERPs filtered in various frequency bands were displayed on the graphics display of the computer. The maximal peak to peak amplitudes of the filtered responses in a predefined time window can be obtained automatically or by marking the peaks manually by means of graphic cursors. The obtained maximal amplitude values of different frequency components of the averaged EP or ERP were statistically processed, and the medians and 95% confidence intervals of the data of all subjects were obtained. As statistical parameters, medians and 95% confidence intervals were selected because the normality of the distribution of the data cannot be tested on a sample of 10–20 observations.

WHAT DOES A HISTOGRAM SHOW? The median amplitudes with 95% confidence intervals of various frequency components of transient EPs or ERPs are displayed in the histograms sorted according to the experimental conditions and recording sites. The histogram presentation allows a simple visualization and comparison of the amplitudes, together with the scalp distribution of various frequency components, under different experimental conditions.

STATISTICAL EVALUATION OF THE RESULTS. The maximal peak-to-peak amplitude values of various frequency components of responses obtained in auditory EPs omitted stimulus, and oddball experiments are tested for the significance of differences by means of a nonparametric test because they do not appear to be normally distributed. For this purpose, the Wilcoxon–Wilcox test (Sachs, 1974) is used. The significance values less than .05 are presented in the tables and figures of this chapter.

Results

From the omitted stimulus paradigm, the response to the stimulus directly preceding the omitted one was included in the analysis because all subjects reported that they had attended to its onset time to be able to mark mentally the virtual onset of the omitted stimulus (see Demiralp and Başar, 1992, for details). From the oddball paradigm, only responses to the nonfrequent task-relevant target stimuli were selected for the analysis.

In Figure 3 the grand averages in time domain and the amplitude frequency characteristics (AFCs) calculated from the grand averages are shown. On the grounds that Demiralp and Başar (1992) found consistent changes only in the theta frequency band, we focused our attention mainly to the low-frequency components of evoked as well as event-related potentials.

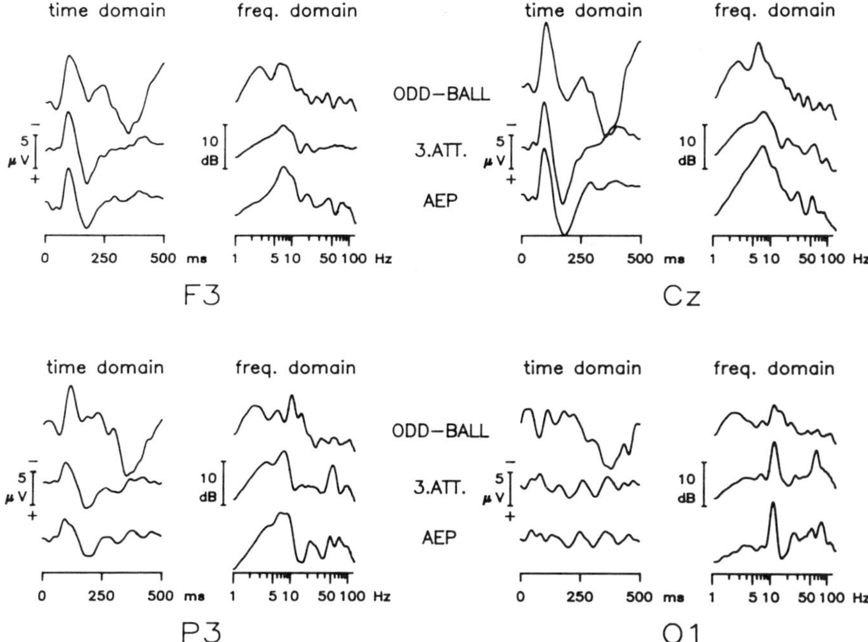

Figure 3.   Time-domain and frequency-domain representations of grand averages of auditory evoked potentials, responses to 3rd attended tones in omitted stimulus paradigm (3.ATT) and responses to nonfrequent target tones in oddball paradigm (oddball) obtained in frontal, vertex, parietal, and occipital (F3, Cz, P3, O1) recording sites.

Differences Between AFCs of Auditory EPs Recorded at Different Locations

The comparison of AFCs computed from auditory EPs elicited in different recording sites revealed differences of the frequency contents. As described in previous studies, the vertex-auditory EP (Cz) showed a peak at 7 Hz with a shoulder at 10 Hz whereas in the parietal region these frequencies were at the same level with an additional side peak occurring at 4 Hz. Frontal response (F3) showed characteristics similar to vertex in alpha range, although it had a more concave form in the subalpha band. In the occipital area (O1), a residue of ongoing alpha activity and a smooth theta peak were detectable. Demiralp and Başar (1992) interpreted these differences as possible manifestations of a distributed processing of the stimuli in the brain. Different brain structures might respond in different frequency bands corresponding to the changing quality and function of the neural networks in these structures.

Differences Between Time Domain Grand Averages and AFCs of Responses in Three Paradigms

The responses to the third attended stimuli in F3, Cz, and P3 locations showed marked increases in the amplitudes of N100–P200 complexes compared with the standard auditory EPs. These increases in the amplitudes were accompanied by increases of theta band (3–6 Hz) amplitudes in AFCs in the frequency domain. In the AFC of the occipital recording, there was an increase in the theta band accompanied by a decrease in the alpha peak, although no evident response could be detected in the time domain in this location.

In the oddball experiments, the target responses showed the characteristic late P300 complexes in all recording sites, including the occipital area where the earlier components were not clearly identifiable. The P300 waves in the time domain were accompanied by additional prominent delta peaks with a center frequency of 2 Hz in the frequency domain. A similar change occurred also in the AFC of the occipital area.

The comparison of frequency domain representations of responses obtained in all four recording sites to standard auditory stimulation third to third attended stimuli, and to oddball tones revealed a progressive increase of amplitudes of subalpha frequency components and a progressive decrease of dominant center frequency of the activity in this frequency band.

Adaptive Filtering of the Responses

For further analysis, the signals were filtered by using digital filters, which caused no phase shift. The bandpass limits of the filters were selected according to the peaks in AFCs. We differentiated, in the subalpha frequency range, two frequency bands that were shared commonly by all conditions with some changes in center frequencies. They consisted of a delta band between 1 and 3 Hz and a theta band between 3 and 6 Hz. Figure 4 shows the frequency components of grand averages obtained in all three experiments filtered in these frequency bands.

Statistical Analysis of Filtered Waveforms

For statistical testing of the changes of these frequency components under different conditions in cognitive performance, we filtered individual subject responses with corresponding bandpass filters. The maximum peak-to-peak amplitudes of filtered frequency components in relevant time windows were measured and tested for significance of differences between the three experimental paradigms by means of the Wilcoxon–Wilcox test. For the delta frequency band, a single time window between 0 and 500 msec has been used, which is approximately equal to the period of a single delta oscillation in

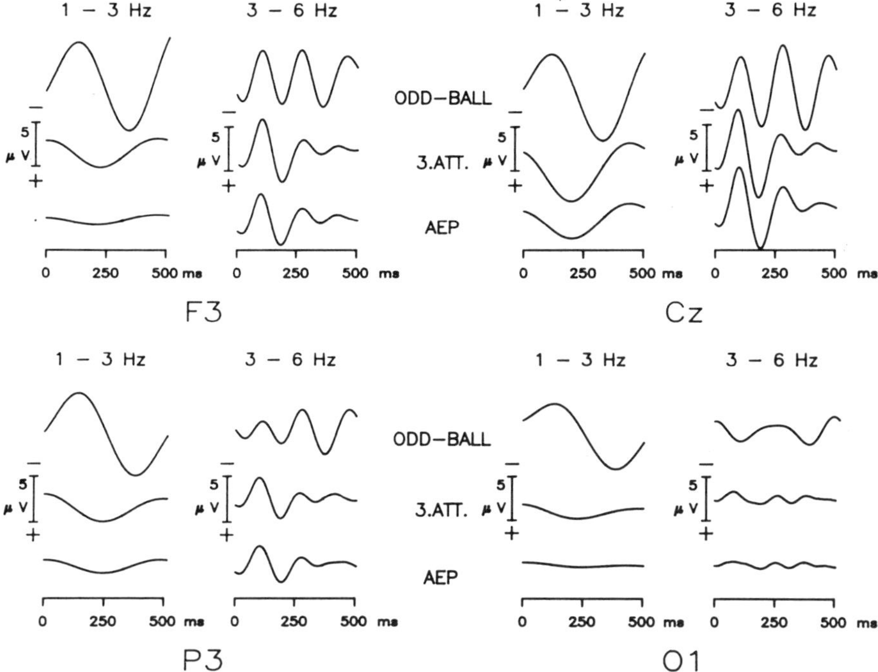

Figure 4.   Delta- and theta-frequency components of grand averages of auditory evoked potentials (AEP), responses to 3rd attended tones in omitted stimulus paradigm (3.ATT), and responses to nonfrequent target tones in oddball paradigm (oddball) obtained in frontal, vertex, parietal, and occipital (F3, Cz, P3, O1) recording sites.

the frequency band 1–3 Hz (center frequency at 2 Hz). For theta oscillations, two different time windows were used to identify the prolonged oscillatory activity with smaller damping factors or delayed enhancements. Previous studies of our group Başar and Stampfer, 1985; Stampfer and Başar, 1985) reported prolonged and enhanced theta oscillations in P300 responses. The first time window covered the N100–P200 complex of auditory evoked responses (0–250 msec), and the second time window included the P300 complex (250–500 msec). The medians of the amplitudes of delta, theta, and alpha components and 95% confidence intervals obtained in three paradigms are shown in Figure 5. The significant differences are marked with symbols representing the significance levels (see also Table 1).

The responses to the third attended stimuli in the omitted stimulus paradigm depicted, in all recording sites, nonsignificant increases in the delta frequency band compared with the standard auditory EPs. In the oddball paradigm, there were further increases in the amplitudes of the delta components, which were statistically significant compared with the responses

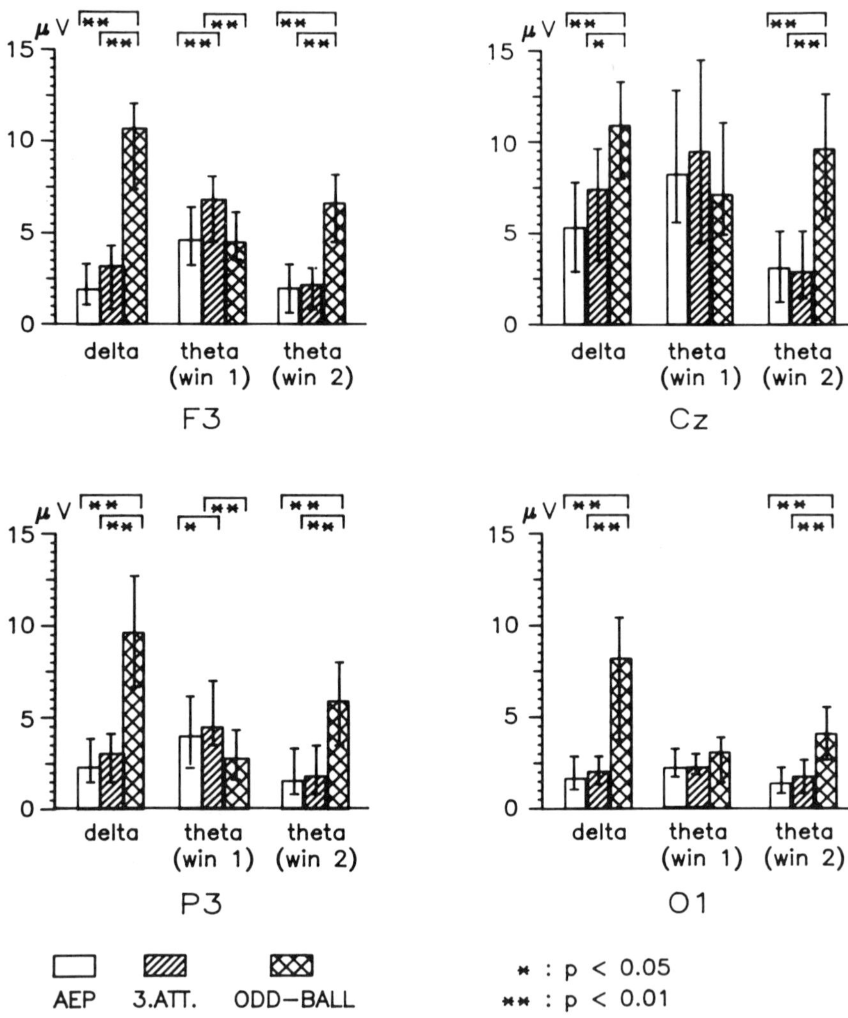

Figure 5.   Medians and 95% confidence intervals of maximum amplitudes of delta-, theta-, and alpha-frequency components of auditory evoked potentials (AEP), responses to 3rd attended stimuli in omitted stimulus paradigm (3.ATT), and responses to nonfrequent target tones in oddball paradigm (oddball) obtained in frontal, vertex, parietal, and occipital (F3, Cz, P3, O1) recording sites. Statistically significant differences are marked with symbols representing significance levels.

to third attended stimuli ($p < .01$ at F3, P3, O1; $p < .05$ at Cz) and with the auditory EPs ($p < .01$ at all locations).

In responses to third attended stimuli in the omitted stimulus paradigm, statistically significant increases of theta oscillations occurred in the early part of the ERPs (window 1: 0–250 msec) in frontal (F3) and parietal (P3)

Table 1. Medians of maximum amplitudes of delta-, theta-, and alpha-frequency components of auditory evoked poentials (AEP), responses to third attended stimuli in omitted stimulus paradigm (3. ATT), and responses to nonfrequent target tones in oddball paradigm (ODDBALL) obtained in frontal (F3), vertex (Cz), parietal (P3), and occipital (O1) recording sites. Percent changes of Amplitudes in (3. ATT) and (ODDBALL) conditions as the percent of Standard AEP Amplitudes Are Given in Parentheses.[a]

| | | | Theta (3–6 Hz) | |
		Delta (1–3 Hz)	window 1 (0–250 ms)	window 2 (250–500 ms)
F3	AEP	1.8	4.7	2.0
	3.ATT	3.0 (66%)	6.7 (43%)	2.0 (0%)
	P300	10.6 (489%)	4.5 (−4%)	6.5 (225%)**
Cz	AEP	5.3	8.2	3.0
	3.ATT	7.3 (37%)	9.5 (16%)	2.9 (−3%)
	P300	10.9 (106%)**	7.2 (−12%)	9.7 (223%)**
P3	AEP	2.4	4.0	1.6
	3.ATT	3.2 (33%)	4.4 (10%)	1.8 (13%)
	P300	9.7 (304%)**	2.7 (−33%)	5.8 (263%)**
O1	AEP	1.6	2.3	1.3
	3.ATT	1.9 (19%)	2.3 (0%)	1.5 (15%)
	P300	8.2 (413%)**	2.8 (22%)	3.9 (200%)**

[a] Statistically significant differences are marked with symbols representing the significance levels (*, $p < .05$; **, $p < .01$.)

locations compared with the standard AEPs ($p < .01$ and $p < .05$, respectively) and compared with the oddball responses ($p < .01$ in both locations). In the late part of ERPs, no significant theta change occurred in responses to third attended stimuli. Further, in oddball responses significant theta increases were registered only in the late part of the responses (window 2: 250–500 msec). The theta changes in oddball responses were more widely distributed in comparison to those in responses to third attended tones, which were localized in frontal and parietal regions. The theta increases could be observed in all four recording sites and were statistically significant in comparison to both the auditory EP and third attended tone responses ($p < .01$ at all locations).

Review and Discussion

A Comparison of Different ERP Paradigms Focused on Delta and Theta Responses

Demiralp and Başar (1992) have described the differences between frequency components of the auditory and visual evoked responses elicited in a standard evoked potential paradigm and those elicited by the stimulus preceding the omitted one in an omitted stimulus paradigm. In this paradigm, the subjects had to mark mentally the virtual onset time of the omitted stimulus. The expectancy and increased attention directed toward the third stimulus (predecessor of the omitted stimulus) induced increases in theta frequency components of the evoked responses.

In this comparative study, we used the auditory oddball responses for comparison. The delta and theta components of ERPs in both paradigms revealed significant differences. In responses to third attended tone, there was a significant increase in theta frequency band at frontal and parietal locations only in the early part of the ERP immediately after the stimulation (window 1: 0–250 msec). Similar changes were also observed in visual modality, but with a more pronounced theta increase in parietal region in comparison to the auditory modality (Demiralp and Başar, 1992). In the delta frequency band there was no significant change in third attended tone responses.

In oddball responses, a prolongation of theta oscillations (a slower damping) was observed. Significant increases in the theta frequency band were recorded only in the late part of the ERP (window 2: 250–500 msec). The theta increases in oddball responses were distributed in all locations like the significant delta increases (F3, P3, Cz, O1).

In the following sections, we comment on whether there are 'pure sensory' or 'pure cognitive' paradigms in ERP research; discuss the results presented previously and those of the Demiralp and Başar study (1992); and tentatively develop a working hypothesis for which a survey of ERP measurements including intracranial recordings is necessary.

'Pure Sensory' or 'Pure Cognitive' Paradigms in ERP Research?

Differences in frequency contents of EPs obtained from different brain locations suggest that responses recorded on different areas of the brain might originate from different neuron populations with possible different functional meanings.

Başar et al. (1991) assumed that it was impossible to design a pure sensory or pure cognitive paradigm in EP research. Even under the minimal conditions necessary for sensation, the higher cognitive functions of the brain are not to be totally rejected. On the other hand, a well-defined cognitive performance needs to be controlled by certain physical events as an interface between the internally running events in the subject's brain and the experimenter. Therefore, it may be expected during the standard EPs that various cognitive processes come into play in addition to the sensory processing. The studies of Posner and Petersen (1990) emphasized the topographical characteristics of cognitive processing. Goldman-Rakic (1988) showed in a neuroanatomic study the parallel distributed networks in primate association cortex. The assumption of Başar et al. (1991) together with the results of the psychological and neuroanatomic studies just mentioned (Goldman-Rakic, 1988; Posner and Petersen, 1990) suggested a distributed sensory-cognitive parallel processing system in the brain. In such a system the primary sensory processes and various associative or cognitive functions might be coactivating in different brain structures during the perception of a physical stimulus. This type of distributed parallel processing could be responsible for the differences of frequency contents of responses obtained in different locations.

Increased Theta Response Elicited by Stimuli Preceding the Omitted Stimulus

If it is hard or even impossible to design pure sensory or pure cognitive paradigms in ERP research, how can "exogenous" and "endogenous" ERP components be differentiated? This question is particularly important for ERP components faster than the P300 wave: They probably cannot be detected in most paradigms used, because they overlap with early exogenous components (Başar et al., 1991; Desmedt et al., 1983; Näätänen, 1988; Picton and Stuss, 1980). Many solutions have been proposed to isolate spatio-temporally overlapping components (for review, see Picton and Stuss, 1980). According to the authors, isolation through experimental manipulation and factor analysis have been the most commonly used methods that yielded objective results. However these methods lack the physiological interpretability of the isolated components.

Frequency domain analysis may provide important tools to solve this problem, as was demonstrated by Demiralp and Başar (1992): They used two types of ERP paradigms, standard EP paradigms and omitted stimulus paradigms, with both auditory and visual stimuli. They compared responses to stimuli in standard EP recording sessions and responses to stimuli preceding the omitted one (third attended stimuli). They found increased peak-to-peak amplitudes for the main peaks in responses to third attended stimuli—N100–P200 for auditory stimuli and N140–P200 for visual stimuli. Frequency domain analysis showed that this increase in amplitude was mainly caused by the selective enhancement of the theta components. The same theta increase is seen in both auditory and visual modalities. This change in low frequency components of ERP is modality independent.

Başar and coworkers (Başar, 1980, 1988; Başar et al., 1975a, 1975b, 1975c; Başar and Özesmi, 1972) have shown that sensory EPs might be considered as a superposition of wave packets in various frequencies with varying degrees of frequency stabilization, enhancement, and time-locking within conventional frequency bands of the ongoing EEG activity. Further, the authors showed that these phenomena can occur interindependently in various frequency bands in separate applications of the stimulus (sweeps) along an ERP recording: "These variations support the hypothesis that different neural or psychophysiological mechanisms come into operation following stimulation" (Stampfer and Başar, 1985).

Başar et al. (1991) have emphasized that the response of a primary sensory area consisted mainly of an enhancement in alpha (8–13 Hz) frequency band to adequate stimuli whereas its response to inadequate stimulation was dominated by the theta activity. The authors found a parallelism between these findings and the studies that showed that primary sensory stimuli elicit impulses or volleys converging over thalamic centers to primary sensory areas, whereas the "sensory stimulation of second order" usually reaches the cortex over association areas (Shepherd, 1988). In this framework the theta dominance during inadequate stimulation of a primary sensory area was interpreted as a possible manifestation of responsiveness of various brain areas in cases of association processes involved in global associative cognitive performance. Mizuki et al. (1980) showed that midline prefrontal region of the cortex generated regular theta rhythms during the performance of simple repetitive mental arithmetic tasks.

The increases in the theta components of the ERPs in comparison to the standard EPs is in agreement with the results of Başar et al. (1991) and Mizuki et al. (1983) mentioned earlier. Our findings supplement these results in terms of a probable functional assignment to the theta band activity. Considering the theta dominance in responses of primary sensory areas to inadequate stimulation and the appearance of theta rhythms in EEG recorded on associated areas during mental tasks together with our results, we incline to explain that neural circuits which perform associative functions share a common information channel that operates in the theta frequency range.

Further, the *topographical differences* in weights of increases of theta components in ERPs suggest an association between the investigated cognitive function, the frequency content of the ERP, and the topography of the frequency components.

INCREASE OF THETA COMPONENTS IS HIGHEST IN FRONTAL RECORDINGS During cognitive performance, Demiralp and Başar (1992) measured the highest—statistically significant—theta increases in frontal and parietal recording sites. In auditory modality, the theta increase was absolutely dominant in the frontal area (44% increase) whereas in visual modality the theta increase in frontal recording site was slightly higher than that in parietal recording site (48% versus 45%). This selectivity has a parallelism with the results of Fuster (1991). The studies of Fuster are based on single unit recordings in the prefrontal cortex of monkeys, which showed a high anticipatory activation level of frontal neurons in time delay tasks. Because the cognitive task in our study was also mainly based on anticipation to an expected stimulus, it is not surprising that the greatest changes are in frontal regions.

Our findings showing the strong participation of the frontal cortex in fulfilling a cognitive task are also in accordance with the results of Knight et al. (1981) on patients with frontal cortex lesions. Results of these studies indicated that the frontal lobes exhibited a modulating influence on the endogenous negativity of ERPs produced in selective attention tasks.

In Visual Modality, the Secondary Dominant Theta Increase Occurs in the Parietal Recordings. In the Demiralp and Başar study (1992), the parietal area was found to be the secondary dominant theta center; in visual modality there was a percent theta increase slightly less than that obtained in the frontal area (45%) whereas in auditory modality the increase of parietal theta activity was not so prominent (10%). This property of visual ERP is in accordance with the specific functions of parietal cortex in visual information processing, as shown by means of cellular measurements on macaque monkeys (Lynch et al., 1977; Mountcastle et al., 1975). According to Mountcastle, 50% of the investigated neurons in the parietal association area 7 of the inferior parietal lobe are visual fixation cells that are active as the animal looks at visual targets which are linked by a strong motivational drive; the rest of the neuronal elements are light sensitive with large and bilateral receptive fields (Robinson, Goldberg, and Stanton, 1978; Yin and Mountcastle, 1977). Lynch et al. (1977) suggested that the neurons in posterior parietal cortex are involved in selective visual attention processes. Mountcastle et al. (1981, 1984) showed that the enhanced responsiveness of light-sensitive neurons in the inferior parietal lobe does not merely occur with changes in general arousal but is more specifically related to the visual attention directed to the target light. Petersen et al. (1988) showed similar effects in the parietal cortex of normal humans by means of positron emission tomography.

The specific parietal theta increase we observed during visual perception in an attentive state with high expectation supports the view of Başar and Stampfer (1985) that the activity of neuronal populations involved in specific stages or parts of perceptual processes use different frequency channels, and hence these specific activities can be differentiated by means of the frequency analysis applied to the surface-recorded evoked responses.

ERPs as Reflections Induced Rhythmicities: A Review of Sources and Psychological Correlates

A Survey of Intracranial Sources of the ERPs. In previous publications (Başar-Eroglu and Başar, 1991; Başar-Eroglu et al., 1991a, 1991b), we have discussed ERP experiments that were pertinent to questions which directly arose from our own experiments on the cat brain. We now extend our discussion by including results from other studies on humans (mostly epileptic patients) as well as on monkeys together with the most recent results using ablation techniques. Despite the interest among psychologists and neurologists, the neural origin of the human event-related potentials is not known. Because P300 is an example of an ERP that is widely investigated, we first focus our attention to its electrogenesis.

Because of the cognitive correlates of the P300, many researchers have assumed that ERPs are generated in neocortex. Neocortical generators postulated for the human P300 have included the frontal cortex (Courchesne, 1978; Desmedt and Debecker, 1979; Knight et al., 1981; Wood and McCarthy, 1985). Some other workers have described the centroparietal or temporoparietal association cortex as the site of generators (Goff, Allison, and Vaughan, 1978; Pritchard, 1981; Simson, Vaughan, and Ritter, 1977; Vaughan and Ritter, 1970) because the P300 amplitude is the largest in these areas. Wood et al. (1980) proposed multiple subcortical sites as generators. The studies of Halgren et al. (1986) suggested involvement of the hippocampal formation and the limbic system in P300 generation. However, according to Wood et al. (1982) and Smith et al. (1990), the surface topography of the P300 is unchanged after unilateral temporal lobectomy that includes the hippocampus, and hippocampal P300s may not be volume conducted to the surface. The analysis of Başar-Eroglu et al. (1991b) also excludes the possibility of a volume conduction from CA3 layer of the hippocampus to cortex; however, a facilitation of the neural signal via hippocampocortical system is not excluded by the description of these authors.

Many authors postulated that generator systems can be better explored in an animal P300 model than in the human brain. Such studies have provided evidence for the neocortical or limbic system involvement. The marginal gyrus, suprasylvian gyrus, and hippocampus have all been postulated as generators of the P300 in the cat (Buchwald and Squires, 1982; O'Connor and Starr, 1985). Frontal cortex (Boyd, Boyd, and Brown, 1976) or locus coeruleus (Pineda, Foote, and Neville, 1987) have been postulated as necessary components for P300 generation in the monkey. A P300-like potential remains after bilateral ablation of the hippocampus. Harrison et al. (1990) have shown that the cat P300 is present after the ablation of the association cortex. These authors also suggested multiple generators of the cat P300, such as marginal and suprasylvian gyri as well as the hippocampal region.

Studies of P300 following unilateral anterior lobectomy, which includes anterior regions of the hippocampus in humans (Stapleton and Halgren, 1987; Wood et al., 1982), or bilateral hippocampal lesions in monkey (Paller et al., 1988) have failed to identify any related change in P300 amplitude. On the other hand, physiological events in the hippocampus have been demonstrated to contribute to scalp-recorded activity (Altafullah et al., 1986).

Smith et al. (1990) tried to isolate the anatomic locus of neural activity most important for directly generating the scalp P300, and they suggested that the critical locus is the lateral neocortex of the inferior parietal lobule. Further, Smith et al. assumed that activity in the hippocampus and frontal lobe may only make minor contributions to scalp recordings. They suggest that the scalp-recordable P300 might best be conceptualized as only the most readily observable aspect of synchronous activity occurring across widely distributed yet highly integrated cognitive activities.

The analysis of Katayama, Sukiyama, and Subokawa (1985) indicated thalamic negativity associated with the endogenous late positivity (or P300) by using conditioning in humans and cats.

Başar-Eroglu and Başar (1991) and Başar-Eroglu et al. (1991a, 1991b) have published a series of results from experiments with freely moving cats and by using a passive P300 paradigm and assumed that P300-like potentials have multiple cortical and subcortical generator sites including reticular formation of the brain stem, hippocampus, and auditory cortex. According to these most recent results of our group, the P300 potential is the most significant, stable, and has the largest amplitudes in the CA3-layer of the hippocampus of the intact cat brain. Başar-Eroglu et al. (1991b) further showed that the hippocampal P300 manifests an enhancement of the theta activity of the field potentials or a type of resonance phenomenon in the theta frequency range. Figure 6 shows (only for discussion purposes) the comparison of spectral activity of spontaneous field potentials and the P300 response in the hippocampus (compare the hypothesis of Miller, 1991, following).

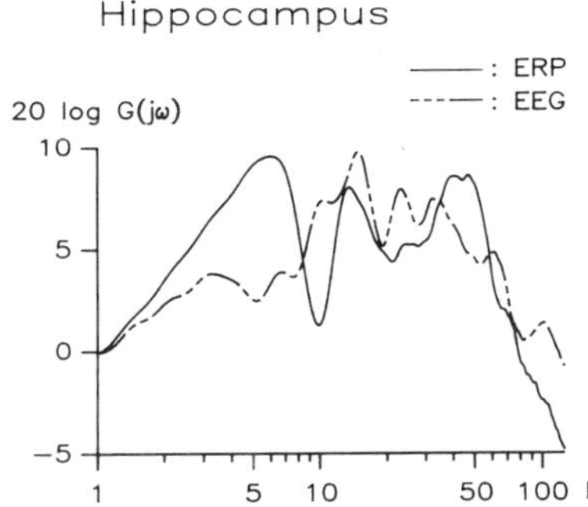

Figure 6.   Comparison of spectral activity of spontaneous field potentials and P300 response in the hippocampus. ERP, event–related potential. (From Başar-Eroglu et al., 1991b.)

THETA ACTIVITY AND LEARNING IN HUMANS: DOES A CORTICOHIPPOCAMPAL INTERPLAY EXIST? In the spontaneous human EEG, theta activity is masked by the alpha rhythm and cannot easily be detected. In earlier studies, Rémond and Lesèvre (1957) reported on a predominance of the theta rhythm in the frontal central region whereas Mundy-Castle (1951) described more pronounced theta activity at temporal regions. The results of Westphal et al. (1990), which showed that the theta amplitude is highest over the anterior midline (Fz and Cz locations), are in accordance with mapping findings of Walter et al. (1984) and Mizuki et al. (1983).

Miller (1991) reviewed extensively that theta activity recorded from the hippocampus has been difficult to find in human subjects because of the difficulties of human central electrophysiology. Some recent evidence that the midline prefrontal region of the cortex can generate theta activity was reported by Mizuki et al. (1980) in certain cognitive states: EEG rhythms of the S- to 5.5-Hz frequency appears with some regularity during the performance of simple repetitive mental arithmetic tasks.

Lang et al. (1987) also used, like Westphal et al. (1990), spectral analysis of frontal EEG and showed that theta frequencies (3–7 Hz) were increased during motor or verbal learning tasks. Miller (1991) stated that the data of the groups of Lang and of Mizuki are compatible with the hypothesis that theta activity in frontal regions is associated with theta activity in hippocampus.

Could the prefrontal cortex show theta activity at the same time as the hippocampus? Aleksanov, Vainstein and Preobrashenskaya (1986) showed that the coherence between hippocampus and prefrontal cortex increased during the course of alimentary conditioning and at the time of presentation of the conditioned stimulus (CS). This increase of coherence was observed especially in the delta and theta frequency ranges. There is yet no clear evidence that the more distantly related areas of association cortex generate theta potentials reflecting the hippocampal ones. Miller (1991), after his strong survey of electrophysiological and functional analysis on the cortico-hippocampal interplay, concludes: Despite the lack of evidence at presence, the central prediction still made is that there are cortical nodes with consistent phase relations to hippocampal theta activity.

THE 'DIFFUSE THETA-RESPONSE SYSTEM IN THE BRAIN' AND THE CORTICO-HIPPOCAMPAL INTERACTION. According to a number of neurophysiological, psychological, and biophysical approaches, the oscillatory phenomena in neural tissues now merit considerable attention in understanding cognitive functions of the central nervous system (Başar, 1992; Bullock, 1992; Galambos, 1992; Gray et al., 1992; Petsche and Rappelsberger, 1992; Pfurtscheller and Klimesch, 1992; Tononi, Sporns, and Edelman, 1992). According to the results of experiments on the P300 response of the cat brain, we here analyze the responsiveness in the theta-frequency range in an extended manner. Tentative interpretations of the results of our group have led us to postulate the existence of a "Diffuse Alpha-Response System" and a "Diffuse Theta-Response System" in the brain. Başar et al. (1991) tentatively assumed that the brain theta response (or the theta component of the evoked potentials) or slower responses might reflect the responsiveness of various brain areas in processes involved with global associative-cognitive performance. We have observed that the theta components in frontal and parietal recordings are significantly enhanced and that the hippocampal P300 response of the brain also shows a dominant theta component (Fig. 6), which emphasize that some special anatomic structures or preferential structures are more involved in generation of theta resonances.

The resonant theta response of the hippocampus to auditory and visual stimuli was explained in details in the report concerning a component analysis of hippocampal EPs (Başar and Ungan, 1973). The concept of theta resonance was extensively analyzed by Miller (1991), who described the corticohippocampal interaction as a basic resonance phenomenon in the theta-frequency range. In the well-documented analysis by Miller (1991), it is envisaged that Hebbian processes of synaptic strengthening select patterns of loops passing from hippocampus to cortex and back to the hippocampus. Miller (1991) states: "The total conduction delay time round each one of these patterns of loops is envisaged to correspond to the theta period. Such resonance between hippocampus and cortex is envisaged to have an important functional role in registration and retrieval of information in the cortex. Each pattern of resonant loops will raise the activation of specific collection of cells which are widely dispersed across the cortical mantle."

Miller (1991) further described the connectional basis of phase locked loops. According to the anatomic and physiological evidence, Miller takes the viewpoint that theta-modulated signals are likely to influence limbic and prefrontal areas, and also (directly or indirectly) other areas of (mainly association) cortex.

Our recent findings of the cat hippocampus indicating a strong theta enhancement [or resonance, as Başar-Eroglu et al. (1991b) emphasized] in the hippocampal P300, and the hypothesis by Miller (1991) give relevant support to emphasize the implication of cognitive components by means of a diffuse theta response system. The term diffuse is used to describe a distributed system in the brain: it is not yet possible to describe it with sharp anatomical boundaries. Such a diffuse theta system, which was experimentally based on theta enhancement phenomena in various brain structures including the brain stem, shows seemingly considerable strength (or more resonant loops) in the hippocampocortical system. Such theta enhancement could also be shown in the cat reticular formation (Başar-Eroglu et al.,1991a).

Hebb's rule implies that groups of synapses converging on a single neuron and having the tendency to fire together will become strengthened as a group. This is, as Miller (1991) also describes, a principle of cooperativity: "When an animal is in a particular environment, a collection of widely dispersed neurons throughout the cortical mantle may be set into a tonic state of elevated activation, as a result of signals reaching the cortex via any of its sensory systems." Miller (1991) further assumes that there is a series of recursive loops, with some temporal divergence/convergence in the pathway in each direction, and there is the possibility of regular oscillation of neural activity imposed at the hippocampal end. "The Hebbian mechanism in this case should favour the strengthening of connections which permit resonance." Başar-Eroglu et al. (1991b) showed that the significant theta response at the CA3 pyramidal layer cannot be recorded in the cortex with high amplitude because volume conduction is negligible. Therefore, the significant cognitive theta enhancements (see also Lang et al., 1987; Miller, 1991; Mizuki et al., 1980) in frontal and parietal recordings (see earlier sections of this study) might occur in the sense of Hebbian cooperation mechanism among the neuronal populations of the frontal cortex, the parietal cortex, and of the hippocampus. It should be emphasized that the cooperativity in the Hebbian sense might, in a diffuse theta response system, be one of the causal factors in magnification of the theta component of ERPs also recorded from the scalp. Again, the use of the term diffuse allows descriptions without neuron-to-neuron tracking, without defining directions of signal flow, and finally without exact boundaries of structures or neural populations involved.

GENERAL REMARKS ON FUNCTIONAL CORRELATES OF ERPs: FOCUSED ATTENTION, SIGNAL DETECTION, RECOGNITION, AND DECISION MAKING. The P300 has been found to correlate with many variables, including 'task relevance,' 'meaningfulness,' 'information delivery,' 'resolution of uncertainty,' and 'decision making.' Regan (1989) criticized that some of these physiological terms overlapped whereas others were not sharply defined, and in itself this led to controversy. The need for psychophysiological models in ERP experiments has also been pointed out by Münte and Künkel (1990) with respect to clinical ERP applications in psychiatry. According to Woods (1990), selective attention refers to the preferential detection, identification, and recognition of selected stimuli in an environment containing multiple sources of information. In P300 experimental designs, the eliciting stimulus is task relevant; that is, attention is paid to it. The contradictory claim that P300 can be recorded in the situation where low-probability stimuli are ignored (Roth, 1973) is suspect, because it assumes that subjects are able to follow instructions to ignore irrelevant stimuli while they just sit with eyes shut (Ritter, Vaughan, and Costa, 1968; Squires et al., 1977).

According to Hillyard and Picton (1979), "the P300 seems to index the operation of adaptive brain systems." A rather different line of thought, derived from autonomic psychophysiology, uses the language of innate

In general, a P300 will be produced by task-relevant stimuli (Chapman, 1973; Chapman and Bragdon, 1964) that occur somewhat unexpectedly and require a motor response or cognitive decision (Donchin, 1981; Ritter, Vaughan, and Costa, 1968).

POSSIBLE NEUROPHYSIOLOGICAL CORRELATES OF SIGNIFICANT INCREASES IN DELTA AND THETA FREQUENCY RESPONSES Woods (1990) claimed that studies of the psychological bases of selective attention have little influence on each other. On the one hand, cognitive psychologists have typically studied selective attention in human subjects performing complex tasks and have monitored attention with simple or complex responses. In contrast, cellular neurophysiologists have examined the effects of selective attention on the firing of single neurons in animals responding to simple stimulus significance. Başar et al. (1991), as well as Demiralp and Başar (1992), tried to bridge the gap between these two different approaches. They aimed at linking both types of experiments by including the discussion of a neurophysiological signal that is common in both categories of analyses: the changes in theta-frequency component (especially of the hippocampus and cortex) during learning, association phenomena, memory retrieval, and attentive behavior. We emphasize that during increased attention (the responses to third stimuli) the significant increases in the theta responses were recorded in frontal and parietal locations. Such overall controlling phenomena in the brain were recently surveyed by Başar (1992). The relation of the theta component to single neuron firing is established in a number of relevant studies (see Buzsaki, 1992; Lopes da Silva, 1992).

We aimed in this work, as in a previous study (Demiralp and Başar, 1992), to reduce this type of controversy by discussing a limited number of measurable neurophysiological correlates, that is, the significant changes in the delta- and theta-frequency channels with some evident functional correlates. Various workers describe frequency channels of the EEG as innate response of neural networks or intrinsic frequency channels of single neurons, related to function (Freeman, 1992; Gray et al., 1992; Llinas and Graves, 1988) or induced rhythmicities (Petsche and Rappelsberger, 1992; Pfurtscheller and Klimesch, 1992). In this study we do not explicitly discuss alpha, beta, and gamma (40-Hz) frequency bands that were discussed elsewhere. The topographical distribution of the prolonged alpha response will be soon analyzed in a detailed study. A 40-Hz diffuse (distributed) system was also described by Başar-Eroglu and Başar (1991) and Başar et al. (1992). The significant change of the alpha oscillations consisted especially in the results of oddball paradigm as a prolongation of oscillation, but not significant increase in the amplitude (see also Kolev and Schürmann, 1992).

*Focused Attention, Signal Detection, Matching, and Decision Making:
Possible Psychological Correlates of Increased Delta and Theta Responses?*

The results described were based on two different types of paradigms and included recordings in various scalp locations. Moreover, animal studies with records of limbic and other subcortical structures were considered for neurophysiological analysis.

We now come back to general definitions of 'focused attention,' 'signal detection,' 'signal recognition' and 'decision making' and tentatively try to combine the results of the theta response and delta response with the psychological context discussed earlier. We aim, therefore, to link physiological results to psychological correlates in terms of natural frequencies of the EEG. In conclusion, we summarize the ERP paradigms used, their psychological contents, and the correlated changes in EEG rhythms:

- oddball paradigm: During the application of this paradigm the subjects have at least two different categories of tasks to perform:
    i.   Focused attention and signal detection
    ii. Matching for target recognition and decision making
  Because the target signals were presented in a random order with random interstimulus intervals, the subjects were not able to prepare themselves for signal detection. There is no anticipation to the coming stimulus. Therefore, the signal detection task refers first to the period after the stimulus delivery. In addition, the subject's task to recognize target stimuli between nontarget ones accomplishes matching and decision-making processes. Accordingly, the involved mental processes contain at least two different groups of components as described above. [The matching mechanisms have been extensively analyzed in reviews by Näätänen (1988, 1990).]
- Paradigm with omitted fourth signal and attention to the third signal: In this simpler paradigm, one of the above-mentioned task categories mainly seems to be in play:
    i.   Focused attention and signal detection
  Because the stimulation is applied repetitively with constant interstimulus intervals, a preparation takes place before the target. Attention is focused to the third signal (see Demiralp and Başar, 1992). The signal is anticipated. Therefore, at the time of signal application there is no relevant surprise or novelty. Accordingly, the difficulty that might stem from matching for target recognition and decision making is highly reduced or not at all involved in the procedure.

Common components in ERPs of both paradigms are focused attention and signal detection. Accordingly, we tentatively assumed that the delta response, the most prominent component of the oddball ERP, is mostly involved with the signal matching and decision making following a novel or unexpected signal or partial surprise. The early theta response (window 1) that is most prominent in frontal and parietal locations during the third attended stimulation is probably the result of focused attention which refers to preferential detection of selected stimuli, and not to matching for signal recognition or decision making.

Significant changes in theta and delta responses in different locations under both categories of the paradigms and the possible functional correlates of the applied paradigms are summarized together in the illustrative Table 2. During the paradigm with third attended stimulus, theta response increase was recorded only immediately following the stimulation. Moreover, such increases were recorded only in frontal and parietal locations. Seemingly, for this less complicated task a theta response increase to sound stimuli in frontoparietal location is representative for the brain mechanisms in play. Considering the anticipatory (time estimation) component of the task, the frontal dominance of the theta increase is in accordance with the studies based on the cellular measurements in the prefrontal cortex (Fuster, 1991). Fuster showed a high anticipatory activation level of frontal neurons in time delay tasks. Başar et al. (1989) have shown, during such a paradigm with repetitive stimulation, that regular, ample and quasideterministic alpha rhythms were recorded in several cortical locations prior to the target. Moreover, such alpha rhythms are almost time-locked to the target for a period of 1 sec before the occurrence of the target. It is remarkable that an enhanced theta response is recorded following these regular and time-locked alpha rhythms. When the task additionally involves signal matching, decision making and surprise, the changes in the frequency channels of ERP reach a higher degree of complexity:

- In all locations a marked change of the delta response is recorded.
- Increases in late theta responses (window 2) were also recorded significantly in all locations.

The existence of important delta increase suggests that processes of decision making and surprise are reflected in this slowest EEG-response component.

Increase in the delta response is probably only related to decision making and matching whereas the theta response seems to be involved in several tasks (focused attention, signal detection, anticipation and expectation), appearing either as an early or late component.

Table 2. Significant changes of the theta and delta responses in different recording sites under both categories of paradigms and the possible functional correlates of the applied paradigms[a]

Paradigms	(1) OMITTED STIMULUS (THIRD ATTENDED SIGNALS)				(2) ODDBALL (TARGET TONES)							
Description of the task	Subject focuses his attention to third signal presented repetitively: Probability of target occurrence = 100%, high expectancy, no surprise				Subject focuses his attention to rare oddball tones presented randomly. Probability of targer occurrence = 20%, low expectancy, surprise, decision making							
Functional Correlates	Association, focused attention, signal detection, expectation, anticipation				Association, focused attention, signal detection, matching for target recognition, decision making, surprise							

	THIRD ATTENDED LIGHT				THIRD ATTENDED TONE				ODDBALL (TARGET) TONE			
	F3	Cz	P3	O1	F3	Cz	P3	O1	F3	Cz	P3	O1
Increase of delta frequency response	–	–	–	–	–	–	–	–	**	**	**	**
Increase of early theta response	**	*	*	*	**	–	*	–	–	–	–	–
Increase of late theta response	–	–	–	–	–	–	–	–	**	**	**	**

[a] Symbols: (*, $p < .05$; **, $p < .01$.)

More evidence for the participation of delta response during the procedure of decision making was provided by the experiments at the auditory threshold level. Başar et al. (1992) have shown that at the hearing threshold the evoked potentials of the subjects were reduced to an almost pure delta oscillation, also detectable without frequency analysis or filter application. At the hearing threshold subjects are supposed to be involved in decision making. Accordingly, the results of Başar et al. (1992) and Parnefjord and Başar (1992), attributing a decision-making function at the threshold level to a delta component, are in good accordance with the results of the present study.

In these analyses a special emphasis was made on the increased theta and delta responses during two different paradigms to accentuate the role of the theta response system and its interplay with the hippocampocortical system. However, the approaches used in this study and also earlier results concerning (i) prolonged alpha oscillations, (ii) preparatory alpha rhythms before the application of repetitive signals and (iii) the 40 Hz response suggest that the EEG dynamics is involved in all cognitive processes and in various frequency channels. Several new experimental strategies on new psychological paradigms using the frequency analysis approach of this study might help to identify, in a clearer manner, the functional correlates of theta and delta responses.

Conclusion

On the basis of the results of the current study, on that of Demiralp and Başar (1992), and on the animal and human ERP studies reviewed here, the following tentative conclusions are offered: Comparison of animal experiments suggests a theta activation circuit or a diffuse theta system in which hippocampus might play a key role. Part of the hippocampocortical interaction might be attributed to cooperativity in the Hebbian sense.

To a certain extent, the different cognitive ERP paradigms used permitted us to investigate focused attention and decision making separately: ERP frequency components appear to be associated with certain aspects of cognitive performance.

The analysis of scalp-recorded human ERPs in the frequency domain allows a link between the cellular neurophysiology and cognitive psychology by discussing a neurophysiological signal that is common in both categories.

Acknowledgment. Work supported by Volkswagen-Stiftung grant 1/67 678, DFG-grant Ba 831/5-1, and IBM-Türk Ltd.

References

Aleksanov SN, Vainstein II, Preobrashenskaya LA (1986): Relationship between electrical potentials of the hippocampus, amygdala and neocortex during instrumental conditioned reflexes. *Neurosci Behav Physiol* 16:199–207.

Altafullah I, Halgren E, Stapleton JM, Crandall P (1986): Interictal spike-wave complexes in the human medial temporal lobe: Typical topography and comparison with cognitive potentials. *Electroencephalogr Clin Neurophysiol* 63:503–516.

Başar E (1980): *EEG Brain Dynamics. Relation Between EEG and Brain Evoked Potentials.* Amsterdam: Elsevier.

Başar E (1983): Toward a physical approach to integrative physiology. I. Brain dynamics and physical causality. *Am J Physiol* 254:R510–R533.

Başar E (1988): EEG-dynamics and evoked potentials in sensory and cognitive processing by the brain. In: *Dynamics of Sensory and Cognitive Processing by the Brain*, Başar E, ed. Berlin: Springer.

Başar E (1992): Brain natural frequencies are causal factors for resonances and induced rhythms. In: *Induced Rhythms in the Brain*, Başar E, Bullock TH, eds. Boston: Birkhäuser.

Başar E, Özesmi Ç (1972): The hippocampal EEG-activity and systems-analytical interpretation of averaged evoked potentials of the brain. *Kybernetik* 12:45–54.

Başar E, Stampfer HG (1985): Important associations among EEG-dynamics, event-related potentials, short-term memory and learning. *Int J Neurosci* 26:161–180.

Başar E, Ungan P (1973): A component analysis and principles derived for the understanding of evoked potentials of the brain: A study in the hippocampus. *Kybernetik* 12:133–140.

Başar E, Başar-Eroglu C, Rahn E, Schürmann M (1991): Sensory and cognitive components of brain resonance responses: An analysis of responsiveness in human and cat brain upon visual and auditory stimulation. *Acta Otolaryngol (Stockholm)* Suppl. 491:25–35.

Başar E, Başar-Eroglu C, Röschke J, Schütt A (1989): The EEG is a quasi-deterministic signal anticipating sensory-cognitive tasks. In: *Brain Dynamics*, Başar E, Bullock TH, eds. Berlin: Springer.

Başar E, Başar-Eroglu C, Rosen B, Schütt A (1984): A new approach to endogenous event-related potentials in man: relation between EEG and P300-wave. *Int J Neurosci* 24:1–21.

Başar E, Gönder A, Özesmi Ç, Ungan P (1975a): Dynamics of brain rhythmic and evoked potentials. I. Some computer methods for the analysis of electrical signals from the brain. *Biol Cybern* 20:137–145.

Başar E, Gönder A, Özesmi Ç, Ungan P (1975b): Dynamics of brain rhythmic and evoked potentials. II. Studies in the auditory pathway, reticular formation and hippocampus during the waking stage. *Biol Cybern* 20:145–160.

Başar E, Gönder A, Özesmi Ç, Ungan P (1975c): Dynamics of brain rhythmic and evoked potentials. III. Studies in the auditory pathway, reticular formation and hippocampus during sleep. *Biol Cybern* 20:160–169.

Başar E, Başar-Eroglu C, Parnefjord R, Rahn E, Schürmann M (1992): Evoked potentials: Ensembles of brain induced rhythmicities in the alpha, theta and gamma ranges. In: *Induced Rhythms in the Brain*, Başar E, Bullock TH, eds. Boston: Birkhäuser.

Başar-Eroglu C, Başar E (1987): Endogenous components of event-related potentials in hippocampus: an analysis with freely moving cats. In: *Current Trends in Event-Related Potential Research*, Johnson R, Jr., Rohrbaugh JW, Parasuraman R, eds. Amsterdam: Elsevier.

Başar-Eroglu C, Başar E (1991): A compound P300-40 Hz response of the cat hippocampus. *Int J Neurosci* 60:227–237.

Başar-Eroglu C, Başar E, Demiralp T, Schürmann M (1992): P300 response: Possible psychophysiological correlates in delta and theta frequency channels. *Int J Psychophysiol* 13:161–179.

Başar-Eroglu C, Başar E, Schmielau F (1991a): P300 in freely moving cats with intracranial electrodes. *Int J Neurosci* 60:215–226.

Başar-Eroglu C, Schmielau F, Schramm U, Schult J (1991b): P300 response of hippocampus with multielectrodes in cats. *Int J Neurosci* 60:239–248.

Bendat JS, Pierson AG (1968): *Measurement and Analysis of Random Data*. New York: Wiley.

Birbaumer N, Elbert T, Canavan AGM, Rockstroh B (1990): Slow potentials of the cerebral cortex and behavior. *Physiol Rev* 70:1–41.

Boyd EH, Boyd ES, Brown LE (1976): Long latency evoked responses in squirrel monkey frontal cortex. *Exp Neurol* 515:22–40.

Buchwald JS, Squires NS (1982): Endogenous auditory potentials in the cat. In: *Conditioning: Representation of Involved Neural Function*, Woody CD, ed. New York: Plenum Press.

Bullock TH (1992): Introduction to induced rhythms: A widespread, heterogenous class of oscillations. In: *Induced Rhythms in the Brain*, Başar E, Bullock TH, eds. Boston: Birkhäuser.

Buzsaki G (1992): Network properties of the thalamic clock: Role of oscillatory behavior in mood disorders. In: *Induced Rhythms in the Brain*, Başar E, Bullock TH, eds. Boston: Birkhäuser.

Chapman RM (1973): Evoked potentials of the brain related to thinking. In: *Psychophysiology of Thinking: Studies of Covert Processes*, McGuigan FJ, Schoonover RA, eds. New York: Academic Press.

Chapman RM, Bragdon HR (1964): Evoked responses to numerical and non-numerical visual stimuli while problem solving. *Nature (London)* 203:1155–1157.

Courchesne E (1978): Changes in P3 waves with event repetition: long-term effects on scalp distribution and amplitude. *Electroencephalogr Clin Neurophysiol* 45: 754–766.

Demiralp T, Başar E (1992): Theta rhythmicities following expected visual and auditory targets. *Int J Psychophysiol* 13:147–160.

Desmedt JE, Debecker J (1979): Wave form and neural mechanism of the decision P350 elicited without pre-stimulus CNV or readiness potential in random sequences of near threshold auditory clicks and finger stimuli. *Electroencephalogr Clin Neurophysiol*, 47:648–670.

Desmedt JE, Juy NT, Bourguet M (1983): The cognitive P40, N60 and P100 components of somatosensory evoked potentials and the earliest electrical signs of sensory processing in man. *Electroencephalogr Clin Neurophysiol* 56:272–282.

Donchin E (1981): Surprise!... Surprise! *Psychophysiology* 18:493–513.

Donchin E, Heffley E, Hillyard SA, Loveless N, Maltzman I, Ohman A, Rosler F, Ruchkin D, Siddle D (1984): Cognition and event-related potentials. II. The orienting reflex and P300. *Ann NY Acad Sci*, 425:39–57.

Freeman W (1992): Predictions on neocortical dynamics derived from studies in paleocortex. In: *Induced Rhythms in the Brain*, Başar E, Bullock TH, eds. Boston: Birkhäuser.

Fuster JM (1991): *The Prefrontal Cortex. Anatomy, Physiology and Neuropsychology of the Frontal Lobe*. New York: Raven Press.

Galambos R (1992): A comparison of certain gamma band (40-Hz) brain rhythms in cat and man. In: *Induced Rhythms in the Brain*, Başar E, Bullock TH, eds. Boston: Birkhäuser.

Goff ER, Allison T, Vaughan HG, Jr. (1978): The functional neuroanatomy of event-related potentials. In: *Event-Related Potentials in Man*, Tuetin EP, Koslow SH, eds. New York: Academic Press.

Goldman-Rakic PS (1988): Topography of cognition: Parallel distributed networks in primate association cortex. *Annu Rev Neurosci* 11:37–156.

Gray CM, Engel AK, König P, Singer W (1992): Mechanisms underlying the generation of neuronal oscillations in cat visual cortex. In: *Induced Rhythms in the Brain*, Başar E, Bullock TH, eds. Boston: Birkhäuser.

364   E. Başar et al.

Halgren E, Stapleton JM, Smith M, Altafullah I (1986): Generators of the human scalp P3. In: *Evoked Potentials*, Cracco RQ, Bodis-Wollner I, eds. New York: Alan R. Liss.

Harrison JB, Dickerson LW, Song S, Buchwald JS (1990): Cat P300 present after association cortex ablation. *Brain Res Bull* 24:551–560.

Heinze HJ, Luck SJ, Mangun GR, Hillyard SA (1990): Visual event-related potentials index focused attention within bilateral stimulus arrays. I. Evidence for early selection. *Electroencephalogr Clin Neurophysiol* 75:511–527.

Hillyard SA, Picton TW (1979): Event-related brain potentials and selective information processing in man. In: *Progress in Clinical Neurophysiology*, Vol. 6, Desmedt JE, ed. Basel: Karger.

Johnson R, Jr. (1988): Scalp recorded P300 activity in patients following unilateral temporal lobectomy. *Brain* 111:1517–1529.

Katayama Y, Sukiyama T, Subokawa T (1985): Thalamic negativity associated with the endogenous late positive component of cerebral evoked potentials (P300): Recording using discriminative aversive conditioning in human and cats. *Brain Res Bull* 14:223–226.

Knight RT, Hillyard SA, Woods DL, Neville HJ (1981): The effects of frontal cortex lesions on event-related potentials during auditory selective attention. *Electroencephalogr Clin Neurophysiol* 52:571–582.

Kolev V, Schürmann M (1992): Event-related prolongation of induced EEG rhythmicities in experiments with a cognitive task. *Int J Neurosci* 67:199–213.

Lang M, Lang W, Diekmann V, Kornhuber HH (1987): The frontal theta rhythm indicating motor and cognitive learning. In: *Current Trends in Event-Related Potential Research, Electroencephalogr Clin Neurophysiol* Suppl 40. Johnson R, Jr, Rohrbaugh JW, Parasuraman R, eds.

Llinas RR, Graves A (1988): The intrinsic electrophysiological properties of mammalian neurons: Insights into central nervous system function. *Science* 242:1654–1664.

Lopes da Silva FH (1992): The rhythmic slow activity (theta) of the limbic cortex: An oscillation in search of a function. In: *Induced Rhythms in the Brain*, Başar E, Bullock TH, eds. Boston: Birkhäuser.

Lynch JC, Mountcastle VB, Talbot WH, Yin TCT (1977): Parietal lobe mechanisms for directed visual attention. *J Neurophysiol* 40:362–389.

Miller R (1991): *Cortico-Hippocampal Interplay and the Representation of Contexts in the Brain*. Berlin: Springer.

Mizuki Y, Takii O, Nishijima H, Inanaga K (1983): The relationship between the appearance of frontal midline theta activity (FmO) and memory function. *Electroencephalogr Clin Neurophysiol.*

Mizuki Y, Masotoshi T, Isozaki H, Nishijima H, Inanaga K (1980): Periodic appearance of theta rhythm in the frontal midline area during performance of a mental task. *Electroencephalogr Clin Neurophysiol* 49:345–351.

Mountcastle VB (1992): Prologue. In: *Induced Rhythms in the Brain*, Başar E, Bullock TH, eds. Boston: Birkhäuser.

Mountcastle VB, Andersen RA, Motter BC (1981): The influence of attentive fixation upon the excitability of the light-sensitive neurons of the posterior parietal cortex. *J Neurosci* 1:1218–1235.

Mountcastle VB, Motter BC, Steinmetz MA, Duffy CJ (1984): Looking and seeing: The visual functions of the parietal lobe. In: *Dynamic Aspects of Neocortical Function*, Edelman GM, Gall WE, Cowan WM, eds. New York: Wiley.

Mountcastle VB, Lynch JC, Georgopoulos A, Sakata H, Acuna C (1975): The posterior parietal association cortex of the monkey: Command functions for operations within extrapersonal space. *J Neurophysiol* 38:871–908.

Mundy-Castle AC (1951): Theta and beta rhythm in the electroencephalograms of normal adults. *Electroencephalogr Clin Neurophysiol* 3:477–486.

Münte TF, Künkel H (1990): Ereigniskorrelierte Potentiale in der Psychiatrie— Methodische Grundlagen. *Psychiatr Neurol Med Psychol* 42:649–659.

Näätänen R (1988): Implications of ERP data for psychological theories of attention. In: *Event-Related Potential Investigations of Cognition*, Renault B, Kutas M, Coles MGH, Gaillard AWK, eds. Amsterdam: Elsevier.

Näätänen R (1990): The role of attention in auditory information processing as revealed by event-related potentials and other brain measures of cognitive function. *Behav Brain Sci* 13:201–288.

O'Connor TA, Starr A (1985): Intracranial potentials correlated with an event related potential, P300, in the cat. *Brain Res* 339:27–38.

Paller KA, Zola-Morgan S, Squire LR, Hillyard SA (1988): P3-like brain waves in normal monkeys and monkeys with medial temporal lesions. *Behav Neurosci* 102:714–725.

Parnefjord R, Başar E (1992): Delta-Oscillationen als Korrelat für akustische Wahrnehmungen an der Hörschwelle. 37. Jahrestagung der Deutschen EEG-Gesellschaft, Magdeburg.

Petsche H, Rappelsberger P (1992): Is there any message hidden in the human EEG? In: *Induced Rhythms in the Brain*, Başar E, Bullock TH, eds. Boston: Birkhäuser.

Petersen SE, Fox PT, Miezin FM, Raichle ME (1988): Modulation of cortical visual responses by direction of spatial attention measured by PET. *Assoc Res Vision Ophthal*, p. 22 (abstr.).

Pfurtscheller G, Klimesch W (1992): Event-related synchronization and desynchronization of alpha and beta waves in a cognitive task. In: *Induced Rhythms in the Brain*, Başar E, Bullock TH, eds. Boston: Birkhäuser.

Picton TW, Hillyard SA (1974): Human auditory evoked potentials. II: Effects of attention. *Electroencephalogr Clin Neurophysiol* 36:193–199.

Picton TW, Stuss DT (1980): The component structure of the human event-related potentials. In: *Motivation, Motor and Sensory Processes of the Brain: Electrical Potentials, Behavior and Clinical Use*, Kornhuber HH, Deecke L, eds. Amsterdam: Elsevier.

Pineda JA, Foote SL, Neville HJ (1987): Long latency event-related potentials in squirrel monkeys: further characterization of wave form morphology, topographic and functional properties. *Electroencephalogr Clin Neurophysiol* 67:77–90.

Posner MI, Petersen SE (1990): The attention system of the human brain. *Annu Rev Neurosci* 13:25–42.

Pritchard WS (1981): Psychophysiology of P300. *Psychol Bull* 89:506–540.

Regan D (1989): *Human Brain Electrophysiology: Evoked Potentials and Evoked Magnetic Fields in Science and Medicine*. Amsterdam: Elsevier.

Rémond A, Lesévre N (1957): Remarques sur l'activité cérébrale des sujets normaux. In: *Conditionnement et réactivité en électroencéphalographie (Electroencephalogr Clin Neurophysiol, Suppl 6)*, Fischgold H, Gastaut H, eds. Amsterdam: Elsevier.

Ritter W, Vaughan HG, Costa LD (1968): Orienting and habituation to auditory stimuli: A study of short term changes in averaged evoked responses. *Electroencephalogr Clin Neurophysiol* 25:550–556.

Robinson DL, Goldberg ME, Staunton GB (1978): Parietal association cortex in the primate: Sensory mechanisms and behavioral modulations. *J Neurophysiol* 41:910–932.

Rösler F (1982): *Hirnelektrische Korrelate kognitiver Prozesse*. Berlin: Springer.

Roth WT (1973): Auditory evoked responses to unpredictable stimuli. *Psychophysiology* 10:125–137.

Sachs L (1974): *Angewandte Statistik*. Heidelberg: Springer.

Shepherd GM (1988): *Neurobiology*. Oxford: Oxford University Press.

Simson R, Vaughan HG, Jr, Ritter W (1977): The scalp topography of potentials in auditory and visual go/no go tasks. *Electroencephalogr Clin Neurophysiol* 43:864–875.

Smith ME, Halgren E, Sokolik M, Bauden P, Musolino A, Liégeois-Chauvel C, Chauvel P (1990): The intracranial topography of the P3 event-related potential elicited during auditory oddball. *Electroencephalogr Clin Neurophysiol* 76:235–248.

Solodovnikov VV (1960): *Introduction to the Statistical Dynamics of Automatic Control Systems*. New York: Dover.

Squires NK, Donchin E, Squires K, Grossberg S (1977): Biosensory stimulation: Inferring decision-related processes from P300 component. *J Exp Psychol* 3:299–315.

Stampfer HG, Başar E (1985): Does frequency analysis lead to a better understanding of human event-related potentials? *Int J Neurosci* 26:181–196.

Stapleton JM, Halgren E (1987): Endogenous potentials evoked in simple cognitive tasks: depth components and task correlates. *Electroencephalogr Clin Neurophysiol* 67:44–52.

Sutton S, Tueting P, Zubin J, John ER (1967): Information delivery and the sensory evoked potential. *Science* 155:1436–1439.

Tononi G, Sporns O, Edelman GM (1992): The problem of neural integration: Induced rhythms and short-term correlations. In: *Induced Rhythms in the Brain*, Başar E, Bullock TH, eds. Boston: Birkhäuser.

Van der Tweel LH (1961): Some problems in vision regarded with respect to linearity and frequency response. *Ann NY Acad Sci* 89:829–856.

Vaughan HG, Jr, Ritter W (1970): The sources of auditory evoked responses recorded from the human scalp. *Electroencephalogr Clin Neurophysiol* 28:360–367.

Walter DO, Etevenon P, Pidoux B, Tortrat S, Guillou S (1984): Computerized topo-EEG spectral maps: Difficulties and perspectives. *Neuropsychobiology* 11:264–272.

Westphal KP, Grözinger B, Diekmann V, Scherb W, Reeß J, Leibing U, Kornhuber HH (1990): Slower theta activity over the midfrontal cortex in schizophrenic patients. *Acta Psychiatr Scand* 81:132–138.

Wood CC, McCarthy G (1985): A possible frontal lobe contribution to scalp P3. *Soc Neurosci Abstr* 11:879.

Wood CC, Allison T, Goff WR, Williamson PD, Spencer DD (1980): On the neuronal origin of P300 in man. In: *Motivation, Motor and Sensory Processes of the Brain: Electrical Potentials, Behaviour and Clinical Use*, Kornhuber HH, Deecke L, eds. Amsterdam: Elsevier.

Wood CC, McCarthy G, Allison T, Goff W, Williamson PD, Spencer DD (1982): Endogenous event-related potentials following temporal lobe excisions in humans. *Soc Neurosci Abstr* 8:976.

Woods DL (1990): The physiological basis of selective attention: Implication of event-related potential studies. In: *Event-Related Brain Potentials, Basic Issues and Applications*, Rohrbaugh JW, Parasuraman R, Johnson R, Jr., eds. New York: Oxford University Press.

Yin TCT, Mountcastle VB (1977): Visual input to the visuomotor mechanisms of the monkey's parietal lobe. *Science* 197:1381–1383.

Chapter 16

Magnetoencephalography in the Study of Human Brain Functions

RIITTA HARI

In magnetoencephalography (MEG), magnetic fields are measured with superconducting sensors outside the human head. The signals result mainly from intracellular currents that flow in synchronously active neurons of the fissural cortex. Source and volume conductor models—like a current dipole within a sphere—are commonly used to determine locations of these currents. With multichannel SQUID magnetometers, field patterns can be constructed and source locations determined even with a "single shot" measurement without repositioning the instrument, opening new possibilities in clinical applications as well as studies of the neural basis of cognitive functions. Examples are given here of MEG recordings from auditory and somatosensory cortices.

In biological tissue, as well as in a wire electric currents producemagnetic fields in which the flux direction is determined according to the right-hand rule. Cerebral sources of magnetic fields detected by MEG are intracellular currents that flow in the fissural cortex during postsynaptic potentials. The MEG signals are first picked up by axial or planar flux transformers. The conventional axial gradiometer records the maximum signals at two locations on opposite sides of a current dipole, whereas the planar gradiometer, with a figure-of-eight pickup loop, detects the largest signal just over the dipole where the field gradient is at its maximum. The insert of Figure 1 shows the flux transformer array of the 24-channel planar gradiometer used

Cognitive Electrophysiology
H-J. Heinze, T.F. Münte, and G.R. Mangun, editors
© 1994 Birkhäuser Boston

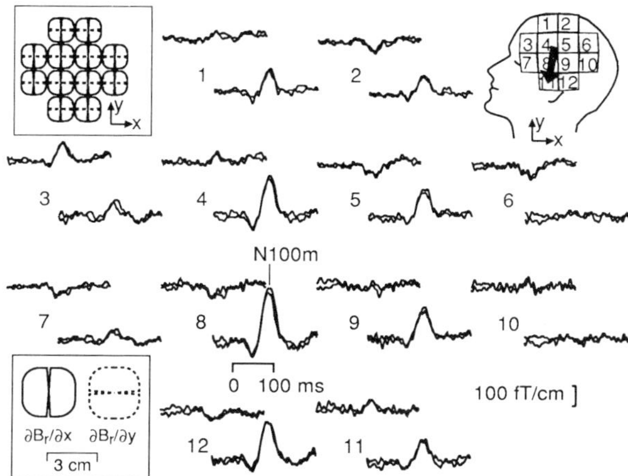

Figure 1.   Auditory evoked magnetic fields to 50-msec 1-kHz square-wave tones repeated once every 1.2 sec; 66 single responses were averaged, and 2 averages are superimposed. Passband of responses is 0.05–90 Hz. Recording locations indicated on schematic head. Upper traces of each response pair illustrate the $\partial B_r/\partial y$ derivative; lower traces, the $\partial B_r/\partial x$ derivative (cf. *insert, below*). Arrow on head shows location and orientation of ECD for N100m. *Insert above*: Schematic illustration of flux transformer array of 24-channel planar gradiometer in Helsinki (Ahonen et al., 1991). *Insert below*: Each sensor unit has two orthogonal figure-of-eight loops to detect the two tangential derivatives of B_r.

in our laboratory during the period 1989–1992. (Ahonen et al., 1991). The two orthogonal figure-of-eight transformers per sensor unit measure the tangential derivatives $\partial B_r/\partial x$ and $\partial B_r/\partial y$ of the radial field component, B_r, simultaneously at 12 locations 3 cm apart. This device often allows the source location to be determined without moving the instrument, thereby increasing the speed of the experiments. After the submission of this manuscript, whole-head neuromagnetometers have emerged.

Interpretation of both MEG and EEG data, in terms of cerebral currents, requires the use of volume conductor and source models because the inverse problem does not have a unique solution. The most common model is a tangential current dipole within a sphere. The success of dipole models derives from the fact that from the typical measurement distance (at least 3 cm from the source), several current configurations appear "dipolar" because the higher order terms, which reflect the complexity of the source, decrease rapidly as a function of distance. The orientation, strength, and three-dimensional location of the equivalent current dipole (ECD), which best explains the measured signal distribution, are determined with a least-squares fit to the data. The locating accuracy is best transversal to dipole orientation and poorest in the direction of depth. A cortical ECD does not rule out the existence of thalamic or other deep triggers of the source area; in fact, the MEG signals often reflect the function of cortical neurons that are driven by thalamocortical fibers and thereby activated in strong synchrony.

When the source areas of the measured signals are approximately known, it is possible to focus on functions of these brain regions. All source estimates can be improved by restrictions based on the known anatomy of the brain as obtained from magnetic resonance imaging (MRI) data, for example. Absolute source locations, with respect to either external landmarks on the head or to brain itself, are not needed in all cases but are important in the determination of epileptic foci. In epileptic patients it is often necessary to determine the *relative* locations of the focus with respect to different sensory projection areas so that these can be spared during surgery.

Some examples are now given of our MEG results from the auditory and somatosensory systems.

Auditory System

Long-Latency Auditory Responses and Spontaneous Auditory Rhythm

A prominent 100-msec deflection, N100m, generated at the supratemporal auditory cortex, is seen in response to various sounds (see Fig. 1). Vascular lesions in the deep mesial auditory cortex abolish N100m unilaterally on the side of the lesion (Mäkelä et al., 1991). N100m is stimulus specific with different source locations, for example, for tones of different frequencies. Several neuronal groups, therefore, seem to give rise to 100-msec responses. The amplitude of N100m decreases by 75% when the interstimulus interval is decreased from 8.8 to 1.1 sec (Hari et al., 1987), probably because of active inhibition. Surprisingly, when 50-msec sounds are presented in pairs with different onset asynchronies, N100m may be larger to the second than to the first sound of the pair at onset asynchronies of 70–150 msec (Loveless et al., 1989); evidently the first sound of the pair modifies the reactivity of the auditory cortex. One possible explanation for the time course of the response enhancement would be an underlying rhythm at about 10 Hz in the auditory cortex.

Evidence of such a 8- to 10-Hz rhythm was recently obtained in MEG recordings over the temporal lobes (see Fig. 3; Tiihonen et al., 1991). The rhythm was damped by auditory stimuli but not by opening of the eyes or by clenching the first, thereby clearly differing from the occipital alpha and the rolandic mu rhythms. The field patterns of the activity could be satisfactorily explained by ECDs situated within 2 cm of the source of the auditory N100m (Fig. 2). Each sensory area, therefore, seems to have its own "idling" rhythm. Steady-state recordings suggest the existence of an additional faster (about 40 Hz) rhythm in the auditory cortex (see following).

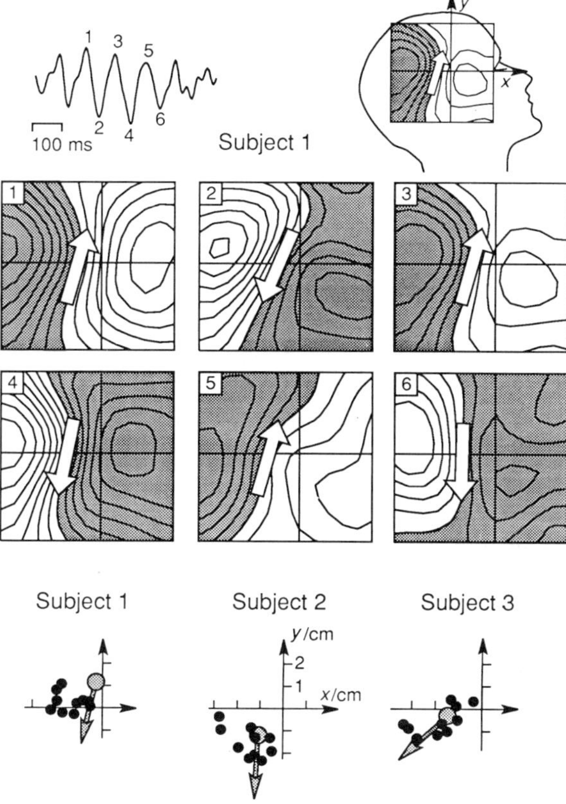

Figure 2.   Upper: Field patterns of radial field component B_r in one subject during six successive peaks (1–6) of representative spindle in 10-Hz spontaneous activity. Insert at top right shows measurement area with respect to head. Shadowed areas indicate magnetic flux out of head; white areas, flux into head; isocontours are separated by 100 fT. *Arrows* show locations and orientations of ECDs. Origin of coordinate system is 7 cm posterior to eye corner; *x*-axis forms angle of 45° with line connecting ear canal to eye corner. Lower: Locations of equivalent dipoles (*filled circles*) for 10 deflections of spontaneous 8- to 10-Hz oscillation for three subjects. *Shadowed circles* and *arrows* show dipole locations and orientations for auditory N100m responses. (Adapted from Tiihonen et al., 1991.)

The activity of the supratemporal auditory cortex would be reflected in frontocentral scalp EEG derivations. It is, therefore, interesting that vigilance decrease has been traditionally considered to be associated with 'alpha spread' from the occipital areas toward the more anterior scalp regions. Instead of a real alpha spread and the movement of generators, a more plausible explanation is that the anterior rhythmic signal may reflect activity of the supratemporal cortex.

Responses to Changes in Auditory Environment

In experiments using "oddball paradigms," infrequent 'deviants' are randomly interspersed among monotonously repeated 'standard' sounds. Different types of deviations evoke strong magnetic responses, or 'mismatch fields' (MMFs). The ECD of MMF is typically 1 cm anterior to that of N100m, indicating that these two responses include activation of different cytoarchitectonic areas (Hari et al., 1992b; Sams et al., 1991b). MMFs are not present in responses to standards nor in those to deviants delivered in the absence of standards. Thus MMF depends not only on the physical quality of the sound but also on the position of the sound in the sequence. In agreement with its electric counterpart, the mismatch negativity (Näätänen and Picton, 1987), MMF has been assumed to be associated with short-term sensory (echoic) memory: to be able to react to a change and to generate an MMF, the cortical neuronal network evidently has to store the "trace" of the previous stimulus for a short period. The duration of the human echoic memory, as reflected by the generation of MMF, is about 10 sec and agrees with behavioral observations (Sams et al., 1993).

Assuming that this sensory–memory connection exists, it is of interest that MMFs can also be elicited by periodicity pitch stimuli in the absence of spectral cues. We recently presented the subject with 100-msec noise bursts, amplitude modulated (depth, 100%) by 80-Hz or 240-Hz square waves (Fig. 3). These stimuli elicited different pitch sensations in spite of their identical flat frequency spectra. Processing of the sounds must thus occur in the time domain.

With these stimuli, MMFs could be evoked at interstimulus intervals (ISIs) as long as 5 sec. Because time-locked activity continuing for such extended periods after the stimulus is improbable, the time code associated with periodicity pitch might be converted to a place code for short-time storage of the stimulus features. The change could take place in neural structures analogous to the coincidence detection systems containing delay lines for detection of binaural time differences (Konishi et al., 1988).

Attentional Modification of Cortical Activity

In a recent study, Rif et al. (1991) presented a monaural randomized sequence of 1- and 3-kHz tones at a constant ISI of 405 ms (Fig. 4). The task of the subject was to count infrequent 150-msec targets of designated frequency among 50-msec standards, and only responses to standards were analyzed.

Figure 4 shows that standard tones, both when attended and when ignored, elicited prominent P50m and N100m deflections. The responses to attended tones differed after N100m from those to the same tones when ignored; this magnetic difference (Md) deflection is shadowed in Figure 4. Md peaked at about 200 msec and its ECD was 8–10 mm anterior to the source of N100m, very close to the generator site of P200m to ignored tones. When

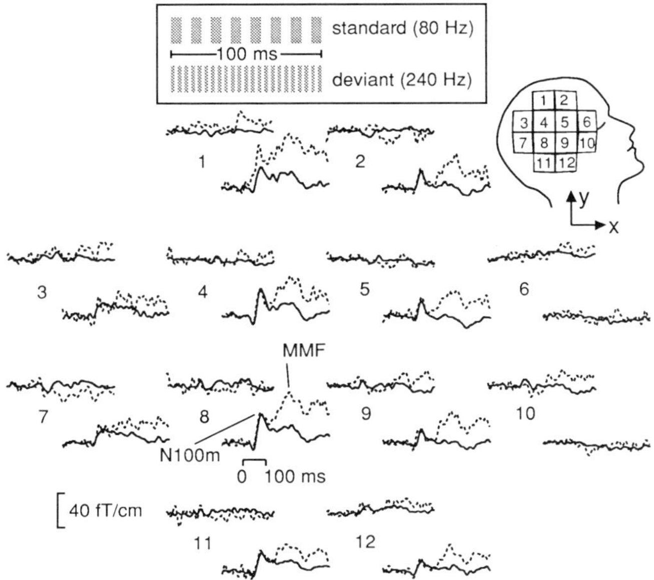

Figure 3.   Periodicity pitch experiment. Responses of one subject to 80-Hz standards (*continuous lines*) and to 240-Hz deviants (*dashed lines*), respectively. Stimuli (gated noise bursts) are shown in insert; recording locations are indicated on head. Other details as in Figure 1.

the ISI was shortened to 240–300 msec and the tones of different pitches were delivered to opposite ears, Md appeared at 30–40 msec, and the amplitude of N100m was larger for attended than for ignored tones. These results indicated that attention is associated with modification of activity of two separate areas in the supratemporal auditory cortex; the time course of the effect depends on the experimental setup. These events can be explained in terms of both gain control and attentional priming mechanisms.

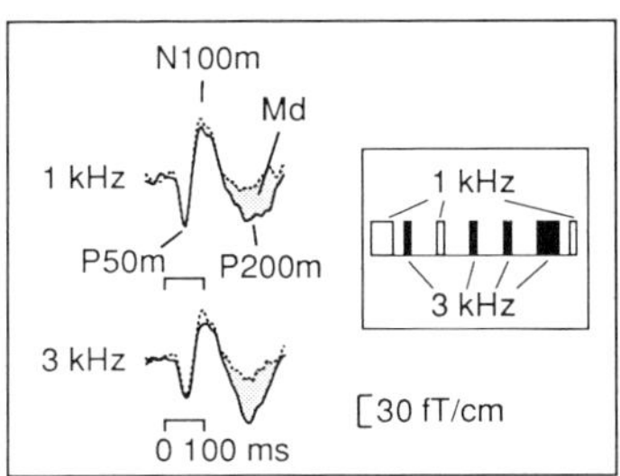

Figure 4.   Auditory evoked magnetic responses from one subject to identical 1- or 3-kHz tones when the subject ignored them (*solid lines*) or attended to them (*dashed lines*). Tones were presented to left ear; responses were recorded over right auditory cortex. Traces are averages of about 150 single responses; signals were digitally low-pass filtered at 45 Hz. Insert illustrates stimulus sequence with tones of two frequencies randomly mixed. (Modified from Rif et al., 1991.)

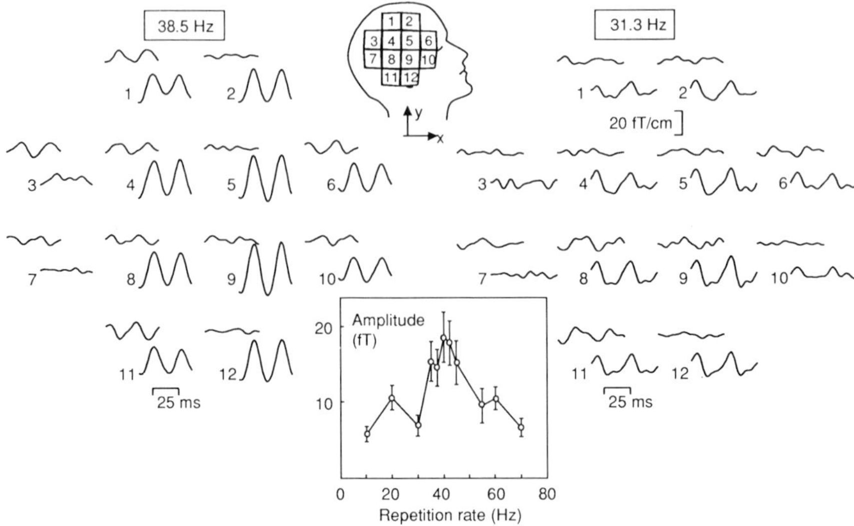

Figure 5.   Steady-state responses to clicks at two stimulus repetition rates (38.5 and 31.3 Hz). Two cycles have been averaged; passband is 0.05–90 Hz. Other details as in Figure 1. *Insert below*: Mean ($\pm$ SEM; 10 subjects) amplitude of steady-state magnetic response as function of stimulus repetition rate according to Hari, Hämäläinen, and Joutsiniemi (1989).

40-Hz Response

Figure 5 shows that steady-state responses to clicks repeated at 38.5 Hz are clearly larger than those to clicks at 31.3 Hz; in both cases the ECDs were at the supratemporal auditory cortex. The signals resemble the electric 40-Hz response first described by Galambos et al. (1981). The amplitude enhancement at stimulus repetition rates of about 40 Hz (also illustrated in Fig. 5) has attracted much attention. We have shown that it is possible to reconstruct the steady-state 20-Hz and 40-Hz MEG responses, including their waveforms and amplitudes, from the 10-Hz response (Hari et al., 1989). Therefore, the 40-Hz response seems to result from a superposition of successive middle-latency responses. The driving force of the enhancement at 40 Hz is, however, the tendency of the cerebral network to react so that responses to single clicks contain frequencies of about 40 Hz. The frequency content of the evoked response then determines the stimulus repetition rate at which the amplitude enhancement from superposition may be observed.

Functions of the Auditory Cortex as Revealed by MEG

To summarize, the auditory cortex, when studied with MEG, reacts strongly to sound onsets and offsets and is extremely sensitive to all acoustic changes as well as to complex sounds (amplitude and frequency modulations; speech).

Monitoring the source locations as a function of time has indicated both parallel and sequential signal processing. In agreement with anatomic asymmetries, source locations are typically about 15 mm more posterior in the left than in the right hemisphere of right-handed subjects.

The auditory cortex is strongly involved in short-term sensory (echoic) memory and of importance for attention and directional hearing. It also seems to be able to integrate heard and seen speech: the magnetic responses to speech sounds can be modified by visual information from lip movements (Sams et al., 1991a). MEG recordings from the auditory cortex have also shown the existence of feature maps (e.g., a tonotopic map, cf. Pantev et al., 1988, Romani, Williamson, and Kaufman, 1982) and of spontaneous idling rhythms.

Somatosensory System

Figure 6 shows somatosensory magnetic fields (SEFs) to electric stimulation of the median nerve at the wrist. The first cortical magnetic response, N20m, peaks at 20 msec and its distribution suggests a tangential current source at the fissural part (area 3b) of the primary somatosensory cortex SI.

In some diseases, SEFs may be altered. In multiple sclerosis, the 20-msec responses to median and ulnar nerve stimuli are delayed, as is well known from previous electric recordings. In addition, abnormally high-amplitude SEF deflections may occur about 50–80 msec at the hand SI (Karhu et al., 1992). In patients with progressive myoclonus epilepsy, the responses have normal generation sites but abnormally high amplitudes (see Fig. 6). The relative sites of sources elicited by electric stimulation of different body parts, such as the ankle, wrist, fingers, tongue, and lips, agree with the somato-sensory homunculus observed by direct cortical stimulations (Penfield and Jasper, 1954).

The responses generated at distinct and somatotopically organized areas show, however, significant interaction. For example, when rare (10%) median nerve stimuli were presented among frequently (90%) repeated ulnar nerve stimuli (ISI, 1 sec), or vice versa (Huttunen et al., 1987), the 70- to 85-msec response waveforms and amplitudes were similar regardless whether the stimulus was a standard or a deviant, indicating strong interaction between afferent inputs from both nerves to SI.

Similar interaction/convergence was observed when a conditioning ulnar nerve stimulus, presented 40 msec earlier, had a strong suppressive effect on N20m evoked by stimulation of the median nerve of the same limb (Huttunen et al., 1992). This effect lasted for about 150 msec and was as strong as that produced by a conditioning stimulation of the median nerve itself (see Fig. 6). Therefore, the lateral and in-field inhibitions seem to have similar time courses. Interestingly, SEF recordings have indicated different behavior at the second somatosensory cortex SII with accurate and in-

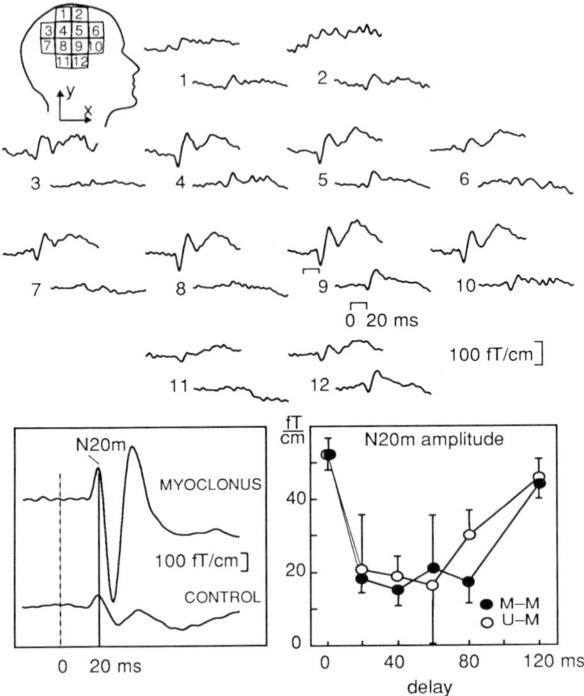

Figure 6.   Somatosensory responses to stimulation of left median nerve at wrist. Recording passband was 0.05–500 Hz; responses were digitized at 2 kHz; traces are averages of about 100 responses. *Insert at left*: Comparison of median nerve SEFs in patient with progressive myoclonus epilepsy and in normal control subject. *Insert at right*: Mean ($\pm$ SEM; six subjects) amplitude of N20m to median nerve stimulation as function of delay after conditioning (median, M; ulnar, U) stimulus. (Insert at right modified from Huttunen, Ahlfors, and Hari, 1992.)

dependent representations for separate fingers: the 100-msec response to middle finger/thumb stimulation was of the same size and latency, regardless whether the stimuli were presented alone or when the intervening stimuli were delivered to the other finger (Hari et al., 1990).

Conclusions

MEG is a method to study noninvasively the activity of the human cerebral cortex, both for increasing our knowledge about functional brain organization and for learning more about the sources of different EEG phenomena and evoked potentials. Instrumental noise is no longer a problem in MEG recordings, and one has to deal with the 'brain noise' itself. It has recently become possible to compare signal processing in the two hemispheres with magnetometers covering the whole head (cf. Ahonen et al., 1992), and thus to focus on complex brain functions associated with simultaneous activity at several cortical areas. Improvements of source models and combination of functional and structural data are necessary for further development of neuromagnetism.

Acknowledgments. This study has been supported by the Academy of Finland, the Körber Foundation (Hamburg, Germany), and by Sigrid Jusélius Foundation. I thank Jari Karhu, Olli V. Lounasmaa, Ritva Paetau, and Mikko Sams for comments on the manuscript.

References

Ahonen A, Hämäläinen M, Kajola M, Knuutila J, Laine P, Lounasmaa OV, Simola J, Tesche C, Vilkman V (1992): A 122-channel magnetometer covering the whole head. In: Dittmar A, J Froment (eds) Proceedings of the Satellite Symposium on Neuroscience and Technology, 14th Annual International Conference of the IEEE Engineering in Medicine and Biology Society. Lyon, France, November 1992, pp 16–20.

Ahonen A, Hämäläinen M, Kajola M, Knuutila J, Lounasmaa OV, Simola J, Tesche C, Vilkman V (1991): Multichannel SQUID systems for brain research. *IEEE Trans Magnet* 27:2786–2792.

Galambos R, Makeig S, Talmachoff PJ (1981): A 40-Hz auditory potential recorded from the human scalp. *Proc NY Acad Sci* 78:2643–2647.

Hari R, Hämäläinen M, Joutsiniemi SL (1989): Neuromagnetic steady-state responses to auditory stimuli. *J Acoust Soc Am* 86:1033–1039.

Hari R, Hämäläinen H, Tiihonen J, Kekoni J, Sams M, Hämäläinen M (1990): Separate finger representations at the human second somatosensory cortex. *Neuroscience* 37:245–249.

Hari R, Pelizzone M, Mäkelä JP, Hällström J, Leinonen L, Lounasamaa OV (1987): Neuromagnetic responses of the human auditory cortex to on- and off-sets of noise bursts. *Audiology* 25:31–43.

Hari R, Rif J, Tiihonen J, Sams M (1992): Neuromagnetic mismatch fields to single and paired tones. *Electroenceph Clin Neurophysiol* 82:152–154.

Huttunen J, Ahlfors S, Hari R (1992): Interaction of afferent impulses at the human primary sensorimotor cortex. *Electroenceph Clin Neurophysiol* 82:176–181.

Huttunen J, Hari R, Leinonen L (1987): Cerebral magnetic responses to stimulation of ulnar and median nerves. *Electroencephalogr Clin Neurophysiol* 66:391–400.

Karhu J, Hari R, Mäkelä JP, Huttunen J, Knuutila J (1992): Somatosensory evoked magnetic fields in multiple sclerosis *Electroenceph Clin Neurophysiol* 83:192–200.

Konishi M, Takahashi TT, Wagner H, Sullivan WE, Carr CE (1988): Neurophysiological and anatomical substrates of sound localization in the owl. In: *Auditory Function. Neurobiological Bases of Hearing*, Edelman GM, Gall WE, Cowan WM, eds., pp. 721–745. New York: Wiley.

Loveless N, Hari R, Hämäläinen M, Tiihonen J (1989): Evoked responses of human auditory cortex may be enhanced by preceding stimuli. *Electroencephalogr Clin Neurophysiol* 74:217–227.

Mäkelä JP, Hari R, Valanne L, Ahonen A (1991): Auditory evoked magnetic fields after ischemic brain lesions. *Ann Neurol* 30:76–82.

Näätänen R, Picton T (1987): The N1 wave of the human electric and magnetic response to sound: A review and analysis of the component structure. *Psychophysiology* 24:375–425.

Pantev C, Hoke M, Lehnertz K, Lütkenhöner B, Anogianakis G, Wittkowski W (1988): Tonotopic organization of the human auditory cortex revealed by transient auditory evoked magnetic fields. *Electroencephalogr Clin Neurophysiol* 69:160–170.

Penfield W, Jasper H (1954): *Epilepsy and the Functional Anatomy of the Human Brain.* Boston: Little, Brown.

Rif J, Hari R, Hämäläinen M, Sams M (1991): Auditory attention affects two different areas in the human auditory cortex. *Electroencephalogr Clin Neurophysiol* 79:464–472.

Romani GL, Williamson SJ, Kaufman L (1982): Tonotopic organization of the human auditory cortex. *Science* 216:1339–1340.

Sams M, Hari R, Rif J, Knuutila J (1993): The human auditory sensory memory trace persists about 10 s: Neuromagnetic evidence *J Cogn Neuroscience* 5:363–370.

Sams M, Aulanko R, Hämäläinen M, Hari R, Lounasmaa OV, Lu S, Simola J (1991a): Seeing speech: Visual information from lip movements modifies activity in the human auditory cortex. *Neurosci Lett* 127:141–145.

Sams M, Kaukoranta E, Hämäläinen M, Näätänen (1991b): Cortical activity elicited by changes in auditory stimuli: Different sources for the magnetic N100m and mismatch responses. *Psychophysiology* 28:21–28.

Tiihonen J, Hari R, Kajola M, Karhu J, Ahlfors S, Tissari S (1991): Magnetoencephalographic 10-Hz rhythm from the human auditory cortex. *Neurosci Lett* 129:303–305.

Keyword Index

This index was established according to the keywords supplied by the authors. Page numbers refer to the beginning of the chapter.